RNA Technologies

Volume 15

Founding Editors
Jan Barciszewski
Volker A. Erdmann

Series Editors

RNA Technologies provides critical and comprehensive discussions of the most significant areas of RNA research, written by leading international authorities. Each volume in the series reviews both general topics, such as RNA interference, RNA biosynthesis and metabolism, microRNAs and diseases, transcriptome analysis and aptamers, as well as highly topical chapters on very recent findings in the field. Recurring themes of the series include the complexities associated with target recognition, RNA-protein interactions, analysis of genetic networks and the promise of small RNAs as cancer diagnostics and therapeutics in combating cancer.

RNA-based approaches have great potential to revolutionize molecular biology, cell biology, biomedical research and medicine. The nucleic acid-based molecules can be used to regulate the level of gene expression inside the target cells and their potential efficacy against several viruses, including SARS-CoV-2, and other chronic diseases represents a promising research area.

RNA Technologies publishes in both print and electronic format since 2010 and is intended for researchers from academia and industry, as well as graduates that look for a carefully selected collection of high quality review articles on their respective field of expertise.

The series is indexed in SCOPUS.

Jan Barciszewski
Editor

Systems Biology II

Springer

Editor
Jan Barciszewski
Nanobiomedical Center, Adam
Mickiewicz University; Institute
of Bioorganic Chemistry of the
Polish, Academy of Sciences
Poznań, Poland

ISSN 2197-9731 ISSN 2197-9758 (electronic)
RNA Technologies
ISBN 978-3-031-62177-2 ISBN 978-3-031-62178-9 (eBook)
https://doi.org/10.1007/978-3-031-62178-9

This Springer imprint is published by the registered company Springer Nature Switzerland AG
The registered company address is: Gewerbestrasse 11, 6330 Cham, Switzerland

Introduction

Living organisms are complex and dynamic systems that consist of matter, energy, and information. The scientific field that studies how organisms organize matter, use energy, and interpret information is called systems biology. Recent technological advancements have enabled us to determine the molecular composition of living cells comprehensively.

The adequate form and function of a multicellular organism requires the perfect implementation of a complicated temporal and spatial choreography of cellular proliferation, morphogenesis, and differentiation, which are powered by developmentally dynamic programs of gene activity. Developmental regulation of gene expression involves transcriptional and posttranscriptional modes including the regulation of mRNA stability and translatability by non-protein coding RNAs. An integrated multidisciplinary approach is necessary for an understanding of how the cell nucleus with its compartmentalized chromatin architecture accomplishes its functional tasks during expression, replication, and repair of genomes. Some proteins required for these functions are synthesized in the cytoplasm but in the nucleus they form functional machines with other proteins and find their target sites within crowded cellular interior. This involves breaking down the individual components and studying their specific features before reconstructing their relationships with other entities.

While biology has traditionally focused on decoding cellular processes one gene at a time, many of the most pressing biological questions and diseases, such as cancer and heart disease, are related to complex systems involving the interaction of thousands of gene products and other factors. Technology has dramatically accelerated biology research, and computation is a standout example. Adopting a new technology typically involves creating a new method, proving its concept in biology, and developing usable tools.

Systems biology emerged as a result of the collaborative efforts of engineers and scientists from various fields who developed algorithms to analyze large quantities of biological data. The development of user-friendly software that bridges computer science and biology has been crucial in driving biological discoveries and yielding high returns on investment. In the past, biologists were limited to studying one biological item or aspect at a time, but with the advent of systems biology, it became possible to study biological phenomena within their natural contexts. Systems biology is an interdisciplinary field that offers a holistic approach to interpreting biological processes and has been responsible for some of the most important developments in the science of human health and environmental sustainability.

Systems biology brings together various scientific disciplines such as cell biology, developmental biology, evolutionary biology, biochemistry, genetics, epigenetics, molecular biology, computer sciences, and mathematics with the aim of developing a global view of cells, organisms, and ecosystems. It is an integrative field that requires a deep understanding of biological systems by gaining insights into the structure–function relationship of individual cellular components. It aims to decipher how these networks and regulatory circuits interact to form living systems and regulatory mechanisms. Chemical interactions between many of these molecules are known, giving rise to the genome-scale reconstructed biochemical reaction networks underlying cellular functions.

To understand network structure and dynamics, systems biology employs various properties such as measures of centrality, community structure, and hierarchy. The complex systems science approach considers that the phenomena under study are strongly interconnected entities composed of many levels and subjected to reciprocal influences. Systems biology continues to be a driving force behind the development and improvement of experimental techniques and equipment to extract large amounts of qualitative and quantitative information from complex biological samples. The bioinformatic pipelines necessary to obtain meaningful biological insights are also constantly evolving.

Understanding biological systems relies heavily on models, which are inherently incomplete descriptions of reality. The ultimate aim of systems biology is to develop predictive models that allow researchers to anticipate how cells and organisms will change over time or in response to external stimuli or disturbances. Mathematical descriptions of the entirety of these chemical interactions result in genome-scale models that facilitate the computation of physiological functions. These models can be leveraged to address key biological questions and enable predictive biology. Networks, graph-theoretic representations that depict how different parts interact with one another at varying levels of biological organization and time scales, can be used to model systems. Mathematical models, particularly those derived from theories of biochemical reactions like ordinary differential equation models, prove to be useful.

Machine learning and artificial intelligence will be pivotal in developing models that incorporate spatial and temporal information. Predicting the fate of cells, tissues, organs, and organisms would be a significant achievement, as it would not only provide us with a fundamental understanding of how life works at the molecular and cellular levels, but it would also be a significant step forward for personalized medicine. This would enable us to anticipate the progression of human diseases and select the most effective treatments for each patient.

The chapters of this book illustrate how systems biology is instrumental in advancing our knowledge about the principles of cellular and tissue organization. Subjects covered include regulation of gene expression, RNA-binding proteins, RNA–RNA interactions, noncoding RNAs, transcriptomics, epigenomics, metabolomics, posttranscriptional gene regulation, systems biology in health and disease, experimental and computational tools for systems biology research, computational methods for multidimensional data analysis, and integration as well as the deduction of predictive models.

How can one understand the complexity of genes, RNAs, and proteins and the associated regulatory networks? One approach is to look for recurring types of dynamical behavior.

The book chapters introduce systems biology methods to construct an integrated genetic and epigenetic cellular network, which combines next-generation sequencing omics data with gene regulatory networks, microRNAs regulatory networks, protein–protein interaction networks, and epigenetic regulatory networks of methylation using high-throughput data.

For many years systems biology is of interest to philosophers of science and philosophers of biology. In particular many philosophers have been interested in the claims systems biologists have routinely made on the organization of biological systems and the need for mathematically based computational approaches to properly control and predict those systems.

Contents

Constructing an Integrated Genetic and Epigenetic Cellular Network by Systems Biology Method for Investigating Whole Cellular Mechanism Using NGS Omics Data

Bor-Sen Chen

Contents

Abstract Epigenetics is the mechanism that leads to heritable changes in the genetic function without affecting the genome's sequence. Since the appearance of epigenomics, epigenetics has been widely investigated in cancer initiation and development. These epigenetic mechanisms could explain how individuals who have the same genotype may produce phenotypic differences in response to environmental stimuli. Recently, through the accumulation of high-throughput next-generation sequencing (NGS) data, the primary purpose of systems biology is to

B.-S. Chen (✉)
Department of Electrical Engineering, Lab. of Control and Systems Biology, National Tsing Hua University, Hsinchu, Taiwan
e-mail: bschen@ee.nthu.edu.tw

J. Barciszewski (ed.), *Systems Biology II*, RNA Technologies 15,
https://doi.org/10.1007/978-3-031-62178-9_1

construct genetic and epigenetic networks for different cellular levels to investigate whole cellular mechanisms. This chapter introduces a systems biology method to build an integrated genetic and epigenetic cellular network (IGECN) using NGS omics data.

We have successfully constructed IGECNs of several diseases and validated them using the literature search evidence. For example, the dysregulation of mir7 leads to the initiation and progression of inflammation-induced gastric cancer; the dysregulation of mir9 leads to HIV-1 infection to hijack CD4+ T cells via dysfunction of the immune and hormone pathways; the dysregulation of mir139-5p, mirLET7i, and mir10a leads to the HIV-1 integration/replication process; the dysregulation of mir101, mir141, and mir152 contributes to the HIV-1 virus assembly and budding mechanisms; dysregulation of mir302a leads to not only the microvesicle-mediated transfer of miRNAs but also the dysfunction of NF-κB signaling pathway in hepatocarcinogenesis. Hence, the coupling dynamic systems of the whole IGECN can permit us to investigate genetic and epigenetic cellular mechanisms via omics data and extensive database mining and are helpful for further experiments from the perspective of systems and synthetic biology.

Keywords HIV · miRNA · Epigenetic regulation · Integrated genetic and epigenetic cellular network · Next-generation sequencing data · Epigenetic cellular mechanisms

1 Introduction

Due to advances in molecular biology technologies for whole genome sequencing, expression profiling, and high-throughput experiments, we have large amounts of biological data covering various biological levels (Blair and Price 2012; Chatr-Aryamontri et al. 2012; Wang et al. 2009). These amounts of "omics" data cover genetic sequences (genomics), microarray-based genome-wide expression profiles (transcriptomics), protein abundance data (proteomics), and microRNA (miRNA) and methylation data, which can prepare an unprecedented view of cellular components and their corresponding cellular mechanisms in biological systems (Broen et al. 2014). Based on the large amount of accumulated genomic/transcriptomic and epigenetic data, understanding more complex cellular functions in living organisms becomes possible. Computational systems biology techniques are able to integrate and unite these large, heterogeneous genomic/transcriptomic, and epigenetic data sets to investigate the corresponding cellular mechanisms. They will give useful tools to have more systems biology insights into cellular mechanisms under a specific biological condition, for example, cancer cells or infected cells (Ke et al. 2013).

Many articles focus on gene regulatory networks (GRNs) and protein–protein interaction networks (PPINs). For example, classical graph algorithms have been employed to unite cellular networks of protein–protein and protein–DNA interactions. It is well-known that simple and complex regulatory circuits have been identified and many putative regulatory circuits have been under discussion. In Yeger-Lotem et al. (2004), authors have searched for composite network motifs, which consist of both transcription regulation and protein–protein interactions (PPIs) that come up significantly more often than expected in random networks. In Lin et al. (2012), integrated networks are composed of transcriptional and PPI data. From the interspecific analysis, it has been shown that several types of network motifs are not due to any particular evolutionary pressure to maintain the correspondent interaction patterns of critical biological functions. Researchers have also integrated both protein–protein and protein–DNA interactions to imply the corresponding signaling transduction pathways. Still, these researchers focus only on these signal transduction pathways that could explain gene expression changes due to gene knockouts (Herrgård et al. 2006). Recently, an integrated cellular network of transcription regulation and PPIs has been introduced (Przytycka and Kim 2010; Wang and Chen 2010). Authors employ gene expression data at multiple time points to prune and combine candidate gene regulatory and signaling networks constructed by the genome-scale data. Then, an integrated and focused network for a specific condition of interest is gained. The transcriptional network is marked out as a dynamic system in which a target gene's gene expression is considered, a cellular function of the regulatory effect of its corresponding transcription factors (TFs) and mRNA degradation. The system modeling of a signaling/protein interaction network is responsible for the activity of neighboring loci in the network. Genomic/transcriptomic and high-throughput methods have successfully identified many GRNs and PPINs. However, we still need more genomics data, such as epigenetic regulation data, to investigate and explore cellular mechanisms.

In real cellular systems, the expression of protein-coding genes is controlled by a complex network of genetic and epigenetic regulatory interactions. In addition to the above-mentioned genetic regulation mechanisms, epigenetic regulation via DNA methylation is an important regulatory mechanism in several cellular processes. Recently, high-throughput next-generation sequencing (NGS) data containing information on mRNAs, micro-RNAs (miRNAs), and methylation has been produced. Therefore, many famous biological questions focus on how regulations and interactions among genes, proteins, and epigenetic regulators cause specific cellular mechanisms. For this problem, this chapter proposes a systems biology method to construct an integrated genetic and epigenetic cellular network that can explain specific cellular mechanisms under genetic and epigenetic regulations in response to specific biological conditions by stochastic dynamic models. Recently, systems biology and computational biology methods have been widely used to develop stochastic dynamic models to describe biological functions from a dynamic system perspective (Hsu et al. 2009; Chen et al. 2002, 2007; Huang et al. 2009; Chen and Li 2010a, b, c; Li and Chen 2010; Li et al. 2015). Dynamic system models for an integrated genetic and epigenetic cellular network (IGECN) not only give a quantitative description of the integrated cellular network but also provide the cellular mechanism of the integrated cellular network in response to various biological conditions, gene knockouts, treatments with external agents, etc. (Guintivano et al. 2014).

In this chapter, we integrated omics data, including NGS (Mohammadi et al. 2013), mRNA and miRNA expression (Kim et al. 2015), RNA sequencing (RNA-Seq) (Trapnell et al. 2009), PPI (Stark et al. 2006), transcription regulation interaction (Bovolenta et al. 2012; Mootha et al. 2003; Subramanian et al. 2005; Zheng et al. 2008), miRNA–target gene association (Griffiths-Jones 2004; Griffiths-Jones et al. 2007; Kertesz et al. 2007; Kozomara and Griffiths-Jones 2010, 2014), and gene ontology (GO) data (http://geneontology.org/) to construct a candidate IGECN. A schematic diagram of the candidate IGECN is shown in Fig. 1. The candidate IGECN mainly consists of three sub-networks. The first is the candidate PPIN, which contains candidate PPIs in signal transduction pathways and metabolic pathways; the second is the GRN, which describes transcription regulations; the third is the candidate miRNA regulatory network. Epigenetic DNA methylation is also included, which is involved in the modification of DNA to influence on mRNA transcription. The TFs in Fig. 1 play the role at the interface between the PPIN and GRN, and genes also play the role of the interface between the GRN, PPIN, miRNA, and methylation regulatory networks.

Since the candidate IGECN was constructed using omics data and extensive database mining, for data being collected under various experimental conditions (including some that were gathered in vitro), it must contain many false-positive interactions and regulations in the candidate PPIN and GRN, respectively. For this reason, it is necessary to prune these false positives, which may not explain real-world cellular mechanisms of biological organisms. Therefore, it is more attractive to validate these interactions and regulations in the candidate IGECN by examining whole cellular mechanisms of a real organism under specific biological conditions (e.g., HIV infections or cancer) using gene expression profile data, protein expression profile data, and miRNA expression profile data via the system order identification method of dynamic interaction and regulation models. Consequently, false-positive interactions and regulations can be pruned by system order detection method, or the number of system parameters of IGECN can be detected by Akaike information criterion (AIC) based on these real omics data to construct an accurate IGECN for some specific cellular mechanisms.

This chapter proposes three coupling stochastic dynamic models to construct a candidate IGECN as the interplay between the GRN, PPI network, and miRNA network to account for methylation regulation in the GRN. After system parameter estimation for coupling stochastic dynamic models of candidate IGECN by these time-profile data, the Akaike information criterion (AIC) and the *p*-value statistical method (Wang and Chen 2010) are used to detect the system order (the number of interactions and regulations) of coupling stochastic dynamic models and then cut off the false-positive regulations and interactions, which are not significant and out of system order, to obtain the real IGECN for some specific cellular mechanisms in response to some specific biological conditions of diseases. By the interplay of genetic transcription regulations, PPIs, miRNA regulations, and DNA methylations

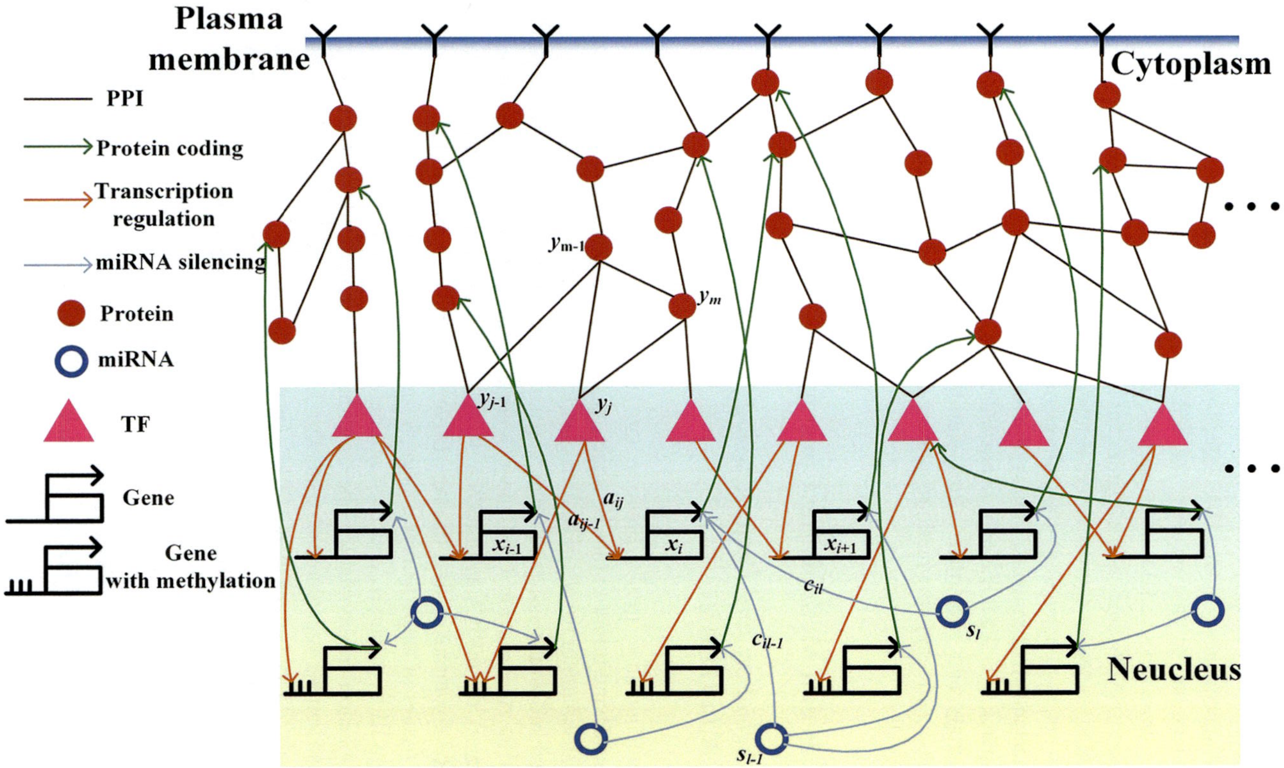

Fig. 1 A figure of candidate integrated genetic and epigenetic cellular networks (IGECN). The candidate IGECN was constructed using omics data and database mining methods. The candidate IGECN mainly consists of three sub-networks: The candidate protein–protein interaction network (PPIN) contains signal transduction pathways and metabolic pathways, the candidate gene regulatory network (GRN) includes candidate transcriptional regulations, and the candidate miRNA regulatory network. The epigenetic regulation of DNA methylation was considered to influence gene transcription. Transcription factors (TFs) represent the interface between PPINs and GRNs, and genes represent the interface between GRNs, miRNA regulatory networks, PPINs, and methylation regulatory networks (Originally published under an Open Access CC BY 4.0 license here https://doi.org/10.1186/s12918-016-0256-5)

of the identified IGECN, the cellular mechanism can be made clearer and understanding.

Investigating host–pathogen interspecies cellular interactions and regulations in the host cells is a critical research topic in order to attempt to understand pathogenic and defensive mechanisms in the HIV infection process and help control clinic pathogenic infections. In this chapter, the IGECN is constructed to investigate host–pathogen cross talks in HIV-infected cells using gene expression profiles and miRNA profiles of NGS data. Some significant cross talk hubs in the interspecies IGECN can be extracted as significant cross talk network markers, from which we could gain and get more insight into the pathogenic mechanism of the HIV virus and the defense mechanism of humans. Finally, the IGECN is also obtained from cancer cells and compared with the IGECN from normal cells to find the significant network marker of therapy and investigate some significant cues of pathogenesis from identifying the genetic and epigenetic regulatory mechanisms using NGS data.

2 Materials and Methods for Constructing IGECN

2.1 Data Selection and Preprocessing for Constructing IGECN in Gastric

Microarray data, including gene expression data and miRNA expression data, 1 pair of data for normal and tumor tissues of stage I gastric cancer in humans, 10 pairs of data for normal and tumor tissues of stage II gastric cancer in humans, and 15 pairs of data for normal and tumor tissues of stage III gastric cancer in humans, are available (Kim et al. 2015). According to The Cancer Genome Atlas (TCGA; https://tcga-data.nci.nih.gov/tcga/), the mRNA and miRNA expression data including 17 for normal tissues and 163 for tumor tissues of stage I liver cancer, 7 for normal tissues and 82 for tumor tissues of stage II liver cancer, 8 for normal tissues and 80 for tumor tissues of stage III liver cancer, and 1 for normal tissues and 6 for tumor tissues of stage IV liver cancer are available. The candidate associations for miRNA–gene regulatory associations are obtained from the TargetScan (Agarwal et al. 2015; Friedman et al. 2009; Garcia et al. 2011; Grimson et al. 2007), and TF–gene regulatory associations are obtained from the Human Transcriptional Regulation interactions database (HTRIdb) (Bovolenta et al. 2012) and integrated transcription.

2.2 Data Selection and Processing for Constructions IGECN in CD4+ T Cells of Humans with HIV-1 Infection

Mohammadi and colleagues (Mohammadi et al. 2013) have reported gene expression data and miRNA expression data in HIV-1-infected cluster of differentiation 4+

(CD4+) T cells and non-infected (normal) cells to analyze host–pathogen interactions at early (or reverse transcription), middle (integration/replication), and late (virus assembly/budding) infection stages.

2.3 *The Construction of the Stochastic System Models in IGECN by Systems Biology Methods*

Our goal is to construct an IGECN to integrate the transcription regulation mechanisms, miRNA regulations, PPIs, and epigenetic regulation via DNA3 methylation and then to investigate whole cellular mechanisms under specific biological conditions. A flowchart of the proposed systems biology method to construct the IGECN is described in Fig. 2. Several kinds of NSG omics data and information gained from databases are combined together, including microarray gene expression data, PPI data, miRNA expression data, and miRNA–target gene association data, to be considered as the input for the proposed IGECN construction method. From these data, the candidate GRN, PPIN, and miRNA regulatory networks using the broad TF–gene regulation pool, PPI pool, and miRNA regulatory pool, respectively, under different biological conditions, were retrieved to construct a candidate IGECN. In this chapter, NGS data with gene expression and miRNA profiles are used to validate these integrated TF–gene regulation data, PPIs, and miRNA regulation data in the candidate IGECN. Therefore, stochastic dynamic system models are employed to identify interactions and regulatory mechanisms in the candidate IGCEN via these gene expression and miRNA expression profiles using NGS data under a specific biological condition of interest, such as cancer or infectious disease.

In the transcriptional regulation network (GRN) of the candidate IGECN, the candidate GRN was depicted as a stochastic dynamic system in which the regulations of TFs and miRNAs are considered as the inputs and gene expression levels of target genes are regarded as outputs (Fig. 1). For target gene i in the candidate GRN, the stochastic dynamic regulatory model of the target gene is described by the following stochastic discrete dynamic equation:

$$x_i(t+1) = x_i(t) + \sum_{j=1}^{L_i} a_{ij} y_j(t) - \sum_{l=1}^{L_L} c_{il} x_i(t) s_l(t) - \lambda_i x_i(t) + k_i + w_i(t) \qquad (1)$$

where $x_i(t + 1)$ is the mRNA expression level of the ith target gene at time $t + 1$, a_{ij} is the regulatory ability of the jth TF to the ith target gene (with a positive or negative value indicating the activation or repression), the regulatory ability a_{ij} is also effected by the DNA methylations of the corresponding gene (Huang et al. 2014), $y_i(t)$ denotes the regulation function of the jth TF binding to the target gene i, λ_i is the degradation rate from the present time t to time $t + 1$, k_i is the basal level (which is also related to the methylation of gene i if k_i becomes large), $s_l(t)$ is the expression level of the lth miRNA, the parameter c_{il} denotes the rate of miRNA–mRNA

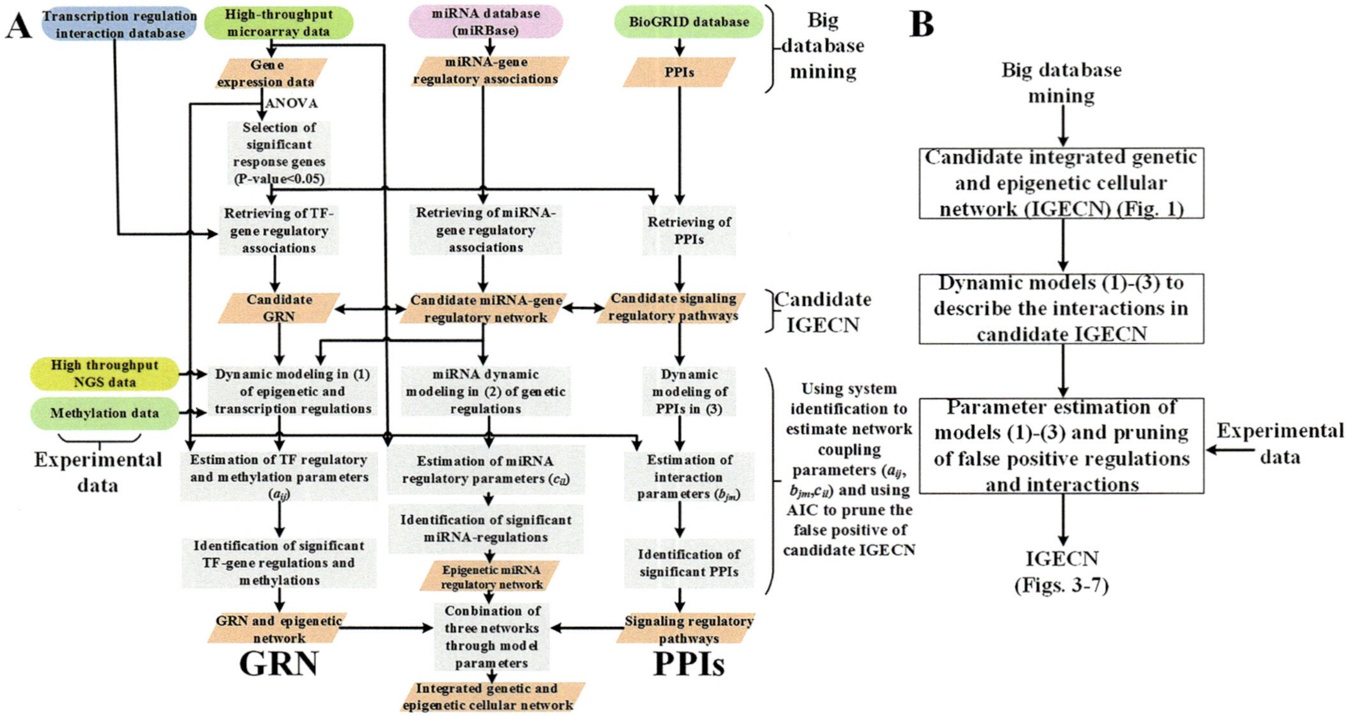

Fig. 2 Flowchart (**a**) and block diagram (**b**) of the proposed systems biology method to integrate genetic and epigenetic cellular networks via big database mining and NGS omics data (Originally published under an Open Access CC BY 4.0 license here https://doi.org/10.1186/s12918-016-0256-5)

coupling and is related to the energy of RNA–RNA binding (Mückstein et al. 2006), and $w_i(t)$ is the stochastic noise due to modeling residue measurement noise and fluctuation in the expression profile of the target gene. The biological interpretation of Eq. (1) from systems biological perspective denotes that the mRNA expression $x_i(t + 1)$ of the target gene i at time $t + 1$ represents a function of the present mRNA expression $x_i(t)$ at time t and mRNA expression due to transcriptional regulation of J_i TFs that bind to the target gene at time t, minus the inhibitory regulation of L_i miRNAs at time t, minus the mRNA due to the self-degradation over time, plus the basal level of mRNA expression due to some unknown regulation, for example, methylation, and stochastic noise due to measurement noise and random fluctuations in the expression profile of target gene i.

In the candidate miRNA regulatory sub-network, the stochastic dynamic regulatory model of the lth miRNA can be described by the following stochastic discrete time dynamic equation (Nandi et al. 2009):

$$s_l(t+1) = s_l(t) - \sum_{i=1}^{I_l} c_l s_l(t) x_i(t) - r_l s_l(t) + k_l + \varpi_l(t) \tag{2}$$

where the degradation of miRNA, i.e., $\sum_{i=1}^{I_l} c_{il} s_l(t) x_i(t)$ is accompanied by mRNA degradation events induced by the miRNA in (1). That is, the lth miRNA induces the degradation of I_l mRNAs at present. r_l represents the self-degradation rate, k_l is the production rate of the lth miRNA, and $\varpi_l(t)$ represents the stochastic noise due to measurement noise and random cellular fluctuation.

In the PPI sub-network, the candidate PPIs can be described as a stochastic dynamic system in which expressions of mRNA and TFs are the input/output of the system (Fig. 1). For a target protein j in the candidate PPI sub-network, the stochastic dynamic model of protein expression level is given as follows:

$$y_j(t+1) = y_j(t) - \sum_{m=1}^{M_j} b_{jm} y_j(t) y_m(t) + \alpha_j x_j(t) - \beta_j y_j(t) + h_j + \zeta_j(t) \tag{3}$$

where $y_i(t)$ denotes the expression level at time t of the target protein, j, b_{jm} represents the interaction ability between the mth interactive protein $y_m(t)$ and the target protein, $y_i(t)$, a_j denotes the effect of protein synthesis from mRNA, $x_j(t)$ represents the mRNA expression level of the jth gene, h_j denotes the basal expression level, and $\zeta_j(t)$ is the stochastic noise. The rate of formation of the protein complex $y_j(t)y_m(t)$ is proportional to the product of the concentration of each protein (Ferreira et al. 2003; Voit 2000), i.e., it is proportional to the probability of two molecular collisions; thus, the protein interaction complex can be considered as a bilinear multiplication scheme.

The biological meaning of Eq. (3) is that the expression level of the target protein j at time $t + 1$ can be considered as a function of the protein expression level at time t, minus the proteins to take part in protein complex interactions with M_j proteins, plus the proteins due to just translation effect from mRNA, minus the present protein degradation, plus the basal protein level from other sources and stochastic interactions. It should be noted that no directionality exists for interacting proteins in the PPI sub-network.

The regulations and interactions between genes, proteins, miRNA, and DNA methylation in the candidate IGECN are described in the following. Some transcriptions $y_j(t)$ of TFs at the end of signal transduction pathways in PPINs can regulate target genes according to the regulation function $a_{ij}y_j(t)$ with the transcription regulatory ability a_{ij} in Eq. (1), and the expression of mRNAs of target genes is also degraded by some miRNAs $s_l(t)$. The regulated genes can influence the corresponding protein expression level via translation from mRNA $x_j(t)$ to protein $\alpha_j x_j(t)$ with the translation rate α_j in (3). An mRNA degradation event, $-c_{il}x_i(t)s_l(t)$ due to the miRNAs $s_l(t)$, an epigenetic regulation described in (1), induced by miRNA $s_l(t)$, also causes the degradation of miRNA, $-c_{il}s_l(t)x_i(t)$, in (2). Epigenetic DNA methylation has also an influence on the regulation parameter a_{ij} and basal level k_i in (1) and the translation rate a_j from mRNA to protein in (3). The genetic and epigenetic regulations in PPIN and GRN interplay among genes, proteins, miRNAs, and DNA methylation in the candidate IGECN can be found in (1)–(3) and Fig. 1, in which TFs are at the interface between PPINs and GRNs and genes are at the interface between GRNs, miRNA, PPINs, and DNA methylation in the candidate IGECN.

From the above stochastic dynamic models (1)–(3) in the candidate IGECN, the candidate GRN can be linked through the transcription regulatory parameters a_{ij} in (1) between genes and their possible regulatory TFs (Fig. 1), and the candidate PPIN can be linked through the protein interaction parameters b_{jm} in (3) between possible interacting proteins and can be linked with candidate GRN through possible regulatory TFs. The candidate GRN and miRNA regulatory networks in the candidate IGEN are linked through with candidate PPIN the miRNA–mRNA coupling parameters c_{il} in (1) and (2). Since the candidate IGECN only indicates potential TF–gene regulatory effects, PPIs, and miRNA–mRNA couplings based on data collected by database mining, they should be verified by the corresponding gene expression profile, protein expression profile, and miRNA expression profile in the real NGS data. The regulatory effects a_{ij} and c_{il} in (1), interaction b_{jm} in (3), and regulation c_{il} in (2), which represent the edges of the candidate IGECN, should be identified and confirmed by empirical gene, protein, and miRNA expression data of cancers or infectious diseases. These regulatory and interactive parameters should be evaluated using temporal data using the constrained least square parameter estimation method in the sequel. Due to the limited number of temporal data points, to avoid overfitting when identifying a large amount of parameters in (1)–(3) of candidate IGECN by the proposed constrained least square parameter estimation method, the cubic spline method (Wang and Chen 2010) was also employed to interpolate extra time points for gene expression data. The gene regulatory parameters a_{ij} and c_{il} in (11) are identified gene by gene (and PPI parameters b_{jm} in (3) protein by protein) so that the candidate IGECN identification process is not limited by the number of genes, proteins, and miRNAs in the candidate IGECN. In other words, the regulatory parameters a_{ij} and c_{il} in (1) are first identified for target gene i and then for target gene $i + 1$, $i + 2$, etc. The models in (1)–(3) only represent a small fraction of real biology. For example, many post-transcriptional regulators (RNA-binding proteins) and epigenetic regulations beyond methylation are also involved. The model uncertainty of a real biological system and the fluctuation of expression data are also involved in the stochastic noises $w_i(t)$ in (1), $\varpi_l(t)$ in (2), and $\zeta_j(t)$ in (3). A model order detection method such as AIC and statistical assessment like Student's t-test (Wang and Chen 2010) is used to cut off the false-positive regulations and interactions and then find out the true interaction and regulation number of each protein, gene, and miRNA in the candidate IGECN. AIC (Wang and Chen 2010; Akaike 1974; Johansson 1993) has been applied to detect the real regulations and interactions by pruning the false-positive regulations and interactions in the candidate GRN in (1), candidate miRNA regulatory network in (2), and the candidate PPI network in (3). AIC, which includes both estimated residual error and model complexity in one statistic, quantifies the relative goodness of the fitness of a model. The minimization achieved in AIC will indicate that the real network is with the true model order. By applying the AIC detection method, the insignificant parameters, a_{ij}, c_{il} *and* b_{jm}, out of network order detected by AIC, are considered false positives and should be pruned from candidate IGECN to obtain true IGECN. We then use Student's t-test to validate the significance of each regulatory parameter in (1)–(3) (Wang and Chen 2010). In this way, the candidate GRN, candidate PPIN, and candidate miRNA regulatory networks based on large-scale data mining are all pruned by the AIC detection method using the temporal NGS omics data, which induces the construction of an accurate GRN, PPIN, and miRNA regulatory network. Based on the interactions and couplings between GRNs and signal transduction pathways (or PPINs) via TFs, and among GRNs, miRNA regulatory networks, and PPINs via coupling genes, these three networks are coupled to consist of the IGECN as shown in Fig. 2.

A general method is proposed in Fig. 2 to construct the IGECN using large-scale database mining and system identification methods for cellular functions. In the following sections, two examples are given to illustrate the application of the IGECN system to investigate cellular mechanisms under different biological conditions. In the first example, an IGECN is constructed to investigate cellular mechanisms for carcinogenesis under cancer conditions, and in the second example, an IGECN is constructed to investigate pathogenic and defensive cellular mechanisms under HIV infection conditions.

2.4 Estimation of the Regulation and Interaction Parameters in IGECN by NGS Data

After constructing the coupling dynamic models (1)–(3) in the candidate IGECN, the regulative and interactive parameters of the coupling dynamic models must be estimated using NGS data. The proposed identification strategy is to identify the IGECN gene by gene, protein by protein, and miRNA by miRNA. Before applying the identification method, we first carefully examine the dynamic models of genetic regulation, protein interaction, and miRNA regulation in IGECN. In (1), the basal level k_i should always be non-negative because the gene expressions are always non-negative. Because the parameters in (1) have certain constraints, the regulatory parameters are always identified by solving the following constrained least square parameter estimation problems.

The mRNA expression model (1) of target genes is rewritten in the following linear regression form:

$$x_i(t+1) = \begin{bmatrix} y_1(t) & \cdots & y_{J_i}(t) & x_i(t) & -s_1(t)x_i(t) & \cdots & -s_{L_i}(t)x_i(t) & 1 \end{bmatrix} \times \begin{bmatrix} a_n \\ \vdots \\ a_{iJ_i} \\ (1-\lambda_i) \\ c_{\text{n}} \\ \vdots \\ c_{\text{IL}} \\ k_i \end{bmatrix} + w_i(t) = \phi_i(t)\theta_i^1 + w_i(t) \tag{4}$$

where $\phi_i(t)$ denotes the regression vector that can be obtained from the processing above and θ_i^1 is the parameter vector of the expression of target gene i to be estimated. To avoid the overfitting in the parameter estimation process, when estimating the regulatory parameters, the cubic spline method (Bar-Joseph et al. 2002; De Boor 1978; Faires and Burden 1998) is also employed to interpolate extra time points (at least double than the number of parameters to be estimated) for $y_i(t)$, $x_i(t)$ and $s_i(t)$ from NGS data.

The mRNA model in (4) at different time points can be arranged as follows:

$$\begin{bmatrix} x_i(t_2) \\ x_i(t_3) \\ \vdots \\ x_i(t_L) \end{bmatrix} = \begin{bmatrix} \phi_i(t_1) \\ \phi_i(t_2) \\ \vdots \\ \phi_i(t_{L-1}) \end{bmatrix} \theta_i^1 + \begin{bmatrix} w_i(t_1) \\ w_i(t_3) \\ \vdots \\ w_i(t_{L-1}) \end{bmatrix} \tag{5}$$

where L is the number of NGS data expression time points after cubic spline interpolation.

We denote the symbols X_i, Φ_i, and W_i to represent (5) as follows for simplicity:

$$X_i = \Phi_i \theta_i^1 + W_i \tag{6}$$

The constrained least square parameter estimation problem θ_i^1 in (6) is proposed as follows:

$$\min_{\theta_i^1} \left\| \Phi_i \theta_i^1 - X_i \right\|_2^2 \quad \text{subject to} \quad A\theta_i^1 \le b \tag{7}$$

where $A = \begin{bmatrix} \overbrace{\begin{matrix} 0 & \cdots & 0 \\ \vdots & \ddots & \vdots \\ 0 & \cdots & 0 \end{matrix}}^{J_i+1} & \overbrace{\begin{matrix} -1 & 0 & 0 \\ 0 & \ddots & 0 \\ 0 & 0 & -1 \end{matrix}}^{L_i+1} \end{bmatrix}$, and $b = [0\ldots0]^T$ that gives the constraints to force the miRNA inhibition $-c_{il}$ to be always non-positive and the basal level k_i to be always non-negative in (1), i.e., $-c_{il}$, ≤ 0 and $k_i \ge 0$. The constrained least square problem in (7) can be solved using the active set method for quadratic programming (Coleman and Hulbert 1989; Gill et al. 1984).

Similarly, the miRNA model in (2) can be rewritten in the following regression form:

$$\begin{aligned} s_l(t+1) &= \begin{bmatrix} s_l(t) & -s_l(t)x_1(t) & \cdots & -s_l(t)x_{I_l}(t) & 1 \end{bmatrix} \begin{bmatrix} 1-n \\ c_{1l} \\ \vdots \\ c_{I_l l} \\ k_l \end{bmatrix} + \varpi_l(t) \\ &= \psi_l(t)\theta_l^2 + \varpi_l(t) \end{aligned} \tag{8}$$

where $\psi_l(t)$ indicates the regression vector and θ_l^2 denotes the parameter vector of miRNA model to be estimated. By cubic spline interpolation method at different time points to avoid overfitting in the parameter procedure, (8) can be rewritten as the following equation:

$$S_l = \psi_l \theta_l^2 + W_l \tag{9}$$

The parameter estimation problem then should be the constrained least square problem as follows:

$$\min_{\theta_l^2} \left\| \Psi_l \theta_l^2 - S_l \right\|_2^2 \quad \text{subject to} \quad B\theta_l^1 \leq b \tag{10}$$

where $B = \begin{bmatrix} 0 & \overbrace{\qquad}^{I_l+1} & -1 & 0 & 0 \\ 0 & & 0 & \ddots & 0 \\ 0 & & 0 & 0 & -1 \end{bmatrix}$ and $b = [0 \ldots 0]^T$ that gives the constraint on the parameter estimation problem in (10) to force the miRNA degradation $-c_{il}$ to be always non-positive and the basal level k_e to be always non-negative in (2), i.e., $-c_{il} \leq 0$ and $k_l \geq 0$. Finally, the protein interaction model in (3) uses the same way as the above to make sure $-b_{jm} \leq 0$ and $\alpha_j h_j \geq 0$. According to the large-scale measurement of protein activities, 73% of the variance in protein abundance can be illustrated by mRNA abundance (Lu et al. 2007). Therefore, mRNA expression levels are frequently used as a substitute for protein expression levels. If the simultaneously measured genome-wide protein expression data and mRNA expression data in each cancer stage are available, the general dynamic models in (1)–(3) can also be employed to identify the real IGECN network in cancer. After the parameter estimation problem for candidate IGECN is solved, we can estimate the IGECN. For example, we estimate the regulatory parameter $a_{RBM5,\ ETS1} = -0.22$ from the TF ETS1 to the target gene *RBM5*, the interaction parameter $b_{PGR,\ NCOA2} = 0.06$ between the two proteins PGR and NCOA2, and the coupling rate $c_{USP15,\ MIR381} = 0.001$ between the miRNA, mir381, and the miRNA USP15 in gastric cancer cells.

By using Student's *t*-test (Seber and Lee 2003), the *p*-values for the estimated parameters, including the regulatory ability a_{ij} in (1), the miRNA–mRNA coupling rate c_{il} in (1) and (2), and the protein interaction ability b_{jm} in (3), are all calculated to determine the significance of the parameters. By the one-way ANOVA method, the *p*-value is calculated for each gene to determine the significance of expression change between normal and tumor cells in cancers and between normal and HIV-infected human cells. The human DNA methylation profiles of liver cancers (including 41 normal and 369 tumor cells) and gastric cancer (including 2 normal and 396 tumor cells) are available in TCGA. A one-way ANOVA has been also used to calculate the *p*-value of each gene and determine the significance of DNA methylation change between normal and tumor cells in each cancer.

3 Results

The construction of IGECN is very important in examining the whole cellular mechanisms of specific biological conditions via big database mining and network identification via NGS data. The following two examples can explain how to construct IGECN to investigate the roles of genetic and epigenetic methylation and miRNA dysregulations in gastric tumorigenesis and pathogenesis in HIV infection. With these identified significant core network biomarkers in IGECN, some potential multiple drug targets are also proposed for the development of therapeutic treatment of the corresponding disease from systems biology perspective.

3.1 *Construction of IGECN for Investigating Significant Cellular Mechanisms in Gastric Tumorigenesis and Hepatocarcinogenesis*

Cancer is a complex and heterogeneous disease with a highly robust ability against therapeutic interventions. It is a constellation of diverse and evolving disorders due to the uncontrolled proliferation of cells that may eventually lead to inevitable cellular dysfunctions. Though some cancer subtypes can be cured by early diagnosis and specific treatment at present, there is no effective treatment being established for many subtypes. Because of the complex, heterogeneous, and evolving nature of cancer, it is appealing to construct an IGECN to investigate the carcinogenesis process via large-scale data mining, systematic genetic modeling, and various types of omics and temporal NGS data.

Since the epigenetic modifications are important in carcinogenesis, IGECNs are constructed by using NGS data collected at various stages of cancer as the temporal information for gastric tumorigenesis (Fig. 3) and hepatocarcinogenesis (Fig. 4) to confirm large differences between miRNA regulations of normal and tumor tissues in gastric tumorigenesis. The results (Fig. 3) show that two cell cycle genes, ubiquitin C (UBC) and ataxia telangiectasia mutated (ATM), and one chromosomal translocation gene, CREB-binding protein (CREBBP), play important roles in gastric tumorigenesis via genetic and epigenetic regulation. ATM is an important cell cycle checkpoint kinase (Abraham 2001), and chromosomal translocation plays an important role in gastric tumorigenesis (McMahon 2010). Seven genes are identified with large basal-level differences between cancer and normal cells, including *RUNX1* (p-value $<1.4 \times 10^{-3}$), *ARHGDIB* (p-value $<1.6 \times 10^{-2}$), *PSME*1 (p-value $<4 \times 10^{-2}$), *VAMP5* (p-value $<<2 \times 10^{-2}$), *POLG* (p-value $<1.3 \times 10^{-2}$), *GZMB* (p-value $<1.44 \times 10^{-2}$), and *RBM5* (p-value $<8.2 \times 10^{-3}$), which have been found to be mainly due to DNA methylation of their corresponding genes. To validate these findings, the epigenomics data of human gastric cancer in TCGA is employed for conformation. Based on DNA methylation profiles of gastric cancer cells and normal gastric cells from TCGA, five corresponding genes are

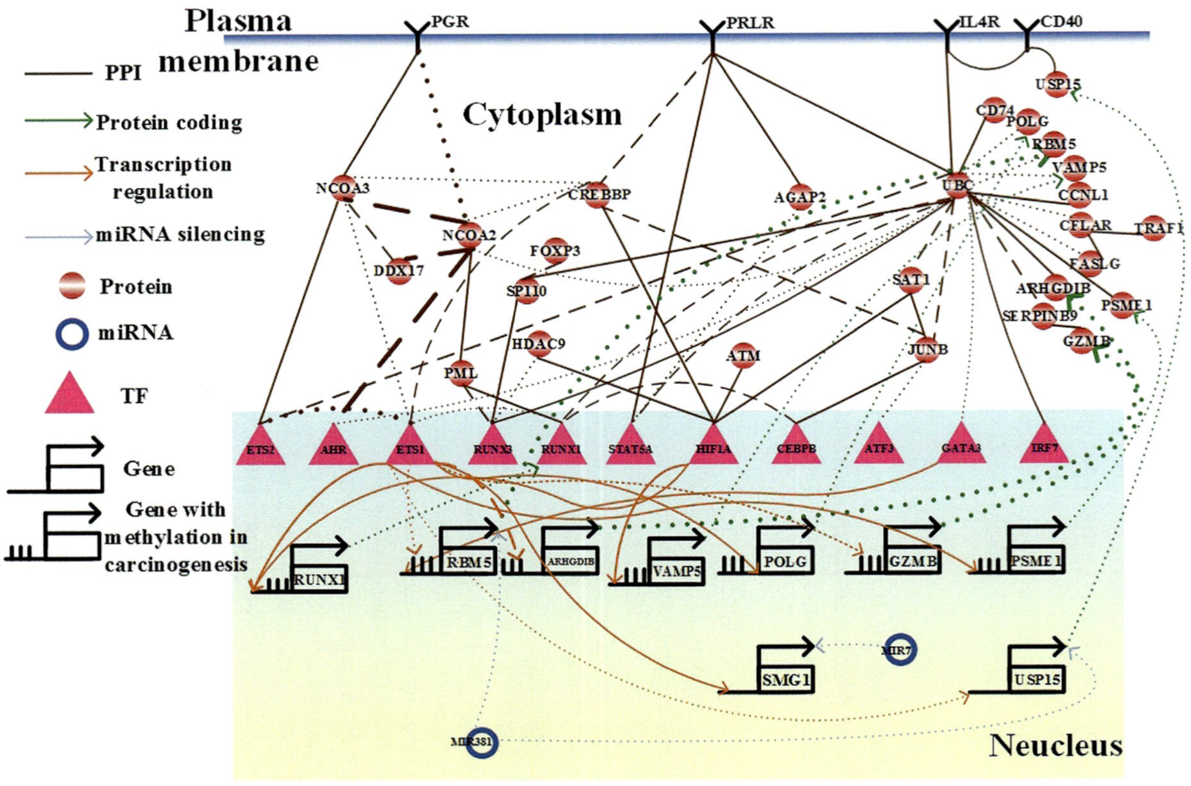

Fig. 3 The integrated genetic and epigenetic cellular network (IGECN) is used to investigate the cellular mechanisms of normal and tumor tissues in gastric cancer. The IGECN includes the regulations and interactions in tumor cells (dotted lines), normal cells (dashed lines), and both (solid lines). The bold lines indicate large genetic and epigenetic regulations or genetic regulation. Dysregulation of mir7 leads to the initiation and progression of inflammation-induced gastric cancer, and PML, a potential biomarker as drug target, can overcome platinum resistance in platinum-based chemotherapy in gastric cancer (Originally published under an Open Access CC BY 4.0 license here https://doi.org/10.1186/s12918-016-0256-5)

found to have methylation change, including *RUNX1* (p-value $<3.56 \times 10^{-1}$), *ARHGDIB* (p-value $<1.81 \times 10^{-1}$), *PSME*1 (p-value $<2.59 \times 10^{-1}$), *GZMB* (p-value $<1.79 \times 10^{-1}$), and *RBM5* (p-value $<3.94 \times 10^{-1}$), and two genes are found to have small methylation change including *VAMP5* (p-value $<8.61 \times 10^{-1}$) and *POLG* (p-value $<8.38 \times 10^{-1}$) between gastric cancer cells and normal gastric cells (Fig. 3).

Further, five genes are also identified to have obvious basal-level differences between hepatocellular carcinomas (HCCs) and normal cells, i.e., *ELAVL1* (p-value $<3.69 \times 10^{-5}$), *JMJD1C* (p-value $<3.59 \times 10^{-1}$), *GADD45A* (p-value $<7.32 \times 10^{-11}$), *EIF2AK3* (p-value $<5 \times 10^{-5}$), and *GSN* (p-value $<1.19 \times 10^{-2}$), which are mainly due to DNA methylation on their corresponding genes (Fig. 4). We have also used the epigenomics data of HCCs in TCGA to confirm these results. From the DNA methylation profiles of HCCs and normal liver cells from TCGA, five genes are also found to have significant methylation changes, i.e., *ELAVL1* (p-value $<5.7 \times 10^{-2}$), *JMJD1C* (p-value $<1.4 \times 10^{-2}$), *GADD45A* (p-value $<9.6 \times 10^{-2}$), *EIF2AK3* (p-value $<1.85 \times 10^{-6}$), and *GSN* (p-value $<8.71 \times 10^{-4}$).

3.2 *Construction of the IGECN to Investigate Pathogenic and Defensive Mechanisms of HIV-1-Infected Cells*

Human HIV was first identified in 1983 and rapidly came out as a virus in response to a pandemic. From the beginning of this century, the HIV epidemic has received the global attention, and therapeutic methods to halt and reverse the AIDS epidemic have become an important United Nations Millennium Declaration topic. Currently, about 35 million people are living with HIV worldwide. With the coming of highly active antiretroviral therapies, HIV can now be managed as a chronic disease. But how to develop a cure for HIV remains a significant challenge. Hence, how investigating host–pathogen interspecific regulatory mechanisms and interactions in cellular genetic and epigenetic systems is a very important research topic to get more insight into pathogenic and defensive mechanisms in the HIV infection process and to help control clinical pathogenic infections. Therefore, through two-sided time-profile HIV-human high-throughput sequence NGS data, RT-PCR data, miRNA data, and other omics data, the interspecific host–pathogen IGECN is constructed for investigating the pathogenic and defensive mechanism in the HIV infection process. To construct this HIV infection model in the following, the coupling dynamic (Eqs. (1)–(3)) should be modified with the pathogen HIV in addition to the host species as follows:

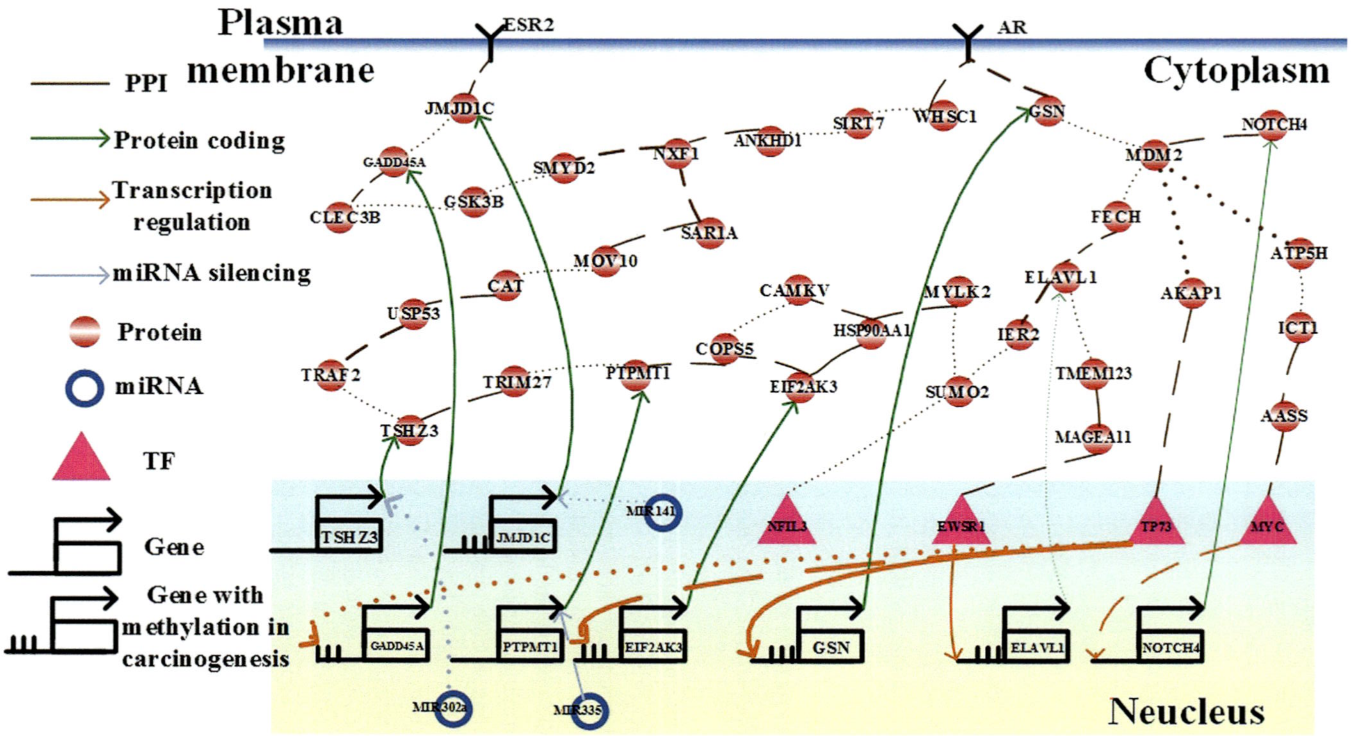

Fig. 4 IGECN is constructed to investigate the cellular mechanisms of normal and tumor tissues in liver cancer. The IGECN includes the regulations and interactions in tumor cells (dotted lines), normal cells (dashed lines), and both (solid lines). The bold lines indicate large genetic and epigenetic regulations or genetic regulations. Dysregulations of mir141 and mir335 contribute to the dysfunction of DNA damage response and cell proliferation in human HCC, respectively (Originally published under an Open Access CC BY 4.0 license here https://doi.org/10.1186/s12918-016-0256-5)

$$
\begin{aligned}
x_i(t+1) &= x_i(t) + \sum_{j=1}^{J_i} a_{ij}y_j(t) - \sum_{l=1}^{L_l} c_{il}x_i(t)s_l(t) + \sum_{o=1}^{O_i} d_{io}h_o(t) - \lambda_i x_i(t) + k_i \\
&\quad + w_i(t), \\
y_j(t+1) &= y_j(t) - \sum_{m=1}^{M_j} b_{jm}y_j(t)y_m(t) - \sum_{o=1}^{O_j} g_{jo}y_j(t)h_o(t) + \alpha_j x_j(t) - \beta_j y_j(t) + h_j \\
&\quad + \zeta_j(t), \\
h_o(t+1) &= h_o(t) - \sum_{j=1}^{J_o} g_{jo}h_o(t)y_j(t) - \sum_{l=1}^{L_o} m_{ol}h_o(t)s_l(t) - K_o h_o(t) + \delta_o + \xi_o(t), \\
s_l(t+1) &= s_l(t) - \sum_{i=1}^{I_l} c_{il}s_l(t)x_i(t) - \sum_{o=1}^{O_l} m_{ol}s_l(t)h_o(t)n - r_l s_l(t) + k_l + \varpi_l(t)
\end{aligned}
\tag{11}
$$

where $h_o(t)$ denotes the expression of the other HIV proteins and the cross talk effect of HIV proteins $h_o(t)$ is also considered. After system identification of coupling IGECN between host and HIV in (11) by NGS data at different infection stages, three genes had differences found in their basal levels between infections and normal cells at the early infection stage, *TAF5* (p-value $<2.4 \times 10^{-1}$), *DDX3X* (p-value $<2 \times 10^{-1}$), and *CELF1* (p-value $<9.47 \times 10^{-2}$); two genes are found with basal-level differences between infections and normal cells at the middle stage, *TAF5* (p-value $<5.8 \times 10^{-2}$) and *CELF1* (p-value $<2.07 \times 10^{-7}$); two genes are found with basal-level differences between infections and normal cells at the late stage, *ZNF451* (p-value $<9 \times 10^{-2}$) and *CELF1* (p-value $<1.6 \times 10^{-3}$). These genes with different basal levels at different infection stages are found due to DNA methylation of stage-specific genes. A previous proof for DNA methylation, including *TAF5* (Wilhelm et al. 2012) and *DDX3X* (Sharma and Bhattacharya 2010) methylation in individuals with HIV infection, *ZNF451* methylation in individuals with Epstein–Barr virus (*EBV*) infection (Kang and Kieff 2015), and *CELF1* methylation in individuals with Moloney leukemia virus infection (Liu et al. 2015), can support these findings. However, DNA methylation of genes is found to be highly tissue-specific. If the methylation profiles of genes in HIV-infected and mock cells become available in the future, these findings can be particularly supported.

By applying the NGS data of normal and infected T cells of HIV-1, the IGECNs can be identified at the three infection stages (Figs. 5, 6, and 7). In the following paragraphs, we will discuss the contribution of miRNA dysregulation to the pathogenic mechanisms of gastric cancer, HIV-1 infection process, and liver cancer with respect to biological cellular functions. The results have applications to the identification of biomarkers as drug targets for the drug development of therapeutic treatments.

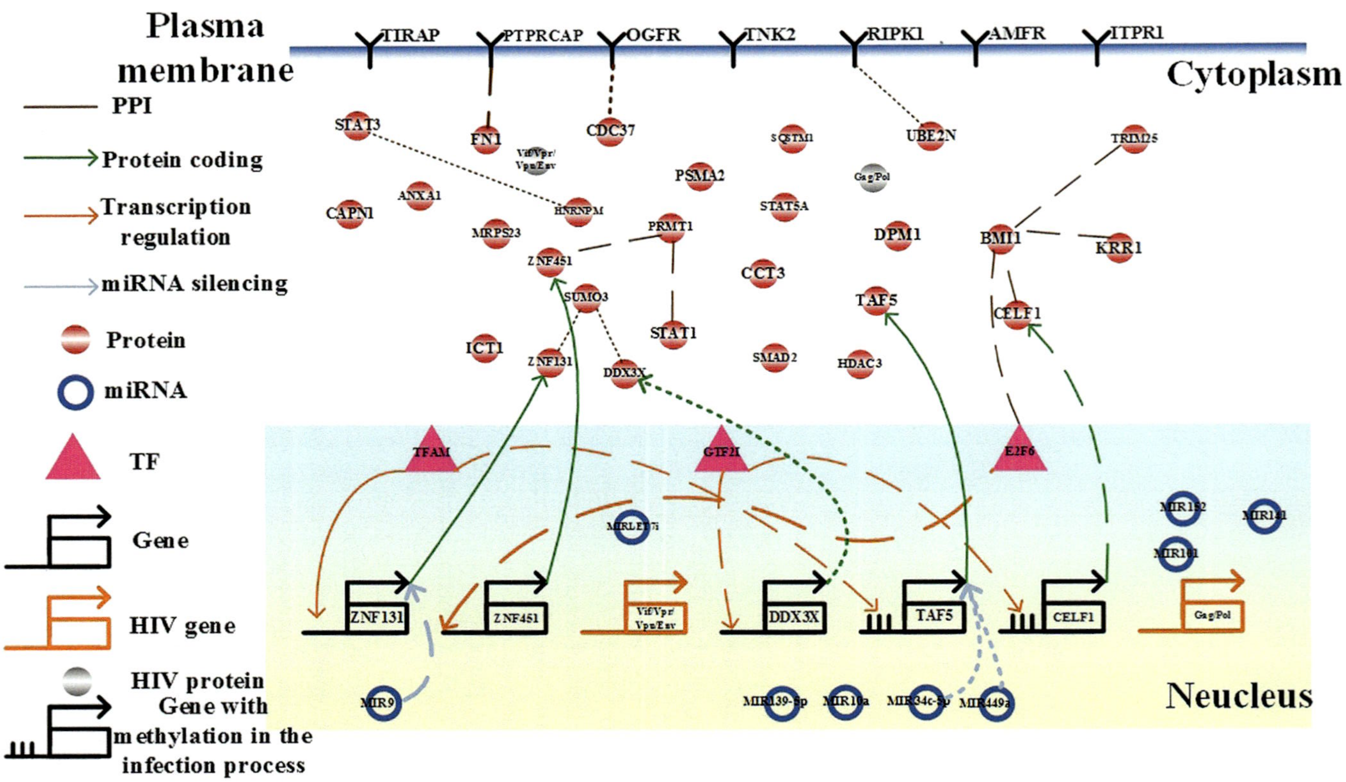

Fig. 5 The integrated genetic and epigenetic cellular network (IGECN) is used to search into the cellular mechanisms of both normal and HIV-infected T cells at the first HIV-1 infection stage, i.e., the reverse transcription stage. The IGECN includes regulations and interactions with HIV-infected cells (dotted lines), normal cells (dashed lines), and both cell types (solid lines). The bold lines denote large genetic and epigenetic regulations or genetic expression. Dysregulation of the signaling cascade from mir9 to *SUMO3* leads to HIV-1 infection, hijacking CD4+ T cells via cellular dysfunction of the immune and hormone pathways at the early HIV-1 infection stage (Originally published under an Open Access CC BY 4.0 license here https://doi.org/10.1186/s12918-016-0256-5)

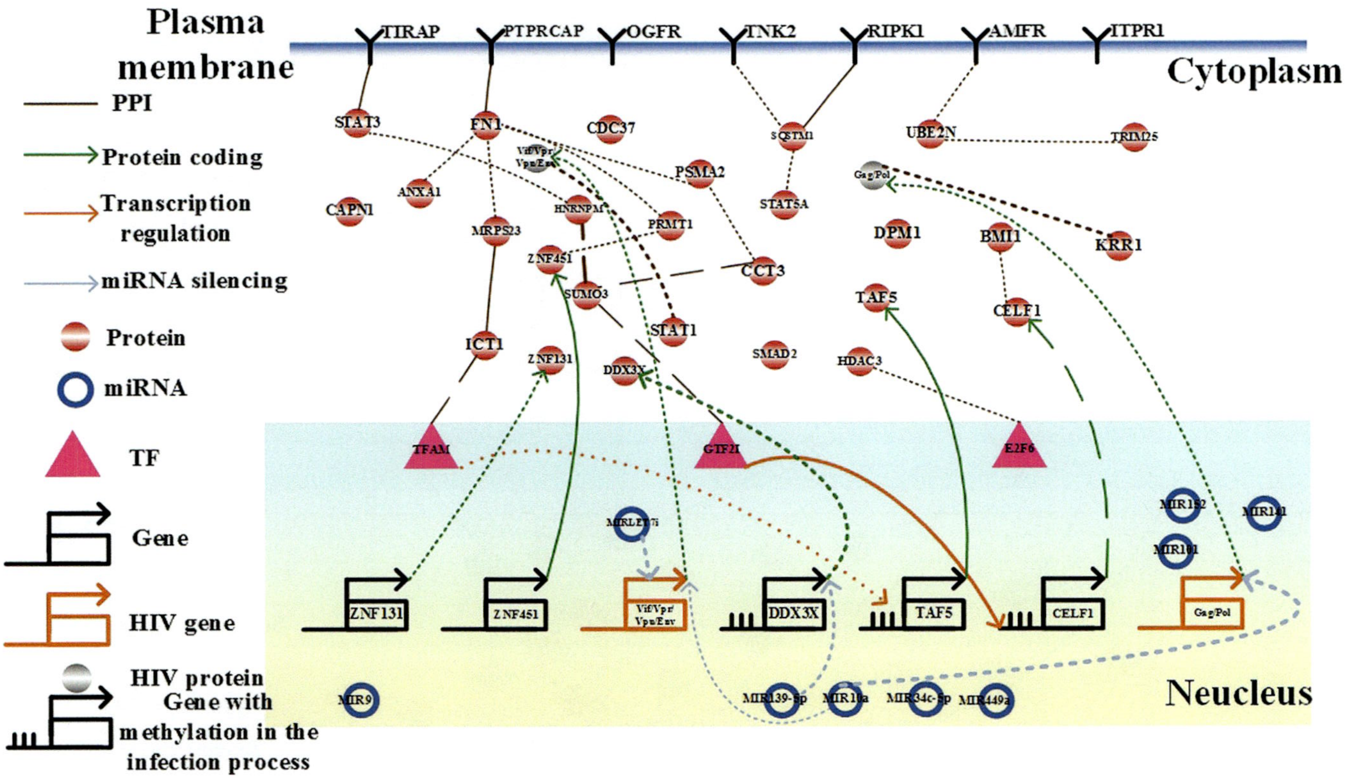

Fig. 6 The integrated genetic and epigenetic cellular network (IGECN) is used to search for the cellular mechanisms of both normal and HIV-infected T cells at the intermediate HIV infection stage, i.e., the integration/replication stage. The IGECN includes those regulations and interactions in HIV-infected cells (dotted lines), normal cells (dashed lines), and both cell types (solid lines). The bold lines denote large genetic and epigenetic regulations. Dysregulations of mir139-5p and mirLET7i lead to the dysfunction of both immune response and viral replication via DNA hypermethylation. Dysregulation of mirLET7i leads to dysfunction of the immune response and viral replication, too (Originally published under an Open Access CC BY 4.0 license here https://doi.org/10.1186/s12918-016-0256-5)

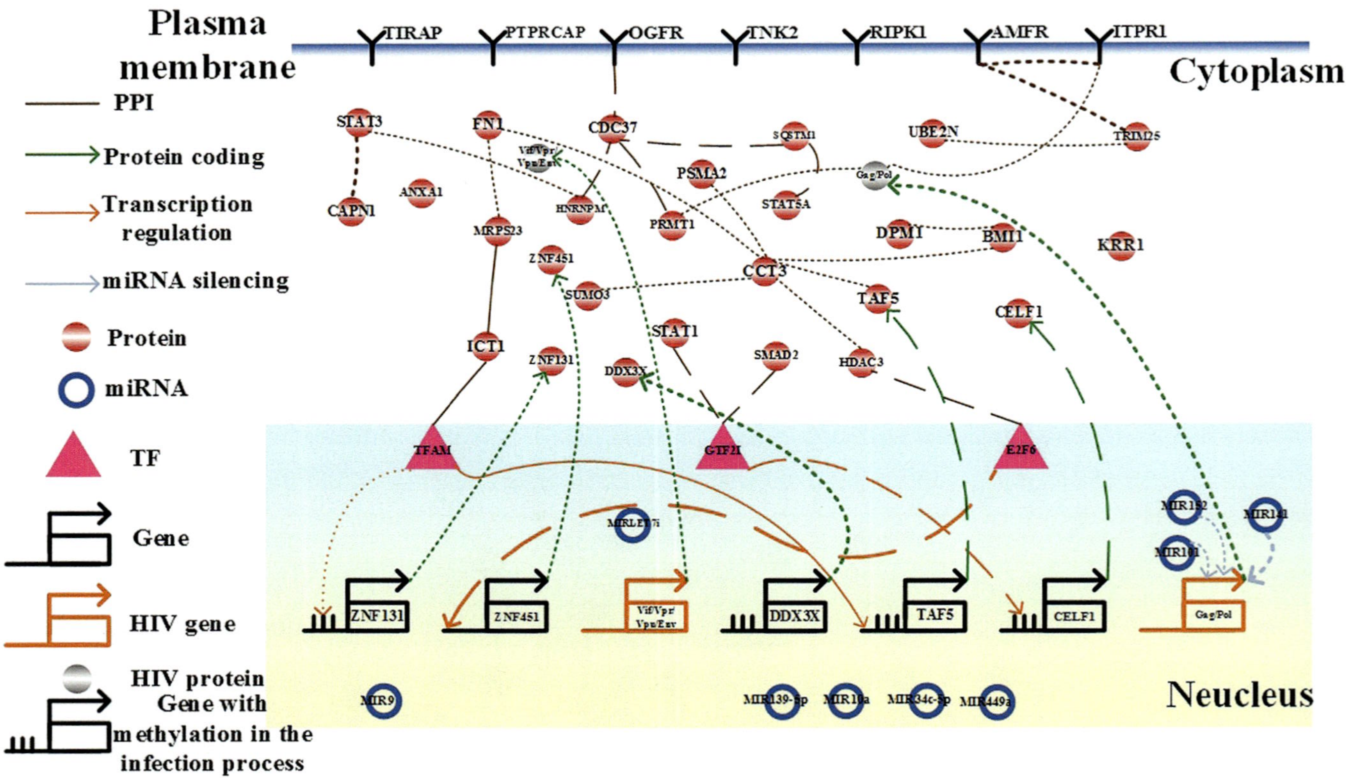

Fig. 7 The integrated genetic and epigenetic cellular network (IGECN) is used to investigate the cellular mechanisms of normal and HIV-infected T cells in the late HIV infection process, i.e., the virus assembly/budding stage. The IGECN includes those in HIV-infected cells (dotted lines), normal cells (dashed lines), and both cell types (solid lines). The bold lines indicate large genetic and epigenetic regulations. Dysregulation of mir101, mir141, and mir152 to the HIV-1 Gag protein leads to HIV-1 budding and release via DNA hypermethylation, ubiquitin transfer, and endoplasmic reticulum-associated degradation at the late infection stage of HIV-1 (Originally published under an Open Access CC BY 4.0 license here https://doi.org/10.1186/s12918-016-0256-5)

Briefly, the dysregulation of mir7 leads to the initiation and progression of inflammation-induced gastric cancer; the dysregulation of mir9 leads to HIV-1 infection to hijack CD4+ T cells through the dysfunction of immune and hormone pathways; the dysregulation of mir139-5p, mirLET7i, and mir10a leads to the HIV-1 integration/replication stage via DNA hypermethylation and immune system dysfunction; the dysregulation of mir101, mir141, and mir152 leads to the HIV-1 virus assembly/budding stage through DNA hypermethylation, ubiquitin transfer, and endoplasmic reticulum-associated degradation; the dysregulation of mir302a leads to both the microvesicle-mediated transfer of miRNAs and the dysfunction of NF-κB signaling pathway in hepatocarcinogenesis. Therefore, these miRNAs have the potential to be selected as drug targets for the corresponding therapeutic treatments in the future.

4 Discussion

4.1 Gastric Cancer

mir7 Leads to the Initiation and Progression of Inflammation-Induced Gastric Cancer

Until now, it has been proposed that miRNAs play important roles in regulating proliferation and apoptosis to promote gastric cancer cells' invasion, metastasis, and angiogenesis (Xie et al. 2014; Zhao et al. 2013; Gao et al. 2015). Even a few connections between genetic and epigenetic regulatory mechanisms in gastric cancer cells have been found independently using biological experiments (Xie et al. 2014; Zhao et al. 2013; Gao et al. 2015; Zheng et al. 2013), and the precise biological functions of miRNAs in gastric cancer cells and miRNAs affecting gastric cancer progression need to be confirmed by comparing the IGECN with that of normal cells.

In this chapter, we have identified the IGECN, including the largest differences in miRNA regulation between normal and cancer cells (Fig. 3). We have also identified that mir7 represses SMG1 (p-value$<10^{-16}$) in cancer cells. It has been reported that both mir7 and SMG1 are involved in the initiation and progression of inflammation-induced gastric cancer (Cao et al. 2015; Roberts et al. 2013). The significant expression changes of mir7 (p-value < 0.151) and SMG1 (p-value < 0.009) between normal gastric cells and gastric cancer cells have also supported the result that mir7 and SMG1 may play an important role in human gastric carcinogenesis.

It is also confirmed that the expression change of mir381 (p-value < 0.83) leads to the expression changes of RNA-binding protein 5 (*RBM5*) (p-value < 0.008) and ubiquitin-specific peptidase 15 (USP15) (p-value $<6 \times 10^{-4}$) through significant miRNA regulations (p-value $<10^{-16}$) in cancer cells. The activities of *RBM5* and USP15 have been also significantly activated in gastric cancer. It has been reported that the reduced *RBM5* and USP15 expressions will appear in tumor cell proliferation (Ishida 2006; Peng et al. 2013) and will induce tumor cell apoptosis (Zou et al. 2014), respectively. It has also been suggested that the dysregulation of two proteins, IL4 and CD40, contributes to the dysfunction of cell proliferation in mantle cell lymphoma (Zou et al. 2014), and the dysregulation of mir381 contributes to the dysfunction of proliferation in esophageal cancer (Zhou et al. 2015). Thus, it has been proposed that the dysregulation of miRNAs in gastric cancer contributes significantly to the dysfunction of proliferation and apoptosis in gastric cancer via the ubiquitin–proteasome system (Mani and Gelmann 2005). TCGA can offer a multilayered insight into genomics and epigenomics data of many human cancer types. Based on DNA methylation profiles of gastric cancer cells and normal gastric cells from TCGA, the methylation changes of *GZMB* (p-value < 0.18), *ARHGDIB* (p-value < 0.18), *PSME*1 (p-value < 0.26), *RUNX1* (p-value < 0.36), *RBM5* (p-value < 0.39), *POLG* (p-value < 0.84), and *VAMP5* (p-value < 0.86) are found by the IGECN. Although the methylation changes of *POLG* and *VAMP5* are insignificant, the gene expression changes of *POLG* (p-value < 0.01) and *VAMP5* (p-value < 0.02) are still significant. Therefore, we could suggest that other regulators may exist to regulate the expressions of *POLG* and *VAMP5* in gastric cancer.

PML, a Potential Drug Target, Conquers Platinum Resistance in Platinum-Based Chemotherapy in Gastric Cancer
Even platinum-based chemotherapy has been employed to treat most malignant diseases, and platinum resistance still limits its efficacy. The nuclear receptor coactivator family (NCOA) and the p160 steroid receptor coactivator family (SRC) in gastric cancer cells have shown to exhibit significant differences in interactions which have been observed in the IGECN than in normal cells. SRC3 has been proposed to play an important role in intrinsic resistance to platinum therapy in cancers (Palmieri et al. 2013). This reveals that the dysregulation of NCOA2 is transduced by signaling cascades via promyelocytic leukemia (PML) and *RUNX1*, which is found to be transcriptionally regulated by v-ets erythroblastosis virus E26 oncogene homolog 1 (ETS1). Therefore, ETS1 has been suggested to play an important role in regulating gastric cancer cell invasion, metastasis, and angiogenesis (Zheng et al. 2013; Nishida et al. 2007). Since *RUNX1* plays an important role in developing human hematopoietic stem cells (Lacaud et al. 2002), it has been shown that PML is considered as a potential drug target that can overcome platinum resistance in the platinum-based chemotherapy process in gastric cancer. It has also been proposed that PML can mediate glioblastoma resistance to the mammalian target in the process of rapamycin-targeted therapies (Iwanami et al. 2013). Consequently, two pharmaceutical drugs, i.e., arsenic trioxide and tretinoin, are considered to potentially target PML to overcome platinum resistance in the therapeutic process of gastric cancer.

4.2 *Hepatocellular Carcinoma (HCC)*

Dysregulations of DNA Methylation and mir141 Can Lead to Dysfunction of DNA Damage Response in HCC
HCC often occurs in chronic disease that can move up DNA damage and chromosomal aberrations (Jemal et al. 2011). Recently, the overexpressed mir141 has been found to lead to extensive DNA damage (Itahana et al. 2013), and *JMJD1C* and *TP73* are the signaling proteins of the DNA damage response pathway with implications for tumorigenesis (Watanabe et al. 2013; Zheng et al. 2010). The accumulated alterations in the DNA damage response pathway are found to improve hepatocarcinogenesis (Lamy et al. 2013; Yang et al. 2014). In this chapter, it has been found that the three genetic and epigenetic regulations, including the gain of the miRNA regulation of mir141 to *JMJD1C* (p-value <0.789) and the genetic regulation of *TP73* to *GADD45A* (p-value <0.037) in cancer cells and DNA methylations of *JMJD1C* and *GADD45A*, can lead to both changes of gene expression profiles of *JMJD1C* (p-value $<7.3 \times 10^{-11}$) and *GADD45A* (p-value $<7.3x \times 10^{-11}$) as well as changes of DNA methylation profiles of *JMJD1C* (p-value < 0.014) and *GADD45A* (p-value < 0.096) between human normal liver cells and liver cancer cells (Fig. 4). These changes can also lead to the gain of PPI between *JMJD1C* and *GADD45A* (p-value < 0.48) in hepatocarcinogenesis. Hence, the dysregulations of DNA methylation and mir141 are suggested to give rise to the dysfunction of DNA damage response in HCC.

Dysregulation of mir335 Leads to Human HCC Cell Proliferation
Recently, we have that the downregulation of *PTPMT1* is sufficient to drive liver cancer cell death (Niemi et al. 2013) and *TRIM27* can regulate cell proliferation in the development process of liver cancer (Hatakeyama 2011). It has been identified that the change in the miRNA expression profile of mir335 (p-value < 0.808) can lead to a significant change in the gene expression profile of *PTPMT1* (p-value $<8.4 \times 10^{-4}$) between normal liver cells and liver cancer cells (Fig. 4). The expressions of *TRIM27* and *PTPMT1* are found to be significantly activated in HCC. The significant genetic changes of *TRIM27* (p-value $<8.4 \times 10^{-4}$) and *PTPMT1* and their interaction (p-value < 0.32) in HCC can improve random proliferation. Consequently, it is suggested that dysregulation of mir335 leads to human HCC cell proliferation.

Dysregulation of mir302a Leads Not Only to the Microvesicle-Mediated Transfer of miRNAs But Also to the Dysfunction of the NF-κB Signaling Pathway in Hepatocarcinogenesis
Microvesicles have been found to transfer miRNA between cells and alter cellular functions by influencing signaling pathways in hepatocarcinogenesis (Li et al. 2014). Microvesicles are found to be enriched with maladjusted miRNAs, which are found to regulate zinc finger proteins. It is identified that the change in the miRNA expression profile of mir302a (p-value < 0.734) contributes to the change in the gene expression profile of teashirt zinc finger homeobox 3 (*TSHZ3*) (p-

value < 0.189), which is also known as zinc finger protein 537 (ZNF537), between normal liver cells and liver cancer cells (Fig. 4). The gene expression change of *TSHZ3* can further contribute to the gene expression profile change of TNF receptor-associated factor 2 (*TRAF2*) (p-value $<1.73 \times 10^{-11}$), to mediate the cellular dysfunction of the nuclear factor-kappa B (NF-κB) signaling pathway in hepatocarcinogenesis. The dysfunction of NF-κB signaling pathway can lead to HCC angiogenesis and metastasis (Li et al. 2014). The suppression of *TRAF2* is identified to inhibit the proliferation of several cancer cells (Shen et al. 2015), and the oncogene in epithelial cancers, *TRAF2*, which can lead to the dysregulation of NF-κB signaling pathway, has also been confirmed (Shen et al. 2015). Consequently, the dysregulation of mir302a can contribute to the microvesicle-mediated transfer of miRNAs and the cellular dysfunction of the NF-κB signaling pathway in hepatocarcinogenesis.

Dysregulation of the Hormone Receptors, *ESR2*, and AR Leads to Hepatocarcinogenesis

The most recent studies have suggested that menopause, which is due to the growing estrogen deficiency and physiological aging, can increase the risk of immune system disorders, mitochondrial dysfunction, cellular senescence, and an imbalance between antioxidant formation and oxidative stress (Brady 2015). These physiologic and biochemical changes in those cellular functions lead to a direct effect on liver function and bring the development of liver disease. In this chapter, two hormone receptors, estrogen receptor 2 (*ESR2*) and androgen receptor (AR), have been identified to be involved in multiple changes in intracellular signaling cascades. The gene expression profile of *ESR2* is found to show the change (p-value < 0.091) between young and old normal liver cells and the change (p-value < 0.063) between young normal liver cells and young liver cancer cells (Fig. 4). We can find that the relatively small change in the gene expression profile of *ESR2* (p-value < 0.45) between old normal liver cells and old liver cancer cells implies that the gene expression change of *ESR2* can be necessary but not sufficient for hepatocarcinogenesis. It is also found that the significant change in the gene expression profile of *AR* not only between young normal liver cells and young liver cancer cells (p-value $<2.46 \times 10^{-6}$) but also between old normal liver cells and old liver cancer cells (p-value $<6.13 \times 10^{-7}$) implies that the gene expression change of AR may be necessary for hepatocarcinogenesis. The result can be confirmed by the most recent studies in hepatocarcinogenesis (Tian et al. 2015).

4.3 HIV-1 Infection

The Signaling Cascade from mir9 to *SUMO3* Leads to HIV-1 Infection to Hijack CD4+ T Cells Through Dysfunction of the Immune and Hormone Pathways at the Early Stage of HIV-1 Infection

Owing to the available HIV-1 mRNA gene expression data at the middle and late infection stages and the lack of HIV-1 protein expression data, the impact of HIV-1 proteins on the host cells cannot be identified directly. In this chapter, we can identify three IGECNs at early, middle, and late infection stages, respectively (Figs. 5, 6, and 7), to investigate the largest differences in the miRNA regulation and HIV-1 protein interactions at the middle and late stages between normal and infected cells. At the early HIV-1 infection stage (Fig. 5), it is found that the expression changes of mir9 (p-value < 0.448) and general transcription factor III (*GTF2I*) (p-value < 0.96) can lead to the expression change of small ubiquitin-like modifier 3 (*SUMO3*) (p-value < 0.016) via the respective regulatory zinc finger protein 131 (ZNF131) (p-value $<9.3 \times 10^{-14}$), DEAD (Asp-Glu-Ala-Asp) box helicase 3, and X-linked (*DDX3X*) (p-value $<1 \times 10^{-16}$). By comparing miRNA expression profiles between the wildtype HIV-1-infected human CEMx174 lymphocytes and Tat RNA silencing suppressor in HIV-1-infected human CEMx174 lymphocytes (Hayes et al. 2011), the Tat RNA silencing suppressor is found to contribute to the expression change of mir9 (p-value < 0.035). Additionally, it has been proposed that Tat can cause mir9 to control inflammatory responses (Yao et al. 2014), and Tat can also regulate gene *GTF2I* expression during HIV infection (Taylor et al. 2011). Further, it has also been reported that SUMO modification of ZNF131 can also suppress estrogen signaling (Oh and Chung 2012), and HIV transcription can be suppressed by inducing estrogen signaling (Szotek et al. 2013). Consequently, it has been suggested that HIV infection leads to the dysregulation of immune and hormone pathways to hijack CD4+ T cells.

Dysregulation of mir34c-5p and mir449 Leads to the Frequency of Latent HIV Infection

Besides, we have also identified that the expression changes of mir34c-5p (p-value < 0.33) and mir449 (p-value < 0.50) lead to the expression changes of TATA-binding protein-associated factor 5 (*TAF5*) (p-value <0.25) via miRNA regulations (p-value $<1 \times 10^{-16}$). It has been also suggested that the genetic and epigenetic regulations in TATA protein binding may control the frequency of latent HIV infection (Miller-Jensen et al. 2013). Hence, we may suggest that genetic and epigenetic regulations in *TAF5* also lead to the control of HIV infection at the early stage.

Dysregulation of mir139-5p Leads to HIV-1 Replication via DNA Hypermethylation

At the intermediate HIV infection stage, we have identified that the expression change of mir139-5p (p-value < 0.23) leads to the expression change of *DDX3X* (p-value < 0.14) via miRNA regulation (p-value $<1 \times 10^{-16}$). The expression of *DDX3X* is also activated at the intermediate HIV infection stage. Therefore, we have suggested that *DDX3X* could regulate HIV replication through mir139-5p regulations (Mahboobi et al. 2015; Yedavalli and Jeang 2011).

Dysregulation of mirLET7i Contributes to Dysfunction of Both the Immune Response and Viral Replication of HIV-1

The HIV-1 Vif/Vpr protein has been reported to be significant for viral replication (Zahoor et al. 2014; Zhang et al. 2000). Therefore, it is determined that the Vif/Vpr protein is regulated by mirLET7i (p-value $<1 \times 10^{-16}$) and interacts with the human STAT1 protein (p-value $<1 \times 10^{-16}$) (Fig. 6). Hence, the expression change of mirLET7i (p-value < 0.079) leads to the expression change of Vif/Vpr protein (p-value < 0.318) during the intermediate HIV infection stage. Therefore, it has been supported by the previous observations that MIR-LET7i has high homology to HIV-1 sequences (90–100%) (Kanak et al. 2010), and the immune response of T cells induced by the Vpr-STAT1 interaction is also involved in regulating HIV-1 replication (Zahoor et al. 2014). Further, by comparing miRNA expression profiles between the wildtype HIV-1-infected human CEMx174 lymphocytes and Vif/Vpr RNA silencing suppressor in HIV-1-infected human CEMx174 lymphocytes (Hayes et al. 2011), Vif/Vpr RNA silencing suppressor is found to contribute to the expression change of mir7i (p-value < 0.107). Consequently, the miRNA regulation of mirLET7i to the Vif/Vpr protein is proposed to be involved in the induction of the immune response and viral replication in the HIV infection process (Fig. 6).

Dysregulation of mir10a Leads to Dysfunction of Immune Response in CD4+ T Cells and Viral Replication in the HIV Infection Process
In the second infection stage, thatmir10a is found to regulate HIV-1 proteins including Gag/Pol (p-value $<1 \times 10^{-16}$), which can significantly interact with small subunit (SSU) processome component, homolog (yeast) (KRR1) (p-value $<1 \times 10^{-16}$), being also known as HIV-1 Rev-binding protein 2 (HRB2) (Fig. 6). Consequently, the expression change of mir10a (p-value < 0.86) is also found to contribute to the expression change of Gag/Pol (p-value < 0.36). *KRR1*, an evolutionarily conserved gene, is found to be essential for viral replication (Panaro et al. 2008). Further, mir10a is also found to be involved in the regulation of immune response in CD4+ T cells (Kelada et al. 2013) and coxsackievirus group B type 3 (CVB3) replication (Tong et al. 2013). Additionally, it has been also shown that the copy number of loci belonging to the MIRLET7 family and mir10a correlates with HIV-1 viral load during the viral infection process (Abdel-Mohsen et al. 2014). Therefore, the cellular dysregulation of mir10a is proposed to be involved in HIV-1 replication via the post-transcriptional regulation of HIV-1.

Dysregulation of mir101, mir141, and mir152 to HIV-1 Budding and Releasing Through DNA Hypermethylation, Ubiquitin Transfer, and Endoplasmic Reticulum-Associated Degradation at the Virus Assembly/Budding Infection Stage of HIV-1
The final mission of HIV-1 replication is mainly virus assembly and budding. HIV-1 Gag protein is proposed to be involved in the regulations of these assembly and budding missions (Grigorov et al. 2006; De Clercq 2000). At the late infection stage, regulations of mir101, mir141, and mir152 to Gag (p-value $<1 \times 10^{-16}$) are found to result in dysfunction of the infected cells through a signaling cascade of five proteins, PRMT1, inositol 1,4,5-trisphosphate receptor type 1 (ITPR1), autocrine motility factor receptor (AMFR), tripartite motif containing 25 (TRIM25), and ubiquitin-conjugating enzyme E2N (UBE2N) (Fig. 7). These four genes have found to be highly expressed in HIV-1-infected T cells too. Consequently, the expression changes of mir101, mir141, and mir152 (p-value < 0.037, p-value < 0.108, and

p-value < 0.126, respectively) have been found to contribute to the expression change of Gag (p-value < 0.090). DNA methylation-associated proteins, including PRMT1 and UBE2N, have been proposed to contribute to virus release from HIV-1-infected cells to dysregulated uninfected cells through the aberrant hypermethylation of host target genes (Luzzi et al. 2014). mir101 (Wei et al. 2013) and mir152 (Huang et al. 2010) have been suggested to induce aberrant DNA hypermethylation in the hepatitis B virus infection. Further, ubiquitin transfer is needed for the efficient HIV-1 release during the final step of viral replication of HIV-1. Consequently, the UBE2N-associated ubiquitin transfer is suggested to be induced by a signaling cascade leading to dysregulation of mir101, mir141, and mir152. Further, ITPR1 is found to be induced to regulate HIV-1 infectivity through endoplasmic reticulum-associated degradation (Casini et al. 2015).

5 Conclusions

In this chapter, a new systems biology method is proposed to construct IGECNs for investigating cellular mechanisms by system dynamic modeling, large-scale database mining, and NGS data. Furthermore, the systems biology method is also applied to identify differences of cellular mechanisms in the IGECNs between normal and cancer cells to investigate carcinogenesis and between HIV-1-infected and non-infected cells to investigate pathogenic and defensive mechanisms at the reverse transcription stage, integration/replication stage, and virus assembly/budding stage by the network identification via omics databases and NGS data. Database mining can provide all possible candidates for genetic and epigenetic regulations and protein interactions in the IGECNs related to cellular functions. We have used AIC and statistical assessment to prune the false-positive regulations and interactions by applying the dynamic regulation and interaction model to NGS data. Finally, we have compared the major differences in IGECNs between different biological conditions to identify cellular mechanistic differences and explore the evolution of cellular mechanisms. Based on these network comparisons between normal and pathogenic IGECN, we can examine how genetic and epigenetic regulation mechanisms affect gastric tumorigenesis, hepatocarcinogenesis, and the progression of HIV-1 infections in CD4+ T cells. In this situation, the proposed IGECN construction method can allow us to identify cellular pathogenic mechanisms from investigating genetic and epigenetic regulatory mechanisms using the system identification method by omics databases and NGS data. A cancer hallmark network framework has been proposed by nine cancer hallmark networks, including angiogenesis-inducing network, mutating network, dedifferentiation network, EMT (epithelial–mesenchymal transition) network, genome duplication network, immune-escaping network, survival network, metabolic network, and stromal network (Wang et al. 2015). By comparing the hallmark network differences between normal and cancer cells, cancer clonal evolution and clinical phenotypes can be predicted to significantly impact diagnosis and personalized treatment.

References

Abdel-Mohsen M, Deng X, Danesh A et al (2014) Role of microRNA modulation in the interferon-α/ribavirin suppression of HIV-1 in vivo. PLoS One 9(10):e109220

Abraham RT (2001) Cell cycle checkpoint signaling through the ATM and ATR kinases. Genes Dev 15(17):2177–2196

Agarwal V, Bell GW, Nam JW et al (2015) Predicting effective microRNA target sites in mammalian mRNAs. elife 4:e05005

Akaike H (1974) A new look at the statistical model identification. IEEE Trans Autom Control 19(6):716–723

Bar-Joseph Z, Gerber G, Gifford DK et al (2002) A new approach to analyzing gene expression time series data. In: Proceedings of the sixth annual international conference on computational biology, pp 39–48

Blair JD, Price EM (2012) Illuminating potential technical artifacts of DNA-methylation array probes. Am J Hum Genet 91(4):760–762

Bovolenta L, Acencio M, Lemke N (2012) HTRIdb: an open-access database for experimentally verified human transcriptional regulation interactions. Nature Precedings, 1–1

Brady CW (2015) Liver disease in menopause. World J Gastroenterol: WJG 21(25):7613

Broen JC, Radstake TR, Rossato M (2014) The role of genetics and epigenetics in the pathogenesis of systemic sclerosis. Nat Rev Rheumatol 10(11):671–681

Cao D, Jiang J, Tsukamoto T et al (2015) Canolol inhibits gastric tumors initiation and progression through COX-2/PGE2 pathway in K19-C2mE transgenic mice. PLoS One 10(3):e0120938

Casini A, Olivieri M, Vecchi L et al (2015) Reduction of HIV-1 infectivity through endoplasmic reticulum-associated degradation-mediated Env depletion. J Virol 89(5):2966–2971

Chatr-Aryamontri A, Breitkreutz BJ, Heinicke S et al (2012) The BioGRID interaction database: 2013 update. Nucleic Acids Res 41(D1):D816–D823

Chen BS, Li CW (2010a) On the interplay between entropy and robustness of gene regulatory networks. Entropy 12(5):1071–1101

Chen BS, Li CW (2010b) On the noise-enhancing ability of stochastic Hodgkin-Huxley neuron systems. Neural Comput 22(7):1737–1763

Chen BS, Li CW (2010c) Robust observer-based tracking control of Hodgkin-Huxley neuron systems under environmental disturbances. Neural Comput 22(12):3143–3178

Chen LZ, Nguang SK, Chen XD (2002) On-line identification and optimization of feed rate profiles for high productivity fed-batch culture of hybridoma cells using genetic algorithms. ISA Trans 41(4):409–419

Chen LZ, Nguang SK, Chen XD (2007) Modelling and optimization of biotechnological processes: artificial intelligence approaches, vol 15. Springer

Coleman TF, Hulbert LA (1989) A direct active set algorithm for large sparse quadratic programs with simple bounds. Math Program 45(1–3):373–406

De Boor C (1978) A practical guide to splines, vol 27. Springer, New York, p 325

De Clercq E (2000) Current lead natural products for the chemotherapy of human immunodeficiency virus (HIV) infection. Med Res Rev 20(5):323–349

Faires JD, Burden RL (1998) Study guide to accompany numerical methods. Brooks/Cole

Ferreira AE, Ponces Freire AM, Voit EO (2003) A quantitative model of the generation of N∈-(carboxymethyl) lysine in the Maillard reaction between collagen and glucose. Biochem J 376(1):109–121

Friedman RC, Farh KKH, Burge CB et al (2009) Most mammalian mRNAs are conserved targets of microRNAs. Genome Res 19(1):92–105

Gao L, Yang Y, Tong D et al (2015) miR-145 mediates the antiproliferative and gene regulatory effects of vitamin D3 by directly targeting E2F3 in gastric cancer cells. Oncotarget 6(10):7675

Garcia DM, Baek D, Shin C et al (2011) Weak seed-pairing stability and high target-site abundance decrease the proficiency of lsy-6 and other microRNAs. Nat Struct Mol Biol 18(10):1139–1146

Gill PE, Murray W, Saunders MA et al (1984) Sparse matrix methods in optimization. SIAM J Sci Stat Comput 5(3):562–589

Griffiths-Jones S (2004) The microRNA registry. Nucleic Acids Res 32(suppl_1):D109–D111

Griffiths-Jones S, Saini HK, Van Dongen S et al (2007) miRBase: tools for microRNA genomics. Nucleic Acids Res 36(suppl_1):D154–D158

Grigorov B, Arcanger F, Roingeard P et al (2006) Assembly of infectious HIV-1 in human epithelial and T-lymphoblastic cell lines. J Mol Biol 359(4):848–862

Grimson A, Farh KKH, Johnston WK et al (2007) MicroRNA targeting specificity in mammals: determinants beyond seed pairing. Mol Cell 27(1):91–105

Guintivano J, Brown T, Newcomer A et al (2014) Identification and replication of a combined epigenetic and genetic biomarker predicting suicide and suicidal behaviors. Am J Psychiatry 171(12):1287–1296

Hatakeyama S (2011) TRIM proteins and cancer. Nat Rev Cancer 11(11):792–804

Hayes AM, Qian S, Yu L et al (2011) Tat RNA silencing suppressor activity contributes to perturbation of lymphocyte miRNA by HIV-1. Retrovirology 8:1–13

Herrgård MJ, Lee BS, Portnoy V et al (2006) Integrated analysis of regulatory and metabolic networks reveals novel regulatory mechanisms in Saccharomyces cerevisiae. Genome Res 16(5):627–635

Hsu HJ, Lee CF, Kaunas R (2009) A dynamic stochastic model of frequency-dependent stress fiber alignment induced by cyclic stretch. PLoS One 4(3):e4853

Huang CH, Jan HY, Lin CL et al (2009) System identification: DNA computing approach. ISA Trans 48(3):254–263

Huang J, Wang Y, Guo Y et al (2010) Down-regulated microRNA-152 induces aberrant DNA methylation in hepatitis B virus–related hepatocellular carcinoma by targeting DNA methyltransferase 1. Hepatology 52(1):60–70

Huang YZ, Sun JJ, Zhang LZ et al (2014) Genome-wide DNA methylation profiles and their relationships with mRNA and the microRNA transcriptome in bovine muscle tissue (Bos taurine). Sci Rep 4(1):6546

Ishida J (2006) Clinicopathological study of the amount of LUCA15 expressed in gastric cancer tissue. [Hokkaido Igaku Zasshi] Hokkaido J Med Sci 81(5):451–459

Itahana Y, Neo SH, Itahana K (2013) miR-141, a new player, joins the senescence orchestra. Cell Cycle 12(23):3586

Iwanami A, Gini B, Zanca C et al (2013) PML mediates glioblastoma resistance to mammalian target of rapamycin (mTOR)-targeted therapies. Proc Natl Acad Sci 110(11):4339–4344

Jemal A, Bray F, Center MM et al (2011) Global cancer statistics. CA Cancer J Clin 61(2):69–90

Johansson R (1993) System modeling and identification. Prentice-hall

Kanak M, Alseiari M, Balasubramanian P et al (2010) Triplex-forming MicroRNAs form stable complexes with HIV-1 provirus and inhibit its replication. Appl Immunohistochem Mol Morphol 18(6):532–545

Kang MS, Kieff E (2015) Epstein–Barr virus latent genes. Exp Mol Med 47(1):e131–e131

Ke X, Cortina-Borja M, Silva BC et al (2013) Integrated analysis of genome-wide genetic and epigenetic association data for identification of disease mechanisms. Epigenetics 8(11): 1236–1244

Kelada S, Sethupathy P, Okoye IS et al (2013) miR-182 and miR-10a are key regulators of Treg specialisation and stability during Schistosome and Leishmania-associated inflammation. PLoS Pathog 9(6):e1003451

Kertesz M, Iovino N, Unnerstall U et al (2007) The role of site accessibility in microRNA target recognition. Nat Genet 39(10):1278–1284

Kim SY, Park C, Kim HJ et al (2015) Deregulation of immune response genes in patients with Epstein-Barr virus-associated gastric cancer and outcomes. Gastroenterology 148(1):137–147

Kozomara A, Griffiths-Jones S (2010) miRBase: integrating microRNA annotation and deep-sequencing data. Nucleic Acids Res 39(suppl_1):D152–D157

Kozomara A, Griffiths-Jones S (2014) miRBase: annotating high confidence microRNAs using deep sequencing data. Nucleic Acids Res 42(D1):D68–D73

Lacaud G, Gore L, Kennedy M et al (2002) Runx1 is essential for hematopoietic commitment at the hemangioblast stage of development in vitro. Blood J Am Soc Hematol 100(2):458–466

Lamy E, Herz C, Lutz-Bonengel S et al (2013) The MAPK pathway signals telomerase modulation in response to isothiocyanate-induced DNA damage of human liver cancer cells. PLoS One 8(1):e53240

Li CW, Chen BS (2010) Identifying functional mechanisms of gene and protein regulatory networks in response to a broader range of environmental stresses. Int J Genomics 2010

Li H, Sun L, Chen X et al (2014) Microvesicle microRNA profiles and functional roles between chronic hepatitis B and hepatocellular carcinoma. Clin Transl Oncol 16:315–321

Li CW, Lo CC, Chen BS (2015) Robust sensorimotor control of human arm model under state-dependent noises, control-dependent noises and additive noises. Neurocomputing 167:61–75

Lin CC, Chen YJ, Chen CY et al (2012) Crosstalk between transcription factors and microRNAs in human protein interaction network. BMC Syst Biol 6(1):1–13

Liu L, Ouyang M, Rao JN et al (2015) Competition between RNA-binding proteins CELF1 and HuR modulates MYC translation and intestinal epithelium renewal. Mol Biol Cell 26(10): 1797–1810

Lu P, Vogel C, Wang R et al (2007) Absolute protein expression profiling estimates the relative contributions of transcriptional and translational regulation. Nat Biotechnol 25(1):117–124

Luzzi A, Morettini F, Gazaneo S et al (2014) HIV-1 Tat induces DNMT over-expression through microRNA dysregulation in HIV-related non Hodgkin lymphomas. Infectious Agents Cancer 9(1):1–18

Mahboobi SH, Javanpour AA, Mofrad MR (2015) The interaction of RNA helicase DDX3 with HIV-1 Rev-CRM1-RanGTP complex during the HIV replication cycle. PLoS One 10(2): e0112969

Mani A, Gelmann EP (2005) The ubiquitin-proteasome pathway and its role in cancer. J Clin Oncol 23(21):4776–4789

McMahon M (2010) RAF translocations expand cancer targets. Nat Med 16(7):749–750

Miller-Jensen K, Skupsky R, Shah PS et al (2013) Genetic selection for context-dependent stochastic phenotypes: Sp1 and TATA mutations increase phenotypic noise in HIV-1 gene expression. PLoS Comput Biol 9(7):e1003135

Mohammadi P, Desfarges S, Bartha I et al (2013) 24 hours in the life of HIV-1 in a T cell line. PLoS Pathog 9(1):e1003161

Mootha VK, Lindgren CM, Eriksson KF et al (2003) PGC-1α-responsive genes involved in oxidative phosphorylation are coordinately downregulated in human diabetes. Nat Genet 34(3):267–273

Mückstein U, Tafer H, Hackermüller J et al (2006) Thermodynamics of RNA–RNA binding. Bioinformatics 22(10):1177–1182

Nandi A, Vaz C, Bhattacharya A et al (2009) miRNA-regulated dynamics in circadian oscillator models. BMC Syst Biol 3(1):1–16

Niemi NM, Lanning NJ, Westrate LM et al (2013) Downregulation of the mitochondrial phosphatase PTPMT1 is sufficient to promote cancer cell death. PLoS One 8(1):e53803

Nishida T, Terashima M, Fukami K et al (2007) PIASy controls ubiquitination-dependent proteasomal degradation of Ets-1. Biochem J 405(3):481–488

Oh Y, Chung KC (2012) Small ubiquitin-like modifier (SUMO) modification of zinc finger protein 131 potentiates its negative effect on estrogen signaling. J Biol Chem 287(21):17517–17529

Palmieri C, Gojis O, Rudraraju B et al (2013) Expression of steroid receptor coactivator 3 in ovarian epithelial cancer is a poor prognostic factor and a marker for platinum resistance. Br J Cancer 108(10):2039–2044

Panaro MA, Mitolo V, Cianciulli A et al (2008) The HIV-1 Rev binding family of proteins: the dog proteins as a study model. Endocrine, Metabolic Immune Disorders-Drug Targets (Formerly Current Drug Targets-Immune Endocrine Metabolic Disorders) 8(1):30–46

Peng J, Valeshabad AK, Li Q et al (2013) Differential expression of RBM5 and KRAS in pancreatic ductal adenocarcinoma and their association with clinicopathological features. Oncol Lett 5(3): 1000–1004

Przytycka TM, Kim YA (2010) Network integration meets network dynamics. BMC Biol 8(1):1–3

Roberts TL, Ho U, Luff J et al (2013) Smg1 haploinsufficiency predisposes to tumor formation and inflammation. Proc Natl Acad Sci 110(4):E285–E294

Seber GA, Lee AJ (2003) Linear regression analysis, vol 330. Wiley

Sharma D, Bhattacharya J (2010) Evolutionary constraints acting on DDX3X protein potentially interferes with rev-mediated nuclear export of HIV-1 RNA. PLoS One 5(3):e9613

Shen RR, Zhou AY, Kim E et al (2015) TRAF2 is an NF-κB-activating oncogene in epithelial cancers. Oncogene 34(2):209–216

Stark C, Breitkreutz BJ, Reguly T et al (2006) BioGRID: a general repository for interaction datasets. Nucleic Acids Res 34(suppl_1):D535–D539

Subramanian A, Tamayo P, Mootha VK et al (2005) Gene set enrichment analysis: a knowledge-based approach for interpreting genome-wide expression profiles. Proc Natl Acad Sci 102(43): 15545–15550

Szotek EL, Narasipura SD, Al-Harthi L (2013) 17β-Estradiol inhibits HIV-1 by inducing a complex formation between β-catenin and estrogen receptor α on the HIV promoter to suppress HIV transcription. Virology 443(2):375–383

Taylor HE, Linde ME, Khatua AK et al (2011) Sterol regulatory element-binding protein 2 couples HIV-1 transcription to cholesterol homeostasis and T cell activation. J Virol 85(15):7699–7709

Tian YE, Xie XU, Lin Y et al (2015) Androgen receptor in hepatocarcinogenesis: recent developments and perspectives. Oncol Lett 9(5):1983–1988

Tong L, Lin L, Wu S et al (2013) MiR-10a* up-regulates coxsackievirus B3 biosynthesis by targeting the 3D-coding sequence. Nucleic Acids Res 41(6):3760–3771

Trapnell C, Pachter L, Salzberg SL (2009) topHat: discovering splice junctions with RNA-Seq. Bioinformatics 25(9):1105–1111

Voit EO (2000) Computational analysis of biochemical systems: a practical guide for biochemists and molecular biologists. Cambridge University Press

Wang YC, Chen BS (2010) Integrated cellular network of transcription regulations and protein-protein interactions. BMC Syst Biol 4(1):1–17

Wang Z, Gerstein M, Snyder M (2009) RNA-Seq: a revolutionary tool for transcriptomics. Nat Rev Genet 10(1):57–63

Wang E, Zaman N, Mcgee S et al (2015) Predictive genomics: a cancer hallmark network framework for predicting tumor clinical phenotypes using genome sequencing data. In: Seminars in cancer biology, vol 30. Academic Press, pp 4–12

Watanabe S, Watanabe K, Akimov V et al (2013) JMJD1C demethylates MDC1 to regulate the RNF8 and BRCA1–mediated chromatin response to DNA breaks. Nat Struct Mol Biol 20(12): 1425–1433

Wei X, Xiang T, Ren G et al (2013) miR-101 is down-regulated by the hepatitis B virus x protein and induces aberrant DNA methylation by targeting DNA methyltransferase 3A. Cell Signal 25(2):439–446

Wilhelm E, Doyle MC, Nzaramba I et al (2012) CTGC motifs within the HIV core promoter specify Tat-responsive pre-initiation complexes. Retrovirology 9(1):1–20

Xie L, Zhang Z, Tan Z et al (2014) MicroRNA-124 inhibits proliferation and induces apoptosis by directly repressing EZH2 in gastric cancer. Mol Cell Biochem 392:153–159

Yang SF, Chang CW, Wei RJ et al (2014) Involvement of DNA damage response pathways in hepatocellular carcinoma. BioMed research international 2014

Yao H, Ma R, Yang L et al (2014) MiR-9 promotes microglial activation by targeting MCPIP1. Nat Commun 5(1):4386

Yedavalli VS, Jeang KT (2011) Rev-ing up post-transcriptional HIV-1 RNA expression. RNA Biol 8(2):195–199

Yeger-Lotem E, Sattath S, Kashtan N et al (2004) Network motifs in integrated cellular networks of transcription–regulation and protein–protein interaction. Proc Natl Acad Sci 101(16): 5934–5939

Zahoor MA, Xue G, Sato H et al (2014) HIV-1 Vpr induces interferon-stimulated genes in human monocyte-derived macrophages. PLoS One 9(8):e106418

Zhang H, Pomerantz RJ, Dornadula G et al (2000) Human immunodeficiency virus type 1 Vif protein is an integral component of an mRNP complex of viral RNA and could be involved in the viral RNA folding and packaging process. J Virol 74(18):8252–8261

Zhao X, Li X, Yuan H (2013) microRNAs in gastric cancer invasion and metastasis. Front Biosci-Landmark 18(3):803–810

Zheng G, Tu K, Yang Q et al (2008) ITFP: an integrated platform of mammalian transcription factors. Bioinformatics 24(20):2416–2417

Zheng T, Wang J, Song X et al (2010) Nutlin-3 cooperates with doxorubicin to induce apoptosis of human hepatocellular carcinoma cells through p53 or p73 signaling pathways. J Cancer Res Clin Oncol 136:1597–1604

Zheng L, Pu J, Qi T et al (2013) miRNA-145 targets v-ets erythroblastosis virus E26 oncogene homolog 1 to suppress the invasion, metastasis, and angiogenesis of gastric cancer cells. Mol Cancer Res 11(2):182–193

Zhou S, Ye W, Ren J et al (2015) MicroRNA-381 increases radiosensitivity in esophageal squamous cell carcinoma. Am J Cancer Res 5(1):267

Zou Q, Jin J, Hu H et al (2014) USP15 stabilizes MDM2 to mediate cancer-cell survival and inhibit antitumor T cell responses. Nat Immunol 15(6):562–570

Fibonacci's Blueprint in Lifespan: Integrating Mathematical Predictions with RNA Dynamics and DNA Heritability in Systems Biology

Robert G. Sacco and John S. Torday

Contents

Abstract Lifespan, characterized by significant interspecies variation, is shaped by an intricate interplay of genetic, environmental, and physiological factors. This study, anchored in systems biology, explores the influence of entropy, fractals, and Fibonacci energy optimization on lifespan evolution. Using the AnAge database, we elucidated the patterns linking the complex mechanics of aging with the Fibonacci sequence. Notably, our findings reveal a significant association between maximum lifespan and Fibonacci age, fostering an intriguing discourse on their connectedness. This connection sparks additional inquiries into the mechanisms and evolutionary

R. G. Sacco · J. S. Torday (✉)
University of California, Los Angeles, CA, USA
e-mail: jtorday@ucla.edu

J. Barciszewski (ed.), *Systems Biology II*, RNA Technologies 15,
https://doi.org/10.1007/978-3-031-62178-9_2

significance of Fibonacci energy optimization in lifespan regulation. By weaving RNA structomics and DNA heritability considerations into a systems biology framework, we aim to unravel the sophisticated interactions bridging genotypes and phenotypes, offering fresh insights into the principles governing lifespan variation. Our study, synergizing systems biology principles with empirical examination, provides valuable insights into the multifaceted nature of lifespan and sets the foundation for future investigations in this fascinating field.

Keywords Energy optimization · Fibonacci sequence · Lifespan · Predictive modeling · RNA structomics · Systems biology

1 Introduction

The evolution of groundbreaking technologies has profoundly reshaped biological sciences, unveiling the intricate processes governing living organisms. A cornerstone of this progression has been deciphering RNA mechanisms, which are pivotal in elucidating the complex relationship between genotypes and phenotypes (Esteller 2011; Lee and Rio 2015; Mattick 2004; Nilsen and Graveley 2010). Investigating diverse RNA forms has markedly advanced our understanding of gene expression, regulation, and mechanics driving biological processes (Alberts et al. 2002; Cech and Steitz 2014; Mattick and Makunin 2006; Sonenberg and Hinnebusch 2009).

Systems biology, a discipline heralding these advancements, leverages an integrative approach, unearthing complex networks, pathways, and interactions steering living organisms (Ideker et al. 2001; Kitano 2002). Employing computational modeling, advanced software, and analytical methods, it inspects biological systems at the genomic, transcriptomic, proteomic, and metabolomic levels, establishing a comprehensive framework to navigate the complexity of life.

A specialized subset of systems biology, RNA structomics, is dedicated to exploring global RNA structure in cellular regulation (Strobel et al. 2016). Investigating RNA structures provides pivotal insights into dynamic cellular processes, expanding our understanding of genotype-phenotype correlations and clarifying the role of RNA structures in gene regulation, alternative splicing, post-transcriptional modifications, and relationships between coding and non-coding RNAs. Understanding RNA regulatory networks and their interactions fosters a deeper understanding of the mechanisms dictating the development, growth, and aging of organisms (Bahar et al. 2006; Lee et al. 2014; Wood et al. 2013). This integrative approach, merging RNA structomics with systems biology, offers a robust framework for deciphering cellular process regulatory networks and deepening our understanding of biological systems' temporal dynamics (Sacco and Torday 2023; Torday and Rehan 2011).

RNA folding, which is pivotal for RNA functionality, transitions the molecule from an unfolded, disordered state to a structured, functional conformation, which is crucial for various cellular activities (Thirumalai et al. 2001; Woodson 2010). Notably, the golden ratio, symbolized by the Greek letter Phi (Φ), has been identified in RNA-folding patterns (Zemora and Waldsich 2010). The Fibonacci sequence and golden ratio, which can be observed in numerous natural phenomena, such as phyllotaxis (arrangement of leaves on a stem) (Douady and Couder 1996), branching patterns within the lung (Goldberger et al. 1985), proportions within DNA (Perez 2010; Larsen 2021), and human lifespan (Sacco and Torday 2023), suggest that intrinsic mathematical principles might be instrumental in regulating lifespan across various species. This convergence of mathematical patterns and biological phenomena could offer a fascinating perspective on the undercurrents that direct the complexities of life and aging.

In this study, we leverage systems biology with a focus on RNA structomics and RNA folding, integrating considerations of DNA heritability, to explore lifespan variation across species. Utilizing the comprehensive AnAge database, a resource rich in maximum longevity data, we explored the Fibonacci age concept and its potential as an evolutionary force alongside entropy, fractals, and Fibonacci energy optimization roles.

We utilized Fibonacci ages, specific values from the Fibonacci sequence, to explore longevity patterns across taxonomic groups, examining whether observed ages align with Fibonacci ages, such as 7.08, 11.45, and 18.53, and understanding how fractal patterns and energy optimization influence lifespan variation.

This study aimed to unravel the complex relationship between RNA structure, DNA heritability, and lifespan and to elucidate the mechanisms and evolutionary forces shaping this interaction. Through this empirical exploration, we aimed to deepen our understanding of the fundamental principles governing longevity and provide a unique perspective on the interplay between genetic information, its expression, and lifespan regulation. The following sections present our findings, discuss their implications, and propose future research directions in this field. By intertwining RNA mechanisms, systems biology, and an interdisciplinary approach, we aimed to enhance our understanding of lifespan and decode the mysteries interwoven into life's intricate tapestry.

2 Lifespan Variability Across Species: Integrating RNA Structomics, DNA Heritability, and Fibonacci Predictive Modeling

This section explores the fascinating diversity in lifespan across different species and examines the factors that drive these variances. Initially, we navigate through the complex interplay between entropy, evolution, and Fibonacci energy optimization, focusing on their role within fractal networks to comprehend lifespan variations. This exploration seamlessly merges the concept of fractal networks with Fibonacci energy optimization, highlighting their interconnectivity and subsequent implications for lifespan.

Moreover, we will integrate insights from RNA structomics and DNA heritability, grounding these molecular perspectives within a comprehensive systems biology framework and Fibonacci mathematical predictions to elucidate lifespan variations. Ultimately, our discussion will traverse through the complexities of accurately predicting lifespan, recognizing the limitations of current models, and emphasizing the need for further refinement and integration of diverse data sources and methodologies.

Together, these discussions present an in-depth perspective on lifespan variation, weaving together crucial concepts that shape our understanding of this complex subject.

2.1 Variability in Lifespan Across Species

Lifespan diversity among species offers a rich and compelling field of study, necessitating a nuanced understanding of the genetic, environmental, and physiological factors that influence longevity. The intricate interplay between RNA mechanisms and DNA heritability sheds light on the processes that drive lifespan variation across diverse organisms.

Genetic factors undeniably cause significant changes in lifespan (Kenyon 2010). RNA mechanisms offer insights into an organism's genetic structure through transcriptome analysis and serve as a conduit for understanding the influence of DNA heritability factors on longevity variations (Djebali et al. 2012; Wang et al. 2009). Using RNA sequencing, researchers can uncover genetic variations, gene expression patterns, and molecular pathways that are intrinsically associated with lifespan (Schlotterer et al. 2014). Examining the transcriptomic profiles between species with different lifespans can illuminate the genes and regulatory networks that contribute to disparities in longevity (Kim et al. 2011; Seim et al. 2013).

Moreover, the study of epigenetic alterations, which are often linked to DNA heritability, can be elucidated through RNA mechanisms. Epigenetic markers such as DNA methylation and histone modifications can dictate gene expression patterns without altering the DNA sequence itself, thereby providing a dynamic yet heritable aspect of gene regulation (Moore et al. 2013; Sen et al. 2016; Torday 2022). Influenced by environmental cues, these alterations can have lasting impacts on gene expression and cellular function.

In parallel, the dynamic interaction of an organism with its environment, subtly integrated with genetic predispositions, also contributes to lifespan variability (Torday 2019). RNA mechanisms help decipher gene expression responses to various environmental demands and stressors, charting adaptive responses and potentially aligning them with predictive mathematical models, such as those based on the Fibonacci sequence, to ascertain their impact on lifespan (Sacco and Torday 2023).

Furthermore, RNA-based methods underscore the intricacy of post-transcriptional regulation, providing a lens through which insights into lifespan variability can be magnified. Crucial components of post-transcriptional regulation, such as alternative splicing, RNA editing, and non-coding RNA molecules, impact gene expression and protein diversity, playing pivotal roles in biological processes and variability (Esteller 2011; Lee and Rio 2015; Nilsen and Graveley 2010; Wood et al. 2013).

In culmination, RNA mechanisms, when studied in concert with aspects of DNA heritability and mathematical predictions, have propelled our understanding of lifespan diversity into new dimensions. Integrating transcriptomic analyses, epigenetic profiling, and insights from post-transcriptional regulation studies illuminates the genetic, epigenetic, and environmental factors that shape lifespan diversity. This knowledge not only amplifies our understanding of the fundamental biology of aging but also supports the flames of curiosity regarding extending a healthy lifespan and mitigating age-related diseases in humans through a synergistic lens that melds RNA dynamics, DNA heritability, and mathematical predictions into a unified narrative.

2.2 *Fractal Networks, Fibonacci Energy Optimization, and RNA Structomics*

In biological systems, the presence of fractal networks, characterized by self-similar patterns and scaling properties, is well established (Sacco and Torday 2023; West and Brown 2005). These networks represent a counterpoint to entropy, synonymous with disorder and aging, by maintaining the structure and organization within living organisms. By optimizing the energy distribution and usage, fractal networks play a role in the overall health and longevity of an organism. This notion of energy optimization is further corroborated by studies on the energetic costs associated with cellular signaling, which have been found to play a crucial role in evolutionary dynamics (Wang et al. 2020).

Fractal networks have been linked to the adaptation of organisms to environmental challenges (Bejan and Zane 2013). The fractal-like organization provides resilience and flexibility, enhancing survival and adaptation to stressors. This is exemplified in mammalian respiratory systems, where the branching structure of the bronchial tree ensures an efficient gas exchange (Goldberger et al. 1985).

A fascinating aspect to consider is the relationship between fractals and Fibonacci numbers explored through Fibonacci energy optimization. Fibonacci numbers, a sequence where each subsequent number is the sum of its two predecessors, pervade numerous natural phenomena, including contributing to efficient energy distribution within biological systems (Prusinkiewicz and Runions 2012; Sacco and Torday 2023).

The intertwined relationship between the structural and functional aspects in organisms coalesces the principles of fractals and Fibonacci energy optimization. Influenced by the Fibonacci sequence, this biological organization amplifies energy utilization efficiency. This hypothesis proposes that the lifespan of an organism is intricately influenced by the synergistic interaction between fractal networks and Fibonacci principles, which collectively maintain homeostasis and regulate physiological processes.

RNA structomics advances the comprehension of fractal networks within biological systems, examining gene expression implicated in branching morphogenesis, extracellular matrix formation, and cellular adhesion to decrypt the molecular fundamentals of fractal network formation (Strobel et al. 2016). This field scrutinizes RNA structures and interactions and elucidates the molecular-level formation of fractal networks.

In addition, RNA structomics helps understand the evolutionary origin and conservation of fractal networks across species (Omland et al. 2008). The identification of conserved genes and pathways involved in the development and maintenance of fractal architecture can be achieved through RNA structomics. This, combined with computational modeling, allows the simulation of fractal network formation and dynamics.

In summary, combining the study of fractals, Fibonacci energy optimization, and RNA structomics offers an intriguing route for exploring lifespan variation mechanisms. These principles provide insights into the energy distribution, resource use, and information processing within biological systems. Advancements in RNA structomics have helped us comprehend the interplay between fractal networks, Fibonacci energy optimization, and the relationship between genotype and phenotype.

2.3 *Challenges in Predicting Lifespan: A Systems Biology Perspective*

The prediction of lifespan, despite advancements in our understanding of its biological basis, remains a multifaceted problem. Numerous complex factors influence an organism's lifespan, such as genetics, epigenetics, the environment, and stochastic events (Houle et al. 2010; Hayflick 2007; López-Otín et al. 2013).

From a systems biology perspective, the challenge is to combine various data types—genomic, transcriptomic, proteomic, and metabolomic—into a comprehensive model that can predict lifespan (Kitano 2002). Each data type provides unique insights into the biology of the organism and the genotype-phenotype relationship. However, integrating these different types of data is a complex task because of the high dimensionality of the data and nonlinear interactions among variables (Ideker et al. 2001).

RNA structomics offers a pathway to navigate this challenge by providing a comprehensive view of dynamic RNA structures and their regulatory roles within cells (Strobel et al. 2016). RNA structure—or, collectively, RNA structomics—significantly influences the determination of phenotypes, including lifespans. Thus, weaving RNA structomics with other data types fosters the creation of a more thorough and accurate lifespan predictive model.

Furthermore, machine learning and computational modeling have emerged as pivotal tools for addressing the complexity of lifespan predictions. These methodologies facilitate the analysis of large-scale data and discernment of complex patterns and relationships, which traditional statistical methods might overlook (Houle et al. 2010). Through machine learning, intricate interactions among various biological factors can be encapsulated within predictive models, thereby enhancing the accuracy of lifespan predictions.

The inherently stochastic nature of biological processes presents an additional hurdle in lifespan prediction, introducing substantial variation, even among genetically identical entities (López-Otín et al. 2013). In a metaphorical parallel, predicting lifespan amidst these chaotic processes resembles deducing a consistent pressure reading amidst the randomness of gas particle collisions within a container. These emergent phenomena, whether pressure or lifespan, ascend not from isolated interactions but from collective outcomes, thus meriting examination at a more encompassing and holistic level.

In summary, lifespan prediction requires an integrative approach that combines diverse data types, advanced computational methods, and an understanding of the stochastic nature of biological processes. By applying systems biology and RNA structomics and leveraging machine learning and computational modeling capabilities, researchers can make a significant progress toward accurate lifespan prediction models.

3 Predicting Lifespan: A Systems Biology Approach

In this section, utilizing a systems biology approach, we endeavor to predict lifespan by introducing the Fibonacci Life Chart Method. This integrative framework combines elements from fractal geometry, Fibonacci sequences, and energy theories to illuminate and predict lifespan across various species. Furthermore, we delve into data-mining techniques employing large-scale datasets and analytics to uncover meaningful patterns with the potential to forecast lifespan. The amalgamation of these approaches facilitates our aim to comprehend the determinants of lifespan and construct predictive models that encapsulate the complexity inherent in biological systems.

3.1 *Predicting Lifespan with the Fibonacci Life Chart Method*

Lifespan variations across species are a topic of ongoing interest in biology. The integration of concepts such as entropy, fractals, and Fibonacci energy optimization into systems biology has illuminated aspects of this variation, while RNA structomics has enhanced our understanding of intricate genotype-phenotype relationships and paved new avenues for lifespan prediction.

An innovative frontier in predicting lifespan is presented by the Fibonacci Life Chart Method (Sacco and Torday 2023), which hinges on Fibonacci energy optimization principles. This suggests that biological systems evolve and optimize energy distribution and utilization through patterns resembling fractals. The Fibonacci sequence, discernible in various biological phenomena, including plant growth, branching patterns, and genetic material, also has potential implications for understanding lifespan (Sacco and Torday 2023).

The Fibonacci Life Chart Method aims to estimate an organism's potential lifespan by mapping key developmental milestones and physiological changes onto a Fibonacci-based timeline. This method acknowledges the interconnectedness and fractal nature of biological systems, implying that phenotypic changes mirror the underlying patterns and processes.

Although this method offers a comprehensive exploration of lifespan via systemic energy optimization principles, the role of DNA heritability remains pivotal. DNA frames an organism's intrinsic lifespan determinants via the genetic blueprint and intertwines significantly with RNA dynamics, manifesting phenotypic expressions relevant to aging and longevity (Kenyon 2010). Understanding how heritable genetic variations might influence observable Fibonacci patterns in an organism's development, and aging is crucial for enhancing predictive accuracy and applicability across various genetic backgrounds by exploring correlations with genetic polymorphisms associated with lifespan.

Incorporating an evolutionary perspective, validating the Fibonacci Life Chart Method may involve examining the interplay between genetic mutations, lineage heritability, and evolutionary pressures with the observed Fibonacci patterns across generations. Deducing conservation or variation in Fibonacci patterns across species and within diverse genetic lineages of a single species can provide deeper insights into the universality and deviations of these mathematical principles in biological contexts.

The predictive power of the Fibonacci Life Chart Method can be enhanced through computational modeling that amalgamates ideas from systems biology. This entails simulating interactions among genes, regulatory networks, and environmental factors, granting a profound understanding of RNA structomics shaping lifespan. Together with RNA structomics, computational modeling enables the integration of voluminous omics data, facilitating in-depth analysis of the molecular processes and networks engaged in aging and lifespan determination.

By infusing systems biology concepts into computational models, the intricacies and feedback loops governing aging processes can be deciphered, refining predictions using the Fibonacci Life Chart Method (Sacco and Torday 2023). This method, which combines fractal geometry, Fibonacci energy optimization, RNA structomics, and computational modeling, encapsulates the complexity and variability of lifespan across species, offering a holistic view of longevity-influencing factors, and synthesizes genetic, environmental, and systemic factors within a singular framework (Sacco and Torday 2023).

In summary, the synergy of the Fibonacci Life Chart Method with RNA structomics and computational modeling holds a tremendous potential for enhancing our understanding of lifespan variation. This multifaceted approach, tapping into the fractal nature of biological systems and harnessing the capabilities of RNA structomics and computational modeling, delves into the molecular complexities that underlie aging and longevity. Ongoing and future research, integrated with additional systems biology insights, will contribute to refining lifespan predictions and deepening our understanding of the fundamental principles of life.

3.2 *Data Mining for Lifespan Prediction*

Data-mining techniques provide powerful tools for extracting valuable insights from large and complex datasets. These techniques involve the exploration and analysis of vast amounts of data to uncover patterns, correlations, and relationships that might otherwise go unnoticed. In the context of lifespan prediction, data-mining techniques can reveal complex relationships between various factors and lifespan outcomes.

Our study employs a data-mining approach grounded in statistical analysis. The strength of this method resides in its ability to examine relationships across multiple variables, control for confounding factors, and quantify the strength of associations. Through the lens of statistical techniques, we delve into a large and diverse dataset. Our primary aim is to validate the accuracy of the Fibonacci Life Chart Method that identifies Fibonacci ages as a significant predictor of lifespan (Sacco and Torday 2023).

Although statistical analysis forms the bedrock of this study, it is worth noting the burgeoning influence of machine learning in the realm of lifespan prediction (Mamoshina et al. 2016; Putin et al. 2016). Machine-learning algorithms can analyze vast amounts of data and identify patterns associated with lifespan outcomes. By training models on historical data that include information on genetic factors, environmental influences, lifestyle choices, and other relevant variables, machine-learning algorithms can predict the lifespan of new individuals or species.

In addition, bridging data mining with systems biology principles offers a potent synergy. Systems biology, with its overarching view of biological component interactions, imbues algorithms with a nuanced understanding of the interplay between genetic, environmental, and physiological elements. This holistic approach magnifies the precision and interpretability of lifespan predictions, fostering an environment in which we can truly appreciate the factors that modulate lifespan variations.

Despite the great potential of data-mining techniques for lifespan prediction, there are challenges. The quality and availability of data, along with the complexity of the underlying biological processes, can present hurdles in accurately predicting the lifespan. However, data mining promises to reveal valuable insights and improve our understanding of lifespan variation. By mining large datasets, researchers can identify novel factors and interactions that influence lifespan, paving the way for targeted interventions and personalized approaches to promote healthy aging.

In summary, our study on lifespan prediction hinges on a robust data-mining strategy accentuated by statistical analysis and enriched by systems biology principles. As we continue to harness advanced data-mining techniques and as datasets evolve in depth and breadth, we remain optimistic about future breakthroughs. These advancements will not only refine our predictions but also enable interventions that promote healthier and more extended lifespans.

4 Method

In the following section, the methods employed in this study to predict lifespan across diverse species from a systems biology perspective are described. This encompasses a rigorous data collection procedure, application of a novel mathematical model, and deployment of advanced statistical analysis techniques. Collectively, these methods create a comprehensive and robust framework that enables in-depth exploration of the complexity of lifespan variation. This approach facilitates examination of the interplay between genetics, environmental stimuli, and molecular mechanisms that give rise to the observed disparities in longevity among different species.

4.1 Data Collection and Integration

The AnAge database (Database of Animal Ageing and Longevity, Build 15, Release date: July 3, 2023) developed by de Magalhães and Costa (2009) served as the primary source of maximum longevity data for this study. This database, with its comprehensive inclusion of over 4,000 species spanning various taxonomic groups, provides a robust foundation for investigating lifespan patterns and the factors influencing longevity.

Our study focused on five specific classes within the phylum Chordata: Amphibia, Aves (birds), Mammalia, Reptilia, and Teleostei (fishes). We also performed a separate analysis of the Primate Order. The choice of these classes is influenced by their diverse lifespan patterns, broad representation in lifespan research, and varying responses to environmental and genetic factors. Furthermore, the Primate Order, which includes species closely related to humans, was particularly valuable in our study because of its physiological similarities and complex behaviors analogous to those of humans, making it a valuable model for understanding longevity patterns (Phillips et al. 2014).

We utilized a set of stringent filters to ensure the reliability and accuracy of our data, which included selecting only entries classified as "acceptable" or "high" quality and excluding those with tiny sample sizes. Age data for both species in the wild and in captivity were included in the analysis, providing a comprehensive representation of longevity across various contexts.

Our final dataset comprised 3603 animal species from five selected classes within the phylum Chordata. The sample sizes for each class were as follows: Amphibia ($n = 144$), Aves ($n = 1275$), Mammalia ($n = 922$), Reptilia ($n = 469$), and Teleostei ($n = 793$). Within the class Mammalia, 135 primate species were included.

4.2 *Mathematical Modeling*

In this study, we employed the Fibonacci Life Chart Method (Sacco and Torday 2023), a model that integrates principles from fractal geometry, Fibonacci sequences, and energy optimization theories, to comprehensively describe and predict variations in lifespan across species. The choice of this method over other potential models is influenced by its unique approach in capturing inherent fractal patterns and energy allocation strategies within biological systems, which provides a distinctive perspective on lifespan variations.

The model incorporates ten key fractal ages, 1.67, 2.70, 4.38, 7.08, 11.45, 18.53, 29.99, 48.52, 78.51, and 127.04, each of which has been strategically selected to align with the observed patterns of energy allocation and utilization in biological systems.

The construction and application of this mathematical model have been extensively detailed in a previous publication (Sacco and Torday 2023). Readers are encouraged to consult this work for a more in-depth understanding of the model's integration of fractal geometry, Fibonacci sequences, and energy optimization theories, as well as their application in predicting the lifespan.

4.3 *Statistical Analysis*

Statistical analysis was conducted to examine the association between the predictive model and maximum longevity data. Descriptive analysis was performed to summarize the variables and assess their characteristics, including calculations of the mean, standard deviation, and interquartile range.

Due to the skewed nature of the data, the interquartile range (IQR) method was employed for outlier removal (Iglewicz and Hoaglin 1993). Extreme values were defined as values below the lower threshold of Q1 and above the upper threshold of Q3. To accommodate the decimal cutoff number obtained from the IQR, we rounded the cutoff number to the nearest whole number to maintain consistency with age-bin intervals.

Continuous age variables were transformed into categorical age bins to facilitate relationship analysis between the predicted age categories and maximum longevity data using the chi-square test (Agresti 2012). The selection of appropriate bin sizes was based on 2-year intervals, resulting in age bins of 10–11.99 years, 12–13.99 years, 14–15.99 years, and so on. The selection of 2-year intervals for age categories was made to strike a balance between granularity and simplicity. This choice enables a detailed analysis of lifespan patterns while keeping the data manageable and interpretable.

A significance level of $p \leq 0.05$ was adopted to determine the statistical significance of the findings. The obtained p-values were compared to this threshold, with results showing p-values less than or equal to 0.05 considered statistically significant. This approach enables a comprehensive assessment of the connection between age predictions and actual longevity patterns, thereby providing valuable insights into the significance of the findings.

5 Results

This section presents the relationship between Fibonacci age and lifespan variation across distinct taxonomic groups. Utilizing filtered datasets that include animal species from the phylum Chordata and span five classes—Amphibia, Aves, Mammalia, Reptilia, and Teleostei—we employed descriptive statistics and the chi-square test to examine the association between predicted and observed maximum longevity categories. This examination highlights the potential influences of fractal patterns and energy optimization mechanisms on lifespan variation.

5.1 Descriptive Statistics

The descriptive statistics detailed in this section provide an exploration of the central tendency, variability, and distribution of the dataset. Of notable significance in the data analysis was the interquartile range (IQR), employed as a robust metric of variability and pivotal in the data-filtering process. Criteria grounded in the first (Q1) and third (Q3) quartiles were applied to exclude outliers, ensuring an accurate representation of the central tendencies and variability of the data, and subsequently, a more reliable analysis of the lifespan patterns within each class.

5.1.1 Overall Sample

Table 1 summarizes the descriptive statistics of the sample, both pre- and post-data filtering, for maximum longevity in the respective chordate classes. The pre-filtering data indicated a sample size of 3603, mean maximum longevity of 18.92 years, and standard deviation of 16.32, among other statistics (Table 1). In contrast, the post-filtering data illustrate changes after the data filtration process, showing a condensed sample size and diminished standard deviation, which suggests a more symmetric and less-peaked distribution in comparison to the pre-filtering data.

Figure 1 shows the main patterns in the histograms of the maximum longevity for the pre- and post-filtering datasets. They provide a clear picture of the distribution changes following the filtration process, indicating a more concentrated pattern of the maximum lifespan.

Table 1 Descriptive statistics for maximum longevity in chordate classes (pre-filtering and post-filtering)

Statistic	Value (pre-filtering)	Value (post-filtering)
Sample size (N)	3603	1808
Mean	18.92	15.57
Standard deviation	16.32	4.38
Median	15.00	15.00
Minimum	0.16	9.00
Maximum	211.00	24.00
Q1	8.90	11.80
Q2	15.00	15.00
Q3	24.00	19.30
Skewness	3.52	0.25
Kurtosis	26.02	1.89

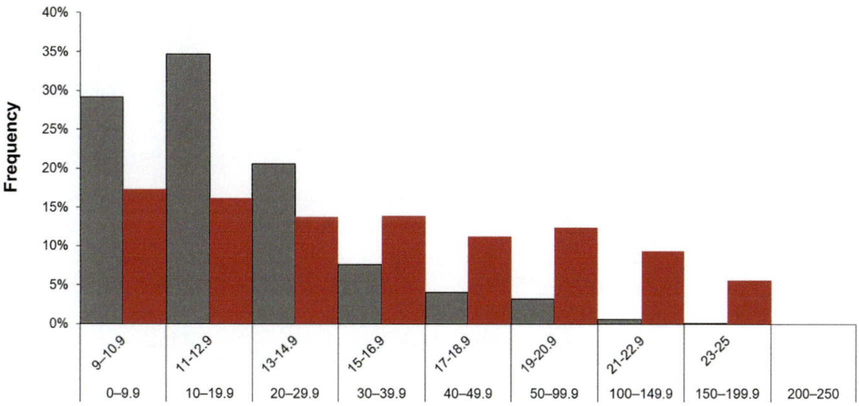

Fig. 1 Histogram of maximum longevity in chordate classes (pre-filtering and post-filtering)

5.1.2 Class and Taxonomic Group Statistics

For each chordate class, namely, Amphibia, Aves, Mammalia, Reptilia, Teleostei, and also the Primate Order, detailed descriptive statistics were computed and are presented in Tables 2, 3, 4, 5, 6, and 7 and histograms (Figs. 2, 3, 4, 5, 6, and 7). These tables show the central tendency, variability, and shape of the distribution within each class and order. Histograms augment these insights by providing visual representations that underscore the concentration of lifespan observations within specified ranges for each category.

A comparative analysis of pre- and post-filtering statistics revealed the impact of outlier removal on data distribution. This insight into the robustness of the data and the reliability of the subsequent analysis is instrumental in revealing unique patterns and variations in the lifespan. It contributes to the understanding of the factors influencing longevity within each taxonomic group, potentially leading to new

Table 2 Descriptive statistics for maximum longevity in Amphibia class (pre-filtering and post-filtering)

Statistic	Value (pre-filtering)	Value (post-filtering)
Sample size (N)	144	75
Mean	15.32	12.64
Standard deviation	11.01	2.56
Median	12.10	11.90
Minimum	4.10	9.00
Maximum	102.00	18.80
Q1	9.05	10.70
Q2	12.10	11.90
Q3	19.15	15.00
Skewness	4.11	0.66
Kurtosis	30.26	2.66

Table 3 Descriptive statistics for maximum longevity in Aves class (pre-filtering and post-filtering)

Statistic	Value (pre-filtering)	Value (post-filtering)
Sample size (N)	1275	646
Mean	18.27	15.68
Standard deviation	11.48	4.13
Median	15.00	15.00
Minimum	2.30	10.00
Maximum	83.00	24.00
Q1	10.00	12.00
Q2	15.00	15.00
Q3	24.00	19.00
Skewness	1.48	0.44
Kurtosis	5.93	1.99

Table 4 Descriptive statistics for maximum longevity in Mammalia class (pre-filtering and post-filtering)

Statistic	Value (pre-filtering)	Value (post-filtering)
Sample size (N)	922	464
Mean	20.16	17.34
Standard deviation	15.99	4.58
Median	17.40	17.30
Minimum	2.10	9.00
Maximum	211.00	26.00
Q1	9.30	13.80
Q2	17.40	17.30
Q3	26.20	21.00
Skewness	3.47	−0.04
Kurtosis	30.03	1.97

Table 5 Descriptive statistics for maximum longevity in Reptilia class (pre-filtering and post-filtering)

Statistic	Value (pre-filtering)	Value (post-filtering)
Sample size (N)	469	240
Mean	22.65	18.87
Standard deviation	18.26	3.71
Median	18.80	18.85
Minimum	1.30	13.00
Maximum	177.00	26.00
Q1	13.00	15.65
Q2	18.80	18.85
Q3	25.90	22.05
Skewness	3.91	0.22
Kurtosis	25.87	1.90

Table 6 Descriptive statistics for maximum longevity in Teleostei class (pre-filtering and post-filtering)

Statistic	Value (pre-filtering)	Value (post-filtering)
Sample size (N)	793	414
Mean	16.98	10.93
Standard deviation	21.59	4.61
Median	10.00	10.00
Minimum	0.16	5.00
Maximum	205.00	20.00
Q1	5.00	7.00
Q2	10.00	10.00
Q3	20.00	15.00
Skewness	3.43	0.50
Kurtosis	19.29	2.15

Table 7 Descriptive statistics for maximum longevity in Primates order (pre-filtering and post-filtering)

Statistic	Value (pre-filtering)	Value (post-filtering)
Sample size (N)	135	74
Mean	31.96	30.83
Standard deviation	13.03	4.12
Median	30.10	30.25
Minimum	13.00	24.00
Maximum	122.50	38.00
Q1	24.00	27.00
Q2	30.10	30.25
Q3	37.50	34.00
Skewness	2.85	0.12
Kurtosis	19.76	2.06

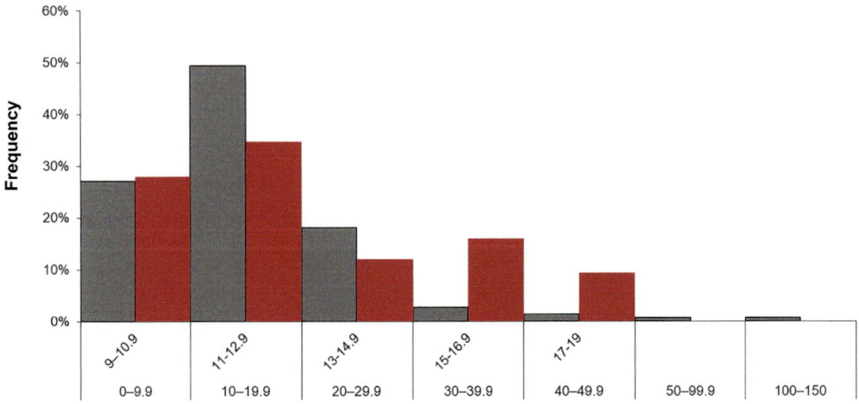

Fig. 2 Histogram of maximum lifespan in class Amphibia (pre-filtering and post-filtering)

hypotheses and research questions, thereby broadening the understanding of longevity in chordates.

5.2 *Chi-Square Test*

This section explores the correlation between predicted age categories and observed maximum longevity data using a chi-square test, transforming continuous age variables into categorical age bins. "Positive" and "negative" associations indicate statistically significant correlations where observed counts are notably higher or lower than expected counts in a particular age bin, respectively. When the p-value does not denote significance, the term "no direction" implies that observed variations

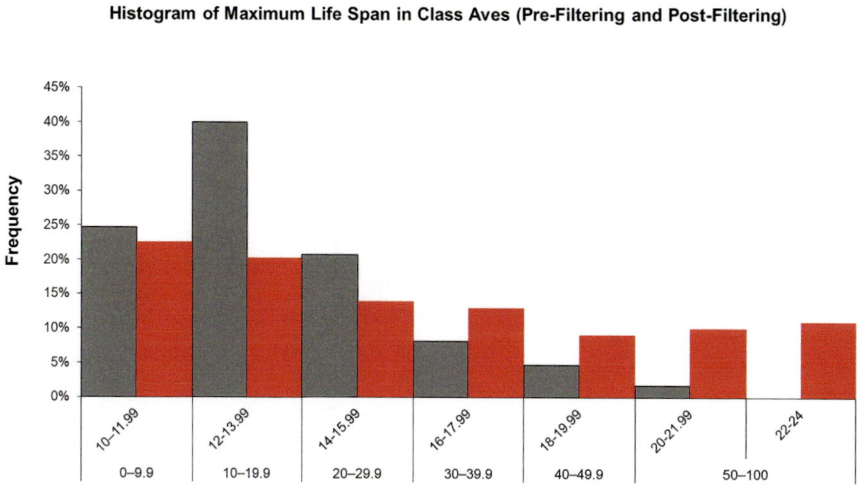

Fig. 3 Histogram of maximum lifespan in class Aves (pre-filtering and post-filtering)

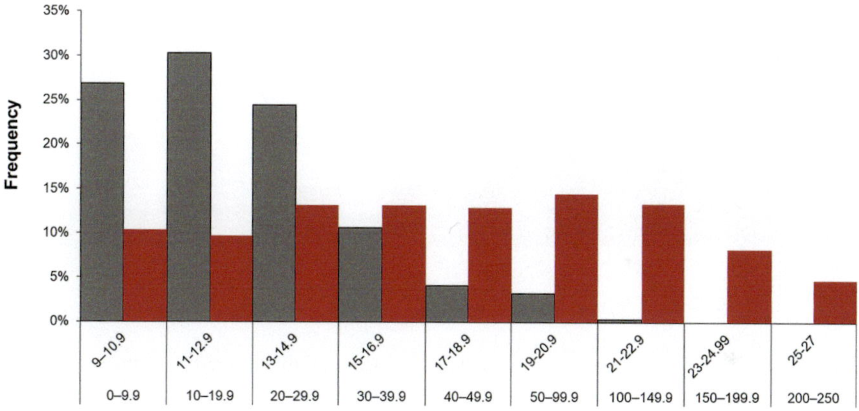

Fig. 4 Histogram of maximum lifespan in class Mammalia (pre-filtering and post-filtering)

may arise from random chance, not providing enough evidence to reject the null hypothesis.

5.2.1 Overall Sample

Table 8 displays the chi-square test results for the entire sample, exploring the correlation between predicted and observed longevities across various age bins. The findings indicated significant positive associations in age bins with average ages near the predicted values, such as 9.81 and 11.76 (average of 10.79). These

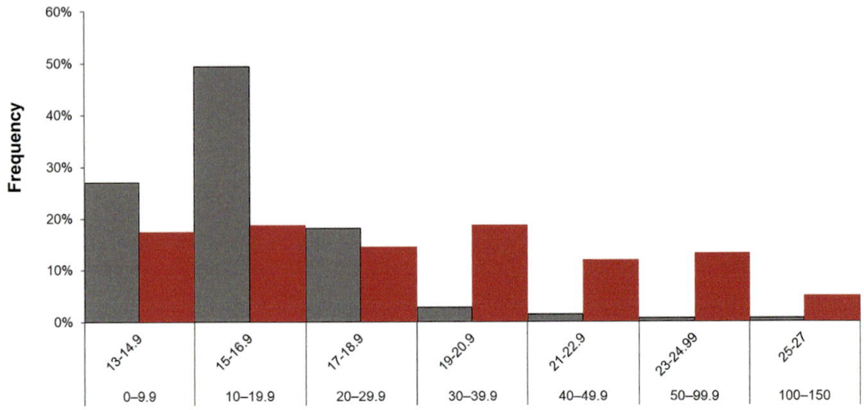

Fig. 5 Histogram of maximum lifespan in class Reptilia (pre-filtering and post-filtering)

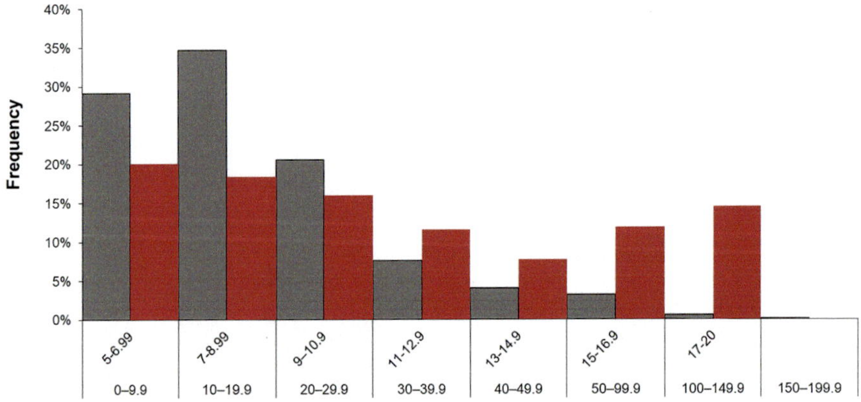

Fig. 6 Histogram of maximum lifespan in class Teleostei (pre-filtering and post-filtering)

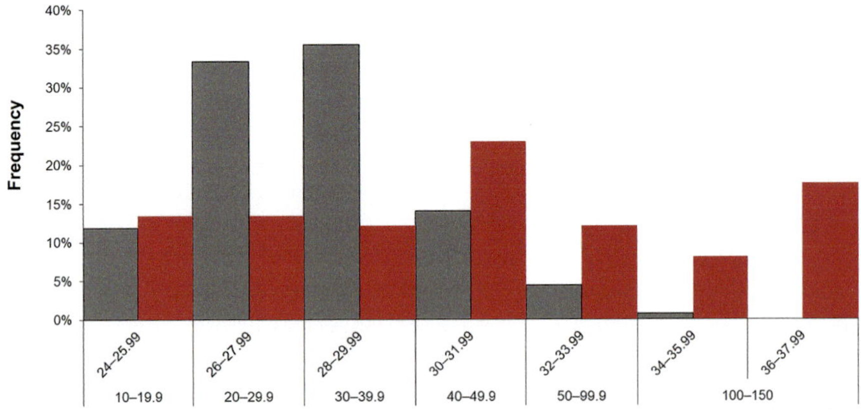

Fig. 7 Histogram of maximum lifespan in order Primates (pre-filtering and post-filtering)

Table 8 Chi-square results for overall sample (post-filtering)

Age bin	Avg. age	Observed	Expected	χ^2	p-Value	Direction
9–10.99	9.81	313	226	38.276	0.0001	Positive
11–12.99	11.76	293	226	22.700	0.0001	Positive
13–14.99	13.83	249	226	2.675	0.1019	No direction
15–16.99	15.70	252	226	3.418	0.0645	No direction
17–18.99	17.79	204	226	2.448	0.1177	No direction
19–20.99	19.92	225	226	0.005	0.9433	No direction
21–22.99	21.91	170	226	5.858	0.0001	Negative
23–24.00	23.55	102	226	77.755	0.0001	Negative

Table 9 Chi-square results for class Amphibia (post-filtering)

Age bin	Avg. age	Observed	Expected	χ^2	p-Value	Direction
9–10.99	10.00	21	15	3.000	0.0833	No direction
11–12.99	11.68	26	15	10.083	0.0015	Positive
13–14.99	13.99	9	15	3.000	0.0833	No direction
15–16.99	15.30	12	15	0.750	0.3865	No direction
17–19.00	17.81	7	15	5.333	0.0209	Negative

results support our hypothesis that the predicted ages function as attractors in the lower threshold range.

Conversely, age bins distant from the predicted values (e.g., 21.91 and 23.55, an average of 22.73) exhibited negative associations. This deviation provides further evidence for the potential relationship between predicted ages and observed longevity patterns, especially suggesting that the predicted age of 18.53 does not function as an attractor for the overall post-filtering sample. However, the predicted age of 11.45 seems to act as an attractor.

These outcomes underscore the importance of the predicted ages as attractors in explaining the observed associations in the lower threshold range while stressing the lack of association in age bins distant from the predicted values.

5.2.2 Class and Taxonomic Group Statistics

We also performed chi-square tests separately for each class and taxonomic group, allowing for a more detailed exploration of the correlations within specific classes or groups.

Tables 9, 10, 11, 12, 13, and 14 present the results for each class. The findings reveal intriguing patterns that could influence our understanding of the hypotheses related to predicted ages. For age bins with averages close to these predicted values, we observed significant positive correlations between the predicted and observed longevity data. However, in age bins distant from these predicted values, the

Table 10 Chi-square results for class Aves (post-filtering)

Age Bbin	Avg. age	Observed	Expected	χ^2	p-Value	Direction
10–11.99	10.88	146	92.29	36.467	0.0001	Positive
12–13.99	12.82	131	92.29	18.943	0.0001	Positive
14–15.99	14.88	90	92.29	0.066	0.7968	No direction
16–17.99	16.97	84	92.29	0.869	0.3513	No direction
18–19.99	18.82	59	92.29	14.010	0.0002	Negative
20–21.99	20.78	65	92.29	9.415	0.0022	Negative
22–24.00	23.03	71	92.29	5.730	0.0167	Negative

Table 11 Chi-square results for class Mammalia (post-filtering)

Age bin	Avg. age	Observed	Expected	χ^2	p-Value	Direction
9–10.99	9.89	48	51.56	0.277	0.5990	No direction
11–12.99	11.86	45	51.56	0.939	0.3325	No direction
13–14.99	13.92	61	51.56	1.944	0.1632	No direction
15–16.99	15.90	61	51.56	1.944	0.1632	No direction
17–18.99	17.78	60	51.56	1.554	0.2125	No direction
19–20.99	19.91	67	51.56	5.202	0.0226	Positive
21–22.99	21.94	62	51.56	2.378	0.1230	No direction
23–24.99	23.62	38	51.56	4.012	0.0452	Negative
25–26.00	25.42	22	51.56	19.066	0.0001	Negative

Table 12 Chi-square results for Primate order (post-filtering)

Age bin	Avg. age	Observed	Expected	χ^2	p-Value	Direction
24–25.99	24.81	10	10.57	0.036	0.8498	No direction
26–27.99	26.57	10	10.57	0.036	0.8498	No direction
28–29.99	28.92	9	10.57	0.272	0.6020	No direction
30–31.99	30.62	17	10.57	4.563	0.0327	Positive
32–33.99	32.86	9	10.57	0.272	0.6020	No direction
34–35.99	35.12	6	10.57	2.305	0.1289	No direction
36–38.00	36.95	13	10.57	0.652	0.4195	No direction

Table 13 Chi-square results for class Reptilia (post-filtering)

Age bin	Avg. age	Observed	Expected	χ^2	p-Value	Direction
13–14.99	13.96	42	34.29	2.023	0.1550	No direction
15–16.99	15.80	45	34.29	3.903	0.0482	Positive
17–18.99	17.76	35	34.29	0.017	0.8958	No direction
19–20.99	19.88	45	34.29	3.903	0.0482	Positive
21–22.99	22.03	29	34.29	0.952	0.3292	No direction
23–24.99	24.08	32	34.29	0.178	0.6727	No direction
25–26.00	25.56	12	34.29	16.905	0.0001	Negative

Table 14 Chi-square results for class Teleostei (post-filtering)

Age bin	Avg. age	Observed	Expected	χ^2	*p*-Value	Direction
5–6.99	5.40	83	51.75	21.567	0.0001	Positive
7–8.99	7.49	76	51.75	12.987	0.0003	Positive
9–10.99	9.59	66	51.75	4.484	0.0342	Positive
11–12.99	11.50	48	51.75	0.311	0.5773	No direction
13–14.99	13.52	32	51.75	8.614	0.0033	Negative
15–16.99	15.29	49	51.75	0.167	0.6828	No direction
17–18.99	17.77	24	51.75	17.006	0.0001	Negative
19–20.00	19.81	36	51.75	5.478	0.0193	Negative

associations were negative, suggesting that the predicted ages may serve as attractors guiding the observed longevity patterns. The inclusion of the average age in each bin served as a reference point for comparing observed and expected values. Age bins with no significant deviation from the expected values denote no association with predicted ages.

Overall, these findings underscore the significance of considering age bins and their correlations with predicted age in understanding longevity patterns in the studied population.

5.3 Discussion of Findings

This study identified several significant positive associations, supporting the Fibonacci age hypothesis. The most compelling evidence was found in the Teleostei class, where the observed ages associated with longevity closely aligned, on average, with the predicted Fibonacci age. The age bin of 7–8.99 years (average age of 7.49), along with the observed ages of 5.40 and 9.59 (average of about 7.50), strongly substantiates the hypothesis for this class.

Likewise, in the Amphibia class, the average age of 11.68 supports the predicted age of 11.45 years, bolstering the evidence of alignment between the observed and predicted ages. For the overall sample, the age bin of 11–12.99 years (average age of 11.76) matches closely with the predicted age of 11.45 years, thereby reinforcing the hypothesis.

In the Aves class, the age bins of 10–11.99 years and 12–13.99 years (average ages of 10.88 and 12.82, respectively) align closely with the predicted age of 11.45 years, lending further support to the hypothesis. Meanwhile, in the Primate Order, the age bin of 30–31.99 years (average age of 30.62) closely aligns with the predicted age of 29.99 years, lending a strong support to the forecasted age in this order.

In contrast, the Reptilia class exhibited potential variation in longevity patterns. This class did not show significant positive associations with the predicted age of 18.53 years in the age bin of 15–16.99 years. However, a positive association was

Table 15 Comparison of the mean age and hypothesized Fibonacci age in different taxonomic groups (pre-filtering)

Class	Mean age	Fibonacci age	% Difference	Sample size
Aves	18.27	18.53	−1.40	1275
Overall	18.92	18.53	2.10	3603
Primates	31.96	29.99	6.57	135
Teleostei	16.98	18.53	−8.36	793
Mammalia	20.16	18.53	8.80	922
Amphibia	15.32	18.53	−17.32	144
Reptilia	22.65	18.53	22.23	469

Table 16 Comparison of the mean age and hypothesized Fibonacci age in different taxonomic groups (post-filtering)

Class	Mean age	Fibonacci age	% Difference	Sample size
Reptilia	18.87	18.53	1.83	240
Primates	30.83	29.99	2.80	74
Teleostei	10.93	11.45	−4.54	414
Mammalia	17.34	18.53	−6.42	464
Amphibia	12.64	11.45	10.39	75
Aves	15.68	18.53	−15.38	646
Overall	15.57	18.53	−15.97	1808

observed in the age bin of 19–20.99 years (average age of 19.88), which was close to the predicted age. This suggests that specific age ranges within Reptilia align more closely with the predicted ages, thus supporting the hypothesis within these age groups.

Interestingly, the data suggest that lower Fibonacci ages, such as 7.08 or 11.45, may act as attractors for certain taxonomic groups. These attractors could account for the observed negative associations for a higher Fibonacci age of 18.53. For instance, in the Teleostei class, a positive association for the attractor age of 7.08 could result in negative associations for the age bins 17–18.99 and 19–20.00. Conversely, the Amphibia and Aves classes displayed negative associations for age bins close to the higher Fibonacci age of 18.53, but positive associations were seen at the lower age of 11.45.

Negative associations for age bins that were not close to the predicted ages in mammals and reptiles can be interpreted as deviations from the predicted ages in the expected direction. These findings underline the potential influence of attractor ages and deviations from expected patterns in understanding variations in longevity among different taxonomic groups.

Further data analysis, detailed in Tables 15 and 16, provided more support for the hypothesis, particularly when allowing a 10% tolerance between the mean age and the predicted Fibonacci age. Remarkably, all considered classes—Reptilia, Primates, Teleostei, Mammalia, Amphibia, and Aves—exhibited adherence to this tolerance in either pre- or post-filtering stages, indicating a potential support for the hypothesis.

For instance, while Amphibia demonstrated a deviation beyond this range (–17.32%) in pre-filtering, it realigned to a 10.39% difference post-filtering.

The overall average percentage differences in the pre- and post-filtered data were 2.34% and –3.90%, respectively. These findings suggest a general alignment with the Fibonacci ages, particularly when considering the law of large numbers. Nonetheless, while the observed alignments across various taxonomic groups lend some support to our hypothesis within the 10% tolerance, it is imperative to note that the distinct species-specific lifespan variations can propel mean age differences outside this range in some instances.

Thus, while Fibonacci age appears to play a role in determining lifespan patterns, it is evident that other factors are also at play and warrant further investigation. By collating and interpreting these key findings, we gain a better understanding of how the results support or challenge our initial hypotheses. This study also lays the groundwork for future research in the fascinating field of longevity studies, providing new perspectives on the factors that shape the lifespan of chordates.

6 General Discussion

Our exploration of lifespan variation intricately links the mathematical precision of Fibonacci sequence and energy optimization with the biological phenomena of RNA dynamics and DNA heritability across species. The manifestation of specific Fibonacci ages, notably 11.45 and 18.53, within various lifespan data not only highlights a plausible synchrony of mathematical logic with biological processes but also opens an avenue to decipher the underlying genetic and molecular frameworks that might be playing a pivotal role in such occurrences.

RNA dynamics, serving as a mediator between the genetic blueprint and phenotypic expression, might be inherently influenced by Fibonacci fractal patterns in guiding cellular processes, particularly those pertaining to development and aging. Our study intertwines the pre-existing knowledge of RNA structomics and new findings in the field of systems biology, providing a novel viewpoint that sees mathematical patterns not merely as coincidental occurrences but potentially as a structured blueprint deeply embedded within life processes.

In addition, DNA heritability and its role in dictating lifespan touch upon intriguing aspects of evolutionary biology. The alignment of Fibonacci numbers with lifespans across a variety of species raises questions regarding the evolutionary pressures and advantages that might be associated with such patterns. This raises the following question: Could these Fibonacci patterns be a heritable trait conserved across various lineages due to some inherent evolutionary advantage?

Furthermore, our findings hint at the relevance of Fibonacci age associations, extending beyond the specific taxonomic groups examined in this study. Examples from the insect world, such as queen ants living up to 18–29 years (Carey 2001) and cicadas with unique life cycles lasting 12 or 18 years (White and Lloyd 1979), suggest an alignment with Fibonacci ages. This indicates the potential wider applicability of Fibonacci age patterns across diverse taxonomic groups.

In dovetailing Fibonacci principles and energy optimization, our study also embraces the concept of entrainment to planetary cycles, shedding light on the potential influence of macroscopic rhythms on life. This exploration propounds the susceptibility of life to planetary cycles, notably those aligned with the Fibonacci sequence (Sacco 2019), and underscores the pivotal linkage between planetary cycles and lifespan variation. Furthermore, it elevates life's role as an entropy exporter in cosmic dynamics, contributing a crucial lens through which to view lifespan regulation across species (Schneider and Sagan 2005).

We acknowledge the limitations of our research, recognizing that the sample size and taxonomic scope may limit the broad applicability of our findings. Further research is needed to examine the specific genetic, molecular, and environmental factors that contribute to lifespan variation.

In summary, our study expands the understanding of lifespan regulation by investigating the relationship between fractal networks, Fibonacci energy optimization, and empirical observations. This integrated approach provides invaluable insights into the complex interplay among genotype, phenotype, and lifespan. The potential universality of Fibonacci age associations across taxonomic groups, along with the examination of entrainment to planetary cycles, paves the way for future research in lifespan regulation.

7 Conclusion

Utilizing a systems biology perspective, our investigation cohesively integrated the mathematical intricacies of the Fibonacci sequence with RNA dynamics and DNA hereditary pathways, thereby constructing a meticulous representation of lifespan discrepancies across a multitude of species.

Employing advanced statistical methods, notably the chi-square test, we illuminated a notable association between Fibonacci ages, especially 11.45 and 18.53, and lifespan. This revelation not only amplifies the concept of an embedded mathematical framework within biological systems, encompassing fractal geometry and energy optimization, but also catalyzes a more profound exploration into the potential encoding or influence of these Fibonacci patterns within genetic material and their expression through RNA dynamics.

However, it is important to acknowledge the limitations of this study. Both sample size and taxonomic scope might impose constraints on the wide-ranging applicability of our findings. Future research endeavors should seek to expand the dataset, incorporate a broader spectrum of organisms, and employ advanced computational models to deepen our insights into the regulatory mechanisms of lifespan. Addressing these limitations and building upon the knowledge gained from this study, we can continue to deepen our understanding of lifespan, paving the way for the formulation of interventions and strategies to not only promote healthier aging but also enhance quality of life.

References

Agresti A (2012) Categorical data analysis. Wiley

Alberts B, Johnson A, Lewis J, Raff M et al (2002) Molecular biology of the cell, 4th edn. Garland Science. https://www.ncbi.nlm.nih.gov/books/NBK21054/

Bahar R, Hartmann CH, Rodriguez et al (2006) Increased cell-to-cell variation in gene expression in ageing mouse heart. Nature 441:1011–1014

Bejan A, Zane JP (2013) Design in nature: How the constructal law governs evolution in biology, physics, technology, and social organizations. Anchor.

Carey JR (2001) Demographic mechanisms for the evolution of long life in social insects. Exp Gerontol 36:713–722

Cech TR, Steitz JA (2014) The noncoding RNA revolution—Trashing old rules to forge new ones. Cell 157:77–94

de Magalhães JP, Costa J (2009) A database of vertebrate longevity records and their relation to other life-history traits. J Evol Biol 22:D537–D543

Djebali S, Davis CA, Merkel A et al (2012) Landscape of transcription in human cells. Nature 489: 101–108

Douady S, Couder Y (1996) Phyllotaxis as a dynamical self-organizing process part I: the spiral modes resulting from time-periodic iterations. J Theor Biol 178:255–274

Esteller M (2011) Non-coding RNAs in human disease. Nat Rev Genet 12:861–874

Goldberger AL, West BJ, Dresselhaus et al (1985) Bronchial asymmetry and Fibonacci scaling. Experientia 41:1537–1538

Hayflick L (2007) Biological aging is no longer an unsolved problem. Ann NY Acad Sci 1100:1–13

Houle D, Govindaraju DR, Omholt S (2010) Phenomics: the next challenge. Nat Rev Genet 11: 855–866

Ideker T, Galitski T, Hood L (2001) A new approach to decoding life: systems biology. Annu Rev Genom Hum Genet 2:343–372

Iglewicz B, Hoaglin DC (1993) How to detect and handle outliers. Quality Press

Kenyon CJ (2010) The genetics of ageing. Nature 464:504–512

Kim EB, Fang X, Fushan AA et al (2011) Genome sequencing reveals insights into physiology and longevity of the naked mole rat. Nature 479:223–227

Kitano H (2002) Systems biology: a brief overview. Science 295:1662–1664

Larsen SH (2021) DNA structure and the Golden ratio revisited. Symmetry 13:1949

Lee Y, Rio DC (2015) Mechanisms and regulation of alternative pre-mRNA splicing. Annu Rev Biochem 84:291–323

Lee MT, Bonneau AR, Giraldez AJ (2014) Zygotic genome activation during the maternal-to-zygotic transition. Annu Rev Cell Dev Biol 30:581–613

López-Otín C, Blasco MA, Partridge L et al (2013) The hallmarks of aging. Cell 153:1194–1217

Mamoshina P, Vieira A, Putin E et al (2016) Applications of deep learning in biomedicine. Mol Pharm 13:1445–1454

Mattick JS (2004) RNA regulation: a new genetics? Nat Rev Genet 5:316–323

Mattick JS, Makunin IV (2006) Non-coding RNA. Hum Mol Genet 15:R17–R29

Moore LD, Le T, Fan G (2013) DNA methylation and its basic function. Neuropsychopharmacol 38:23–38

Nilsen TW, Graveley BR (2010) Expansion of the eukaryotic proteome by alternative splicing. Nature 463:457–463

Omland KE, Cook LG, Crisp MD (2008) Tree thinking for all biology: the problem with reading phylogenies as ladders of progress. BioEssays 30:854–867

Perez JC (2010) Codon populations in single-stranded whole human genome DNA are fractal and fine-tuned by the Golden Ratio 1.618. Interdiscip Sci Comput 2:228–240

Phillips KA, Bales KL, Capitanio JP et al (2014) Why primate models matter. Am J Primatol 76: 801–827

Prusinkiewicz P, Runions A (2012) Computational models of plant development and form. New Phytol 193:549–569
Putin E, Mamoshina P, Aliper A et al (2016) Deep biomarkers of human aging: application of deep neural networks to biomarker development. Aging (Albany NY) 8:1021
Sacco RG (2019) Modeling celestial mechanics using the Fibonacci numbers. Int J Astron 8:8–12
Sacco RG, Torday JS (2023) Systems biology of human aging: a Fibonacci time series model. Prog Biophys Mol Biol 177:24–33
Schlotterer C, Tobler R, Kofler R et al (2014) Sequencing pools of individuals – mining genome-wide polymorphism data without big funding. Nat Rev Genet 15:749–763
Schneider ED, Sagan D (2005) Into the cool: energy flow, thermodynamics, and life. University of Chicago Press
Seim I, Fang X, Xiong Z et al (2013) Genome analysis reveals insights into physiology and longevity of the Brandt's bat Myotis brandtii. Nat Commun 4:1–8
Sen P, Shah PP, Nativio R et al (2016) Epigenetic mechanisms of longevity and aging. Cell 166: 822–839
Sonenberg N, Hinnebusch AG (2009) Regulation of translation initiation in eukaryotes: mechanisms and biological targets. Cell 136:731–745
Strobel EJ, Watters KE, Loughrey D et al (2016) RNA systems biology: uniting functional discoveries and structural tools to understand global roles of RNAs. Curr Opin Biotechnol 39: 182–191
Thirumalai D, Lee N, Woodson SA et al (2001) Early events in RNA folding. Annu Rev Phys Chem 52:751–762
Torday JS (2019) Cell-cell communication predicts aging, senescence and death: an integrated, predictive evolutionary approach. Biomed Rev 30:15
Torday JS (2022) Hormones and reality: epigenetic regulation of the endocrine system. Springer Nature
Torday JS, Rehan VK (2011) A cell-molecular approach predicts vertebrate evolution. Mol Biol Evol 28:2973–2981
Wang Z, Gerstein M, Snyder M (2009) RNA-Seq: a revolutionary tool for transcriptomics. Nat Rev Genet 10(1):57–63
Wang TL, Kuznets-Speck B, Broderick J, et al (2020) The price of a bit: energetic costs and the evolution of cellular signaling. bioRxiv. https://www.biorxiv.org/content/10.1101/2020.10.06.327700v3
West GB, Brown JH (2005) The origin of allometric scaling laws in biology from genomes to ecosystems: towards a quantitative unifying theory of biological structure and organization. J Exp Biol 208:1575–1592
White J, Lloyd M (1979) 17-year cicadas emerging after 18 years: a new brood? Evolution 33: 1193–1199
Wood SH, Craig T, Li Y et al (2013) Whole transcriptome sequencing of the aging rat brain reveals dynamic RNA changes in the dark matter of the genome. Age 35:763–776
Woodson SA (2010) Compact intermediates in RNA folding. Annu Rev Biophys 39:61–77
Zemora G, Waldsich C (2010) RNA folding in living cells. RNA Biol 7:634–664

Refinement of Single-Cell RNA-seq Gene Expression Signatures with Combiroc

Mauro Bombaci, Ivan Ferrari, Saveria Mazzara, and Riccardo L. Rossi

Contents

Abstract In this work we showcase how the combiroc R package can be used to refine and optimize gene signatures generated by high-throughput omics. Our workflow identifies subsets of gene expression signatures from traditional single-cell RNA sequencing experiments and allows distinguishing cells with higher confidence and fewer markers. Using a well-annotated single-cell RNA-seq dataset, we demonstrate that smaller signatures of lower-ranking genes selected with this combinatorial approach exhibit a greater power than top differentially expressed genes: these sub-signatures maintain a descriptive ability in terms of biological

M. Bombaci · I. Ferrari
Istituto Nazionale Genetica Molecolare "Romeo ed Enrica Invernizzi" Via F. Sforza 35, Milan, Italy

S. Mazzara
Department of Computing Sciences and Bocconi Institute for Data Science and Analytics (BIDSA), Bocconi University, Via Rontgen 1, Milan, Italy

R. L. Rossi (✉)
Istituto Nazionale Genetica Molecolare "Romeo ed Enrica Invernizzi" Via F. Sforza 35, Milan, Italy

Revelo Datalabs Srl, Viale Liguria 40, Milan, Italy
e-mail: rossi@revelodatalabs.com

J. Barciszewski (ed.), *Systems Biology II*, RNA Technologies 15,
https://doi.org/10.1007/978-3-031-62178-9_3

ontologies related to the cells to which they belong, defining actual and specific bio-semantic spaces.

Keywords Biomarkers · Gene signature · Single-cell RNA-seq · ROC curves · Combinatorics · R package

1 Introduction

To understand biological systems, we heavily rely on models, which are intrinsically incomplete descriptions of reality. As a common aphorism notes, "all models are wrong, but some are useful" (Box 1976), and the usefulness and usability of models when they are fed quantitative or qualitative data depend on the quality of such data. In biological and medical science today, data that are collected with the goal of building a model are collected at increasingly growing resolution and can be of many types: images, feature quantities, variation frequencies, and semantics. Omics methods (genomics, transcriptomics, proteomics, and many others) mainly deal with cell or tissue-specific gene or protein signatures that are made of many features, often tens to hundreds. Gene signatures as they are designed and obtained, take into account the few top differentially expressed genes compared to pairwise contrasts; applying such hard thresholds can hamper the ability to describe wider and more complex contexts since biological systems often share molecular determinants due to their cooperative nature and functional redundancy, and these determinants can work and do cooperate under a priori hard thresholds. Limiting the investigation to a few features only because they are positioned at the very top of thresholds-imposed ranking is an even more incomplete description of what is actually happening in the system, and either the selected features are used to explain a biological phenomenon or they are used as diagnostic markers.

To address this issue and discover more comprehensive descriptors, we harness the combinatorial power of a method we previously described (Mazzara et al. 2017) as a marker signature refiner. We recently implemented the method in the new combiroc package (Ferrari et al. 2023) and devised a protocol tailored for single-cell RNA-seq (scRNA-seq), one of the prevailing molecular techniques nowadays. scRNA-seq allows for the analysis of gene expression profiles of individual cells within heterogeneous populations; this level of granularity enables researchers to gain a detailed and precise insight into gene expression patterns that traditional bulk RNA sequencing methods cannot provide. Leveraging single-cell transcriptomics allowed for a more comprehensive exploration of phenotypes characterized by cellular heterogeneity and a more accurate depiction of the dynamics of development and differentiation at the single-cell level.

Here, we present a procedure in which we employ the new combiroc R package to refine differential expression gene signatures, using them as input to identify the most efficient gene combinations (in terms of specificity and sensitivity) for discriminating among different clusters of annotated cell types. Over the past few years, numerous scRNA-seq datasets have been published and collected in repositories. Some have been organized as resources in reference databases, and we have leveraged these collections, with a particular emphasis on the annotated immunological single-cell dataset available today (Butler et al. 2022). For our analysis, we opted to use a peripheral blood mononuclear cells (PBMCs) dataset from a healthy donor which provides comprehensive descriptions and annotations of immunological cells at three levels of granularity. This dataset served as an initial reference point to assess the rigor of cell-specific gene signatures as they are currently reported.

2 Data Retrieval, Annotation, and Differential Expression

Well-characterized peripheral blood mononuclear cell (PBMC) datasets, generated through single-cell RNA-seq methods, are readily available today from various sources, including analytical packages in the R ecosystem (Ihaka and Gentleman 1996; R Core Team 2023) and sequence technology providers. In particular, 10x Genomics offers on its website a wide resource section where many different types of datasets can be downloaded from (10x Genomics 2023). At the time of our accession of such resources, more 632 datasets were present, and among them we obtained the dataset “10k PBMCs from a Healthy Donor (v3 chemistry)” (shortly the PBMC-10 K dataset) relative to peripheral blood mononuclear cells from a healthy donor sequenced on Illumina NovaSeq with approximately 54,000 reads per cell (10x Genomics 2018). The HDF5 file format for the filtered feature/cell matrix of the dataset was then read with the Seurat package (Satija et al. 2015) and processed as an object of 20,292 features (genes) across 11,537 samples (cells) and a single RNA assay. We followed the package’s standard single-cell gene expression protocol to identify cell clusters based on their gene expression profiles. After standard pre-processing and automatic clustering, more than 20 different cell clusters were identified. Nevertheless, we used the reference-based system implemented in the Azimuth package (Butler et al. 2022) to categorize all cell clusters into biologically meaningful groups. This method employs a weighted nearest neighbor representation of RNA and protein expression, obtained through CITE-seq experiments, which Azimuth uses to assign cell types to a query dataset. By superimposing the level1 reference for PBMCs from Azimuth onto our dataset, we obtained clusters for the main immunological subpopulations, including B cells, T cells (CD8-T and CD4-T), dendritic cells (DC), NK cells, and mononuclear cells (Mono), as well as other less-characterized clusters (labeled “other” and “other T”). Using this clustering, we performed the standard Seurat (version 4) protocol to find differentially expressed genes within each of the six well-annotated subpopulations versus all others using the *FindMarkers()* command with default parameters. Once the differential expression gene signatures for each cell type were obtained, we carried on the combiroc protocol as detailed in the following paragraphs (Fig. 1).

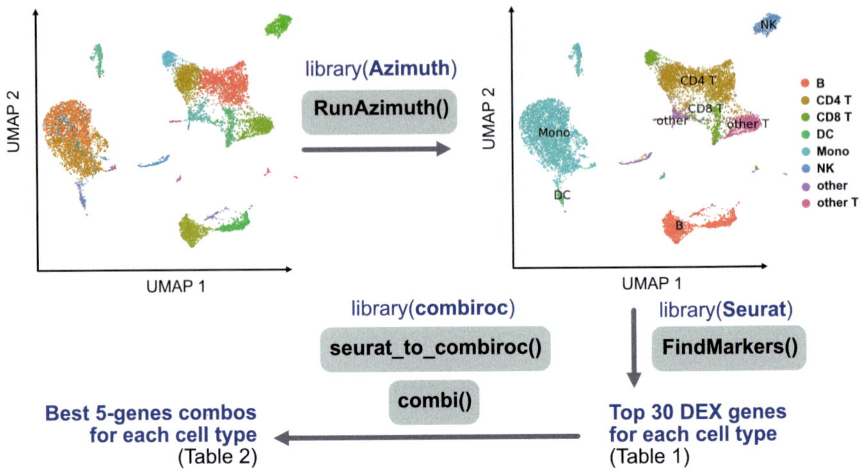

Fig. 1 Workflow of the combiroc R package for scRNA-seq. The PBMC-10 K scRNA-seq dataset was processed with Seurat version 4, and single cells were grouped into 21 different clusters (UMAP plot on the left). These cells were subsequently annotated with the level1 Azimuth reference (UMAP plot on the right) and then funneled into the differential expression Seurat protocol with standard processing; differential gene expression signatures were determined for six main cell types, and the 30 top genes by fold change were selected for each comparison. These lists of 30 genes (see also Table 1) were used as the combiroc input in the computation of all possible combinations made of up to five genes. The best five-gene combinations (see also Table 2) were selected then according to their Youden index ranking

3 Combiroc Workflow

3.1 Search for the Best-Performing Gene Combinations

Gene expression analysis performed with Seurat resulted in a gene signature for each pairwise comparison of a specific cell type versus all other cells. We expressly determined the gene signatures for six cell clusters (NK and B cells, dendritic and mononuclear cells, CD8 and CD4 T cells) made of positive genes only (*only.pos* argument equal to TRUE in the *FindMarker* function), i.e., genes that resulted overexpressed in the specific cell cluster compared to other clusters altogether. All other parameters were the default ones. For each one of these six cell types, we took into consideration only the 30 most overexpressed genes as cell-type-specific gene signatures (Table 1).

These gene lists were then used as input for the combiroc package. Given a list of markers, combiroc assesses performance in terms of specificity (SP), sensitivity (SE), and ROC curve computations for all possible combinations of these markers. By combinations, we mean combinations without repetition, i.e., a subset of k distinct elements from a set of n elements, which can be calculated using binomial coefficients. The computational burden of this process can be significant when dealing with more than ten markers: a list of n markers generates $2^n - 1$

Table 1 Differential expression signatures

Rank	NK	B	DC	CD8T	Mono	CD4T
1	GNLY	IGLC2	HLA-DPA1	CD8B	S100A9	IL7R
2	NKG7	IGKC	HLA-DPB1	CCL5	S100A8	TRAC
3	PRF1	IGHM	HLA-DQA1	CD8A	LYZ	TRBC1
4	GZMA	IGHA1	HLA-DRA	CD3D	S100A12	CD3D
5	KLRD1	IGLC3	HLA-DRB1	GZMH	CTSS	LDHB
6	GZMB	CD79A	HLA-DQB1	TRAC	FCN1	LTB
7	SPON2	MS4A1	CD74	GZMK	VCAN	TRBC2
8	CTSW	IGHD	FCER1A	IL32	MNDA	IL32
9	FGFBP2	CD79B	IRF8	CD3E	FTL	CD3G
10	CST7	TCL1A	CST3	CD3G	PSAP	CD3E
11	KLRF1	JCHAIN	PTGDS	S100B	AIF1	TCF7
12	HOPX	BANK1	GZMB	TRGC2	TYROBP	NOSIP
13	TRDC	LINC00926	HLA-DQA2	GZMM	NEAT1	CD2
14	CLIC3	CD37	HLA-DRB5	CTSW	FOS	BCL11B
15	CCL4	CD22	PLD4	TRBC2	LST1	RPS29
16	CD247	RALGPS2	TCF4	CST7	DUSP1	LCK
17	IL2RB	TNFRSF13C	ITM2C	LCK	TYMP	MAL
18	CD7	CD74	PPP1R14B	CD27	FTH1	RPS27
19	GZMM	HLA-DQA1	CLEC10A	TCF7	S100A6	PIK3IP1
20	CCL5	VPREB3	CCDC50	GZMA	CD14	CD27
21	MYOM2	IGHG1	IRF7	LINC02446	LGALS1	RPS12
22	GZMH	PLPP5	HLA-DMA	NKG7	FGL2	AQP3
23	CMC1	HLA-DQB1	JCHAIN	NELL2	FCER1G	LEF1
24	MATK	BLK	UGCG	TRBC1	CST3	RPSA
25	KLRB1	FCER2	CPVL	CD7	MS4A6A	RCAN3
26	KLRC1	FCRL5	HLA-DMB	CD2	SERPINA1	CD6
27	ADGRG1	BCL11A	CD1C	LYAR	S100A11	EVL
28	ABHD17A	FAM129C	BASP1	STK17A	CYBB	ETS1
29	APOBEC3G	SPIB	LILRA4	BCL11B	SRGN	RPS3
30	IFITM2	SWAP70	BCL11A	CLEC2D	COTL1	RPS25

The table reports the gene signature made of the top 30 differentially expressed genes (overexpressed) in each one of the six cell types taken into account. Genes are ranked according to log fold change, with higher effect size for top markers

combinations, resulting in over a billion possible combinations for $n = 30$. To manage the computational load, we decided to exclude combinations made by more than five genes, reducing the total number of combinations to compute to 174,436.

We applied the standard workflow protocol as described in the combiroc package official documentation either on CRAN (Ferrari et al. 2023) or the GitHub repository to evaluate the specificity and sensitivity of each combination in each cell type (Fig. 1), resulting in a total of over a million gene combinations with their corresponding ROC curves. These gene combinations were then ranked based on

Table 2 The best five-gene combinations selected for each cell type

NK	B	DC	CD8-T	Mono	CD4-T
CLIC3	CD22	BASP1	CD8A	AIF1	CD3G
IL2RB	CD79A	CLEC10A	CD8B	CYBB	CD6
KLRF1	FCRL5	FCER1A	GZMH	FCN1	LEF1
MYOM2	JCHAIN	ITM2C	LINC02446	SERPINA1	MAL
TRDC	MS4A1	LILRA4	TRAC	VCAN	RCAN3

the Youden index J, which represents the additive effect of SP and SE (J = SP + SE-1), and the best-performing five-gene combination was selected for each cell type. The genes belonging to these winning combinations were annotated alongside the top 5 differentially expressed genes (Table 2).

3.2 Ranking of Combiroc Combinations

It's important to note that the five genes belonging to the best combination for each cell type did not occupy the top-ranking positions within the original 30 gene signatures from which they were extracted. In other words, they were not the most overexpressed or the most differentially expressed genes. Matching the position of genes from the combinations with the rankings of differentially expressed genes in the originating gene signature showed that their position is evenly distributed across all positions, without any evident enrichment. In CD8-T cells only four out of five combination genes appeared in the top 10 positions. The overall distribution of genes from the combiroc-selected combinations across the top 30 positions was essentially random (Fig. 2a). This remained true also for sub-positions: even when dividing the rankings into three tiers (first tier from ranking 1 to 10, second from 11 to 20, third from 21 to 30). Genes from the combiroc-selected combinations were evenly distributed across these tiers without any discernible preference or pattern (Fig. 2b). The position of combination genes in the differential expression ranking might seem counterintuitive, as they are selected by combiroc for their combined specificity and sensitivity for each cell cluster. To further investigate, we examined whether the cumulative expression of combination genes correlated with their specificity compared to the cumulative expression of an equal number of top differentially expressed genes.

3.3 Comparison of Combinations and Signature Performance

Overall, markers selected with our combinatorial method proved to be more effective in discriminating cell types compared to the top 5 differentially expressed

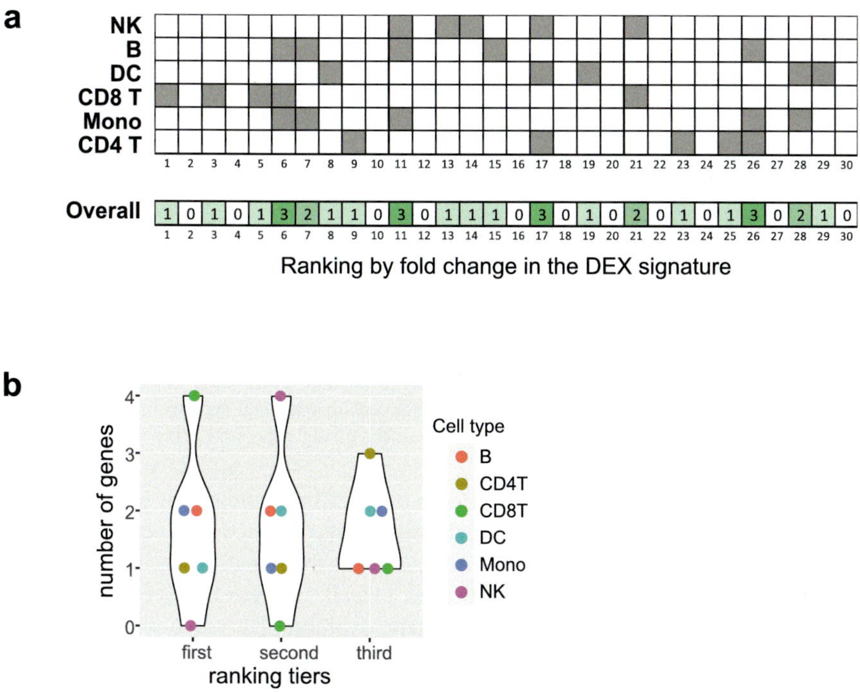

Fig. 2 Genes from combiroc combinations in the context of differential expression ranking: (**a**) Positional distributions of the genes from the five-marker combiroc combinations on the differentially expression ranking (from 1 to 30), for each cell type. A gray box represents one gene from the combination positioned in the differentially expression ranking. The "Overall" line shows the cumulative number of genes (green gradient) present in each ranking position across all cells and along the ranking. (**b**) Overall distribution of genes from combinations divided in three ranking tiers: *first*, upper tier from rank 1 to 10; *second*, middle rank from 11 to 20; and *third*, lower rank from 21 to 30. Dots indicate the number of genes from the cell-specific combination that can be found in any one of the three tiers (same color dots' values sum up to 5)

markers. The primary advantage of combiroc was observed in specificity, rather than in sensitivity. Even in cases where combiroc markers had lower sensitivity, the cumulative Youden index consistently favored combiroc markers (Fig. 3). This demonstrates that the top differentially expressed genes are not necessarily the most cell-type-specific ones.

For each cell type, we also determined the cumulative expression of the five most differentially expressed genes, i.e., the first five genes from each column of Table 1, and compared it to the cumulative expression of the five genes in the cognate combination (genes in Table 2). The cell cluster annotation taken as a reference (Fig. 4a) was obtained from the Azimuth reference-based process previously described, and it corresponds to the cell types used for pairwise comparisons in differential expression analysis (one specific cell type versus all other cells in the dataset). While one might expect the five most overexpressed genes to be the best descriptors for the cells they were selected for, this was not the case: the five genes

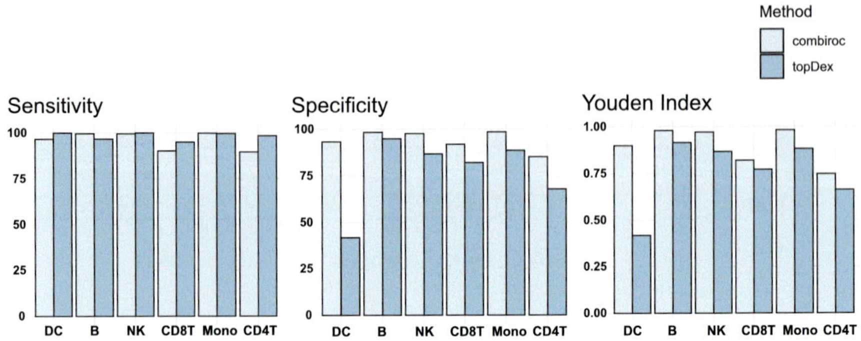

Fig. 3 Performance metrics with combiroc and differential expression. Plots for sensitivity, specificity, and Youden index are shown as determined for each cell type using combiroc five-gene combinations (light blue) and the top 5 most differentially expressed genes (blue)

from the combinations were more effective at discriminating their respective cell clusters, despite not being at the top of the differential expression gene signatures. In dendritic cells (DCs), the cumulative expression of the five differentially expressed genes extended to mononuclear and B cells as well, whereas the five combo genes were confined to the two smaller DC clusters, exhibiting very low residual expression elsewhere, indicating higher specificity (Fig. 4b, DC panels). Markers for B cells obtained through both differential expression and combiroc methods showed significant specificity, but combiroc markers exhibited stronger signals and higher sensitivity, as shown by the higher expression level (Fig. 4b, B panels). Similarly, NK cells were better discriminated by combiroc markers with less contamination from CD8-T cells compared to DEX markers. In T cells (CD8 and CD4) as well as mononuclear cells, combiroc markers also performed slightly better (Fig. 4b).

4 Genes from Combinations Define Specific Semantic Space

4.1 Definition of the Common Semantic Space

Objects, actions, and their categories are organized in a bigger overall structure that describes reality as our brain is able to represent it. The ability to organize such pieces in a smooth and consequential manner defines a de facto continuous space reflecting semantic similarities among these objects (Huth et al. 2012). In a similar way, every gene is an object that can be associated with functions (actions) and categories, while an ensemble of genes and their products is organized in pathways, acting into and through compartments, across tissues and organisms. Such cooperation defines a biological semantic space, or context; given any list of genes or gene products, such space can be described using the vast ontology databases that are currently available (The Gene Ontology Consortium 2018; Ghulam et al. 2020). These knowledge bases are sources of biological information on the functions of

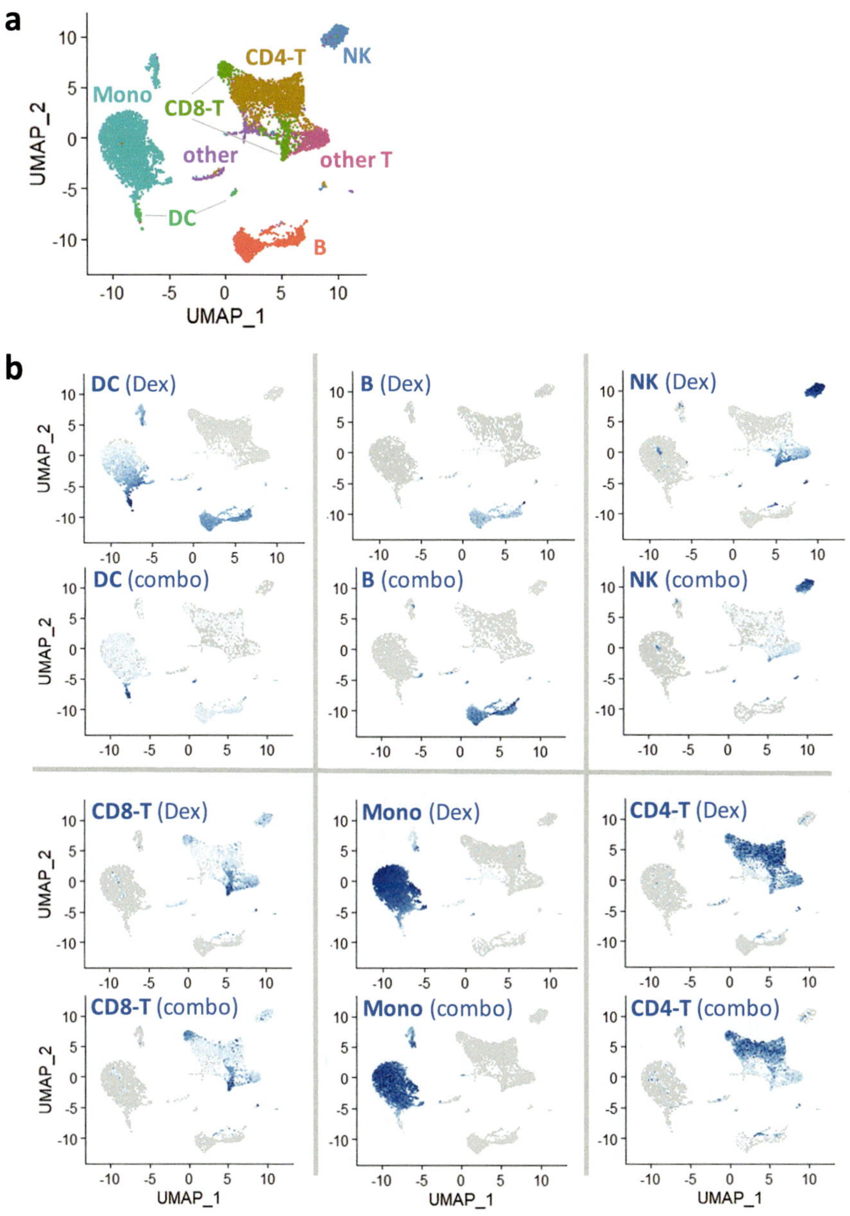

Fig. 4 Combinations exhibit greater specificity than top differentially expressed (DEX) genes. (**a**) UMAP embedding of PBMC-10 K cells clustered with the Azimuth reference cell-type annotation in different colors: dendritic cells (bright green), B (red) and NK (blue) cells, monocytes (turquoise), CD8 (forest green), and CD4 (tan) T cells. (**b**) Cumulative feature plots for the expression of the top 5 differentially expressed genes ("Dex" panels) and for the five genes from the best selected combination ("combo" panels), in each cell type

genes that are both human-readable and machine-readable, and this duality allows for network-like visualizations useful to emphasize both specificity and overlaps of actions, i.e., spaces of specific functional action in a wider biological semantic space relative to the overall context and its functional limits. Using this approach when dealing with gene signatures and their sub-samples is possible to see how the functional domains of a different group of genes intersect and overlap.

While the combinations selected for the six cell types are completely non-overlapping, the original differential expression signatures (made by 30 genes each; see Table 1) are not. The bigger overlap is observed for CD8-T, CD4-T, and NK cells (a total of 20 shared common genes), while a minor overlap exists between B cells and dendritic cells (five genes) and between DCs and mononuclear cells (one gene). We decided to take into account the gene signatures of cell types with the biggest overlap, CD4-T cells, CD8-T cells, and NK cells (Fig. 5a), amounting to a total of 70 unique genes, to build their whole additive semantic space, i.e., the ensemble of most represented biological terms, defined by these 70 unique genes from the three original signatures: on top of such semantic space, we asked which was the informational load (or sub-space) defined by the three combiroc combinations.

The genes from DEX signatures of CD8-T, CD4-T, and NK cells (summing up to a total of 70 unique genes) were used as the input for String dbQuery app (Szklarczyk et al. 2015) in Cytoscape (v.3.9.1) (Shannon et al. 2003) to retrieve the graph of annotated interactions. On this graph we determined the functional enrichment with String Enrichment, and then we used Enrichment Map (v.3.3.6) (Merico et al. 2010) to visualize the whole semantic space defined by the input genes. In this network view, each node represents a biological term, and weighted edges represent the number of genes shared by two nodes. The network was manually annotated with labels for the main terms (Fig. 5b).

4.2 *Combination-Specific Semantic Sub-spaces*

Then we queried this semantic space network separately with the five-gene combinations from NK, CD8-T, and CD4-T cells, highlighting edges containing genes from the combinations themselves (Fig. 6). While all three combinations matched some more general "central" biological functions and terms such as immune system process, cell surface receptor signaling, adaptive immunity, and T cell activation, it was immediately evident that each combination displayed a characteristic topological matching: combination from NK cells was specific for antigen processing, NK receptor-like cell, cytokine signaling, and C-type lectin domain (Fig. 6a); CD8-T cells brought about the enrichment of serine/trypsin protease and peptidase as well as cytotoxic and helper T lymphocyte functions (Fig. 6d); finally, the gene combination from CD4-T cells matched specifically with gamma-delta T cell functions, WNT signaling, and TCR recombination (Fig. 6e).

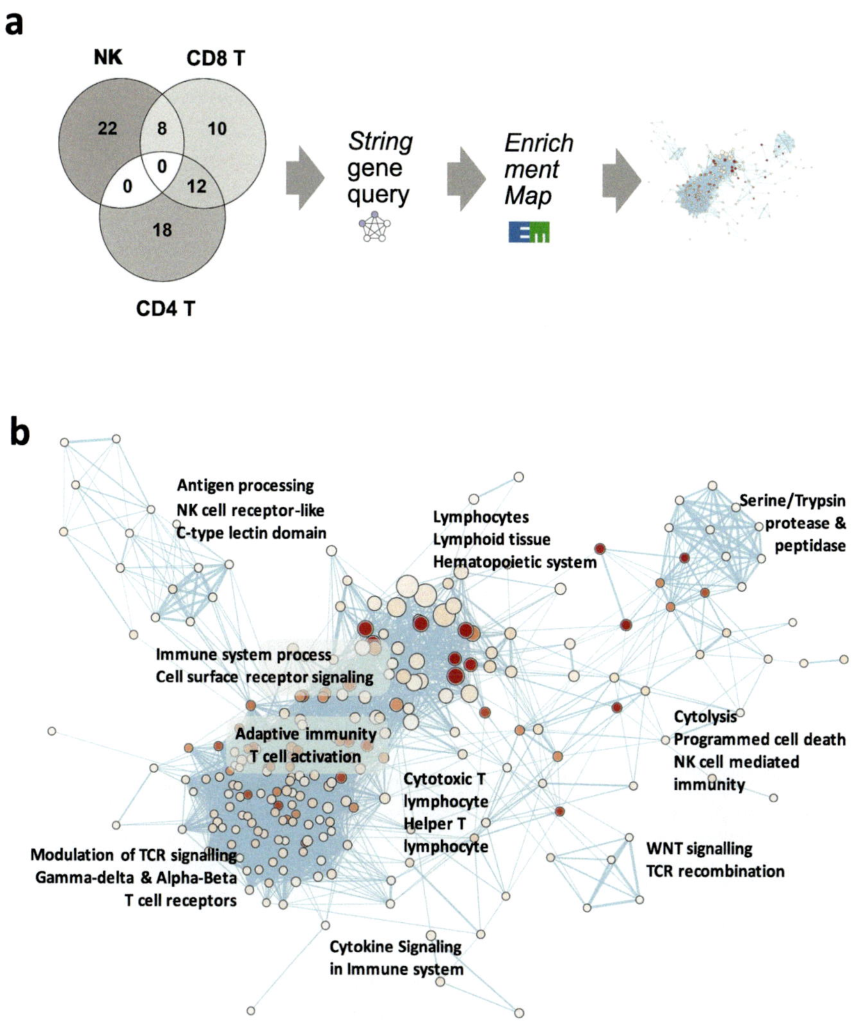

Fig. 5 Gene signatures define a common semantic space. (**a**) Venn plot of the overlap among the 30-gene differential expression signatures of CD8-T, CD4-T, and NK cells. Signatures of NK and CD8-T cells share eight genes, while signatures of CD8-T and CD4-T cells share 12 genes. Seventy unique genes are processed with String database and Enrichment Map to obtain the common semantic space. (**b**) Semantic space in higher details defined by all genes (70 unique genes) from the NK, CD8-T, and CD4-T differential expression signatures. Network nodes are biological terms; edges correspond to genes shared by two terms. The size of nodes is proportional to genesets' size and color proportional to enrichment. Ontologies for enriched pathways are topologically grouped for shared genes and main groups are indicated with text labels

Semantic spaces obtained in this way can be easily browsed term by term (the nodes of the network) looking for genes defining them (the network's edges) with a research-oriented tool such as Cytoscape or any other tool able to digest network

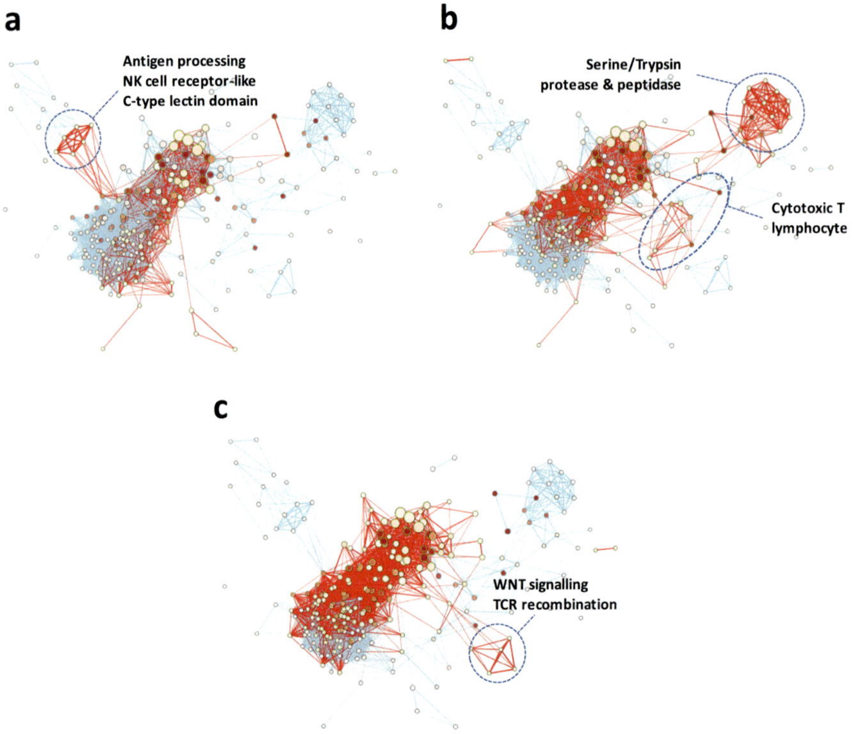

Fig. 6 Combinations from combiroc occupy a specific semantic space. Semantic sub-spaces are highlighted in red as they are defined by the network edges in which genes from the cell-specific combiroc combinations are found, namely, in (**a**) NK cells, (**b**) CD8-T cells, and (**c**) CD4-T cells. Some of the most characteristic groups of terms are labeled for the three cell types

structures. Cell-specific sub-spaces can be used to identify functional specificity of the gene combinations, as well as any other gene signature, helping to build research hypotheses or experimental validations. The specificity that characterizes the gene combination selected by combiroc can not only be devised as a diagnostic tool able to discriminate among cell types but also maintains its validity for functional interpretation of cell clusters, separating common, shared functions from specific molecular or biochemical tasks, which is a very common and more research-oriented purpose.

5 Conclusions

Extracting from the original full gene signatures' smaller subsets (i.e., combinations) of features that are able to completely describe and characterize the cells or tissues, independently to their ranking in the original signatures, is a doable but computationally heavy task, since the number of possible combinations increases exponentially, and from a few tens of genes, a huge number of combinations can be obtained, up to billions. In this work we showed that smaller gene combinations extracted from the original full signature do not impair the discriminatory power of the original signature and sometimes have an even higher cell or tissue identification power.

Differential gene expression per se is designed to find the strongest and most stable effect size, i.e., differential expression signal between two conditions without controlling for sensitivity and/or specificity. Recalling a famous slogan, "Power is Nothing Without Control," we showed how combiroc gives gene expression the missing control allowing to optimize both specificity and sensitivity, thus strongly ameliorating the power of gene expression signatures in the wider context of the single-cell gene expression analyses. Sensitivity and specificity are thus treated as a system property, not solely related to the contingency of a single signature but in the context of all possible, even if smaller, signatures.

Gene expression experiments based on cutoff-driven gene signatures is the golden method to look for specific molecular determinants, and current detection or sequencing technologies facilitated these approaches skyrocketing research into a wealth of data; in this context the downstream data reduction and ability to superimpose biological meaning and categories are key. This is why we advocate for a less dogmatic view on gene signatures, whose system-level validity is most of the time jeopardized by the limited and binary context they are extracted from. Refinement of marker signatures through search of combinations made of only a few gene markers can have a wider value as system-level descriptors if they have overall higher specificity and sensitivity. The ability to identify such combinations from the existing gene signatures or new ones through ad hoc designed transcriptomic experiments should leverage both diagnostic and research purposes. This essentially allows rising above the molecular basal noise of long and ranking-dependent signatures, separating the wheat from the chaff.

Acknowledgments We would like to thank Davide Cittaro and Asia Zonca for their fruitful discussion and comments as well as the members of the Bioinformatics Units at INGM for constructing criticism with a special mention to Andrea Gobbini. SM was funded by the European Research Council (ERC) Consolidator Award MicroC 772970 (to PI F. Buffa).

References

10x Genomics (2018) 10k PBMCs from a Healthy Donor (v3 chemistry). Official 10x Genomics Support. https://support.10xgenomics.com/single-cell-gene-expression/datasets/3.0.0/pbmc_10k_v3

10x Genomics (2023) 10x Datasets Single Cell Gene Expression, Official 10x Genomics Support. https://www.10xgenomics.com/resources/datasets/. Accessed 7 Oct 2023

Box GEP (1976) Science and statistics. J Am Stat Assoc 71:791

Butler A, Darby C, Hao Y et al (2022) Azimuth: a shiny app demonstrating a query-reference mapping algorithm for single-cell data (R package version 0.4.6). Version 0.4.6. GitHub. https://github.com/satijalab/azimuth

Ferrari I, Rossi RL, Mazzara S et al (2023) combiroc: selection and ranking of omics biomarkers combinations made easy. In: CRAN https://CRAN.R-project.org/package=combiroc. Accessed 5 Oct 2023

Ghulam A, Lei X, Guo M, Bian C (2020) A review of pathway databases and related methods analysis. Curr Bioinforma 15:379–395

Huth AG, Nishimoto S, Vu AT et al (2012) A continuous semantic space describes the representation of thousands of object and action categories across the human brain. Neuron 76:1210–1224

Ihaka R, Gentleman R (1996) R: a language for data analysis and graphics. J Comput Graph Stat 5: 299

Mazzara S, Rossi RL, Grifantini R et al (2017) CombiROC: an interactive web tool for selecting accurate marker combinations of omics data. Sci Rep 7:45477

Merico D, Isserlin R, Stueker O et al (2010) Enrichment map: a network-based method for gene-set enrichment visualization and interpretation. PLoS One 5:e13984

R Core Team (2023) R: A language and environment for statistical computing. https://www.R-project.org. Accessed 7 Oct 2023

Satija R, Farrell JA, Gennert D et al (2015) Spatial reconstruction of single-cell gene expression data. Nat Biotechnol 33:495–502

Shannon P, Markiel A, Ozier O et al (2003) Cytoscape: a software environment for integrated models of biomolecular interaction networks. Genome Res 13:2498–2504

Szklarczyk D, Francescini A, Wyder S et al (2015) STRING v10: protein-protein interaction networks, integrated over the tree of life. Nucleic Acids Res 43:D447–D452

The Gene Ontology Consortium (2018) The gene ontology resource: 20 years and still GOing strong. Nucleic Acids Res 47:D330–D338

A Cognitive Study of Modeling in Systems Biology

Miles MacLeod

Contents

Abstract In this chapter we present some results and insights of a 4-year study of cognitive practices in systems biology. Systems biology is of interest to cognitive scientists and philosophers for a number of reasons including its interdisciplinarity and its application of systems-level and mathematical explanations to biology, but one source of interest is the ability of modelers to engage with highly complex nonlinear biological networks. In this study our group tracked model-building processes among modelers in two systems biology labs during the course of their graduate degree projects. We were able to identify the main cognitive strategies by which these modelers were able to keep complexity under control, develop an intrinsic understanding of their systems, and turn this into informative inferences on their systems. They were able to this while keeping within cognitive constraints, such as working memory constraints. Computational simulation and visualization play a large role here. We identify mesoscopic and middle-out modeling as

M. MacLeod (✉)
University of Twente, Enschede, The Netherlands
e-mail: m.a.j.macleod@utwente.nl

J. Barciszewski (ed.), *Systems Biology II*, RNA Technologies 15,
https://doi.org/10.1007/978-3-031-62178-9_4

important general step-by-step strategy when it comes to managing complexity. Overall these observations can help give insights for new modelers on the kinds of day-to-day processes they may need to engage in to build models effectively and provide some theory to experienced modelers on how and why modeling might need to be constrained on cognitive terms.

Keywords Complexity · Mesoscopic modeling · Simulation · Cognition · Ethnographic study

1 Introduction

Systems biology has for many years now been of interest to philosophers of science and philosophers of biology. In particular many philosophers have been interested in the claims systems biologists have routinely made on the organization of biological systems and the need for mathematically based computational approaches to properly control and predict those systems (Green 2022). Others, including myself, have been interested in the practices of systems biologists, specifically how complexity is addressed or responded to in practical decision-making (MacLeod 2020; MacLeod and Nersessian 2018). Particularly we are interested in the capacities of human cognition and the different strategies for coping with such limits. Here I will report on an ethnographic study collected and analyzed over the past 10 years of model-building practices in two systems biology labs. The goal of the study was, among other things, (1) to explore the decisions made in the course of model-building and (2) to analyze the rational basis of those decisions—why and for what purpose they are made—given the constraints modelers are operating under. From this study we were able to give a rich cognitive account of modeling in at least strains of systems biology. In this paper I will report on five aspects of those practices and the results of our own analysis of the roles those practices play in handling complex systems. This analysis itself contributes to certain philosophical perspectives on the roles engineering the study of biological systems—and the training of modelers—but also more importantly to a refinement and rationalization of the strategy underlying much modeling in systems biology, which can be helpful to the field's long-term planning and justification. Overall this analysis should also help systems biologists develop a more theoretical understanding of how practices in their field operate, providing concepts from cognitive science and philosophy to assist in that task.

The paper will proceed as follows. In Sect. 2 I will give an overview of the ethnographic study. In Sect. 3 I will report on the results of that study. In Sect. 4 we look at some of the relevance and ramifications of those results.

2 The Study

The study I am reporting on here was carried out with the support of a NSF Reese grant from 2008 to 2013. The study took the form of an ethnographic study of model-building practices in two systems biology labs, led by Professor Nancy Nersessian. It sought to understand the means by which modelers through modeling processes, and interdisciplinary collaboration with experimenters, were able to compile models of *complex* biological systems. In particular, for instance, our group was interested in how modelers were capable of generating inferences on such systems despite complexity and despite often having no biological training. Ethnographic approaches are particularly suited to providing a fine-grained account of practices and the relations of those practices to the contexts in which they occur. For example, an ethnographic approach can help understand the contextual factors motivating certain types of decision or patterns of behavior and thus can help rationalize why certain say seemingly suboptimal decisions might be made (Nersessian and MacLeod 2022). The study of the rationality of scientific decisions is a part of the philosophy of science. In our case complexity of problems was a significant constraint on practice, as well as often relations with experimental partners or limitations on the existing data. We were able to tease apart different aspects of these, how they were informing modelers' decisions, and the compromises often involved.

As is common in ethnographic studies, our study was organized around different forms of data collection, here chiefly lab observations and interviews. The interviews were often longitudinal. Individual modelers were repeatedly interviewed over the course of a model-building project (often a PhD project). Observations took not only the form of lab observations (particularly in the case of Lab C) but also attendance and note-taking at group meetings. Acronyms are used for the names of the labs and members to preserve anonymity. The two labs we studied were different in character. Lab G was a computational lab. It contained only modelers. The modelers were for the most part from engineering backgrounds. It operated at the time through collaboration with experimental labs and studied a variety of topics concerning metabolic and cell signaling systems. Lab C was a fully equipped wet lab containing experimenters and modelers and also bimodal researchers (modelers also trained in experimental methods). The lab at the time studied in particular reductive oxidation signaling systems. Lab G however was often more engaged with the projects of experimental collaborators and thus had a variety of research subjects. Elsewhere we have written on the abilities of Lab C modelers to combine experiment and quantitative skills and the affordances and limitations of doing so (MacLeod and Nersessian 2013b). We will not focus on that in this paper. We will not discuss experimentation to a substantial extent.

For Lab G 44 interviews were conducted. For Lab C 62 interviews were conducted. We recorded over 20 hours attending lab meetings. We combined this data with analyses of the outputs of these research labs: chiefly publications and presentations. Interview data was coded using grounded theory. Grounded theory is an inductive approach to data analysis (Corbin and Strauss 1990). Analysts identify important categories in the data which represent intentions, word uses, behaviors, emotions, etc. and then observe the connections interviewees make between them. These identifications were done independently to help triangulate objectively on a set of descriptive categories. Results of this project were checked against what could be observed of practices in the scientific literature, at least for publications emerging out of labs using similar practices in systems biology. This helped confirm the result applied beyond the two labs we studied. Further, results were presented to participants in the study, as well as at scientific conferences to help provide further support for the generality of the results. For the most part, feedback did serve to confirm the results.

3 Lab Practices

As mentioned the majority modelers we observed and interviewed in these two labs had an engineering undergraduate degree. Others came from physics, applied mathematics, or another quantitative field. There is certainly a question about why the lab PIs preferred engineers which we will circle back to in Sect. 4. For the most part, these two labs were engaged in the ordinary differential equation modeling to track the dynamics of specific intra-cellular metabolic systems which perform some important function. Their approach is sometimes described in the literature as "bottom-up" systems biology, since modelers aim to build models of systems for which the underlying sequence of metabolic (or gene regulatory) interactions is known, rather than "top-down," in which data is used to reverse-engineer this system structure (O'Malley and Dupré 2005). As we will see, "middle-out modeling" can arguably be taken as a more useful description of what it is bottom-up modelers are doing. The modelers in our labs usually have predictive or control goals in mind, namely, to apply a model in order to help direct system function to a useful end.

Modelers in our labs, whether pure or bimodal modelers (those with experimental skills), usually began their work by assembling a detailed pathway representation of their chosen system from information available in published literature (and through collaboration with experimenters). This can be an intensive process in the frequent case in which that overall cellular pathway has never been constructed before. Modelers would have to pull in information from different sources. At the same time, modelers would try to collect what data they could to describe rates of reactions governing different catalyzed interactions between pathway components. Having assembled this information, modelers would then choose (on advice of a PI) how to represent the network using ordinary differential equations. Different options were available such as using power functions to represent interactions or Michaelis-Menten representations (Voit 2000). Variables in these equations represent the concentrations of each pathway metabolite in a cell (in the case of metabolic modeling). These levels go up and down as elements interact. Parameters represent rates of reactions and other constants.

Having constructed the equations and put in available data modelers would need to estimate unknown parameter values by trying to tune those parameters to ensure the equations fit available data on network performance. The resulting model could then be tested against data reserved for testing the model. If an effective model could not be found through this process, then modelers would need to think about why their model was not working and what information they might need to include in order to produce a better result. Given the complexity of these systems—particularly their nonlinearities—this was a not an easy task in general. Most modelers did not as a result achieve anything like a robust predictive model. But they were able often to come up with creative uses of the models and the modeling process. These uses help suggest alternative framings of the roles of models and the goals of systems biology in general, as we will see below.

4 Results

In this part we will discuss some of the significant findings which emerged from our study which I think would be of some value to a scientific audience. Many of our findings combine observations from the data with philosophical or cognitive theories which help interpret them.

4.1 Detail-Driven Practices

In the first place, our findings recognized that the modeling practices were specifically what could be called *detail-driven*. This is important because historically many modeling practices are not detail-driven but prefer largely simplified or abstract representations in order to extract underlying principles or laws (MacLeod 2021). Modeling in the traditional sense usually begins with some idealized or abstract representation as we see, for instance, in the Lotka-Volterra equations. If these models can reproduce some basic dynamics, it is often taken to indicate that the model captures some underlying property or principles governing the system. For instance, in some stem-cell modeling, models are structured around rarefied systems of a few interacting genes in order to show that such basic mechanisms can reproduce pluri-potentiality. That they can is taken as evidence for the importance of this basic mechanism and the principle it instantiates, despite the otherwise substantial complexity of stem-cell systems (Fagan 2016).

Systems biology modeling of this kind seems different in that respect. Modelers usually work from the starting point that given complexity, particularly nonlinearity, they are not necessarily in a good position to assess what features of a network are important to represent at the outset or whether or not there really are discrete underlying principles for such networks. Rather they operate on the assumption that the more information they can place in a model to start with the better chance of accurate or reliable predictions. As such *high-fidelity* models are needed to make progress. Once models are assembled and working, the model itself can help make decisions about how complexity might be reduced (e.g., through sensitivity analysis). But it is hard to do this reliably given the complexity of these systems without a complex model to begin with. These practices, which we observed in our labs, thus represent a break with traditional mathematical biology and modeling in many other fields, although probably there are similar modern trends across science given the power of modern computation. This detail-driven practice does represent however a distinct modern mathematical tradition which is afforded by powerful computation. And as we will see this work requires different skills from traditional mathematical ones.

Detail-driven practices demand that researchers, such as those in our labs, put a substantial effort into assembling detailed network models. For our modelers this was itself often a difficult process, requiring foraging through sources. It also ensured that their efforts were tied to the quality of the information they had at the outset and whether or not those representations were complete. This observation is pertinent to the next finding.

4.2 Debugging

A significant aspect of the practices we observed among modelers could be considered debugging processes. These were intense processes and sometimes frustrating ones. But they give us I think a more realistic picture of what skills and thought-processes modeling in this strand of systems biology actually requires in order to achieve useful outcomes. Students are often taught the mathematical aspects of modeling—how to do stability analysis, for instance. But they really seem taught or instructed on the kinds of practices and heuristics which are often required to locate errors in models and hypothesize solutions. Yet the real world rarely fits the existing mathematical frameworks perfectly.

As mentioned the models modelers produce at the first instance were rarely complete—missing necessary structural elements or including inaccuracies in parameter values. Modelers almost immediately needed to turn their attention to locating or diagnosing the sources of error. They do not do this solely or even chiefly by checking the correctness of the biological knowledge contained within this models. Often this knowledge is speculative or unknown. Rather modelers relied on the models themselves and pursued a variety of heuristics to help locate errors and improve models through targeted hypotheses on what needed to be included. Our group analyzed these practices in detail based on longitudinal study model-building.

In the first place, modelers have to locate sources of error. A basic heuristic is to run the model and compare it against the known output at various stages or steps in the model (see Fig. 1). If expectation is on track for some elements of the pathway but not others with respect to metabolite concentrations, then this can help localize where an error might be. The computer can be asked to "offline variables" and isolate

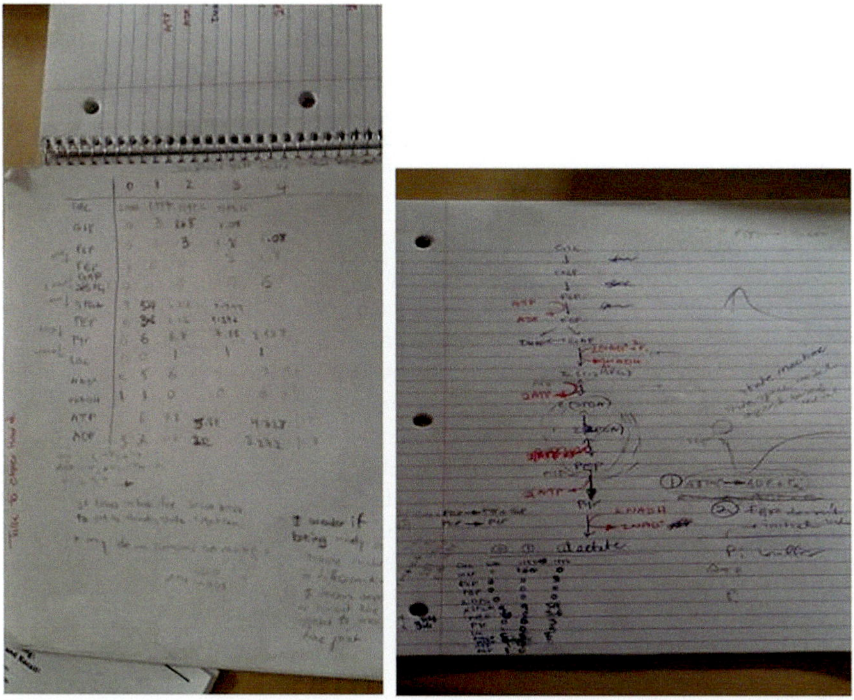

Fig. 1 A modeler "getting a feel" for the model in this case using pen and paper to explore how the model works (Republished from MacLeod and Nersessian (2018, p. 11))

just specific sub-pathways to observe their specific behavior. Through affordances of computation, modelers can have a good control in these respects. They can run simulations frequently and observe a graphical output at any stage in the model they wish too and look for elements that do not meet expectations. Such information will not always localize an error, but doing so helps build a set of hypotheses about where that error might be and what is in fact missing. For instance, if there is too much flux along a particular channel or section of the model, then this should show up in the numbers and will be responsible for a model not settling to equilibrium appropriately (in the case that equilibrium is an expected outcome).

In order to resolve inaccuracies modelers can experiment with different hypotheses—in practice modifications to the models which would remove the error. This is relatively easy to do with a simulation model. Generally modelers can apply a heuristic which moves from small localized changes (such as the addition of an arrow representing a plausible new interaction between the existing elements or adjusting parameters) to changing mathematical representations of interactions or posing new elements in the network de-black-boxing components which stand in for a more complex set of interactions. By running simulations modelers can run through options (MacLeod 2020).

From this information we have an image of modeling practice centered on the ability to diagnose and fix errors. As one of the modelers put it:

> I find glitches in the model, and why is it that, for example. . .And when you look at it there's no way it can get better because it depends on two things, and those two other things, for example, are increasing. So you can never get it decreasing for a period of time from those two. Maybe something else has a role that I haven't taken into account.

What is remarkable about these practices is that diagnosing and resolving errors lead modelers to make inferences on their biological systems, hypothesizing new relations or new elements at work in those systems, despite not having a deep background in biology. In one case, for instance, a modeler was able to postulate the existence of a novel element interacting on his system. He did so by recognizing that a proportionality relation between fluxes leaves the system at one part of the network and the imbalance in the ratio between fluxes moving through two other channels. He proposed a novel metabolic element which would result from the flux leaving the system but which would also regulate the flow in the other two channels. This would provide a mathematical solution. In this case the existence of an element playing these roles was identified experimentally. The modeler in this case did not have a firm idea of what this biological element could be, but from the errors in the model, it was for him the most parsimonious solution to produce a working model.

4.3 Intensive

The third feature I want to identify here is the intensity of the model practices we observed. These model processes are described by the modelers themselves as difficult and intensive. They require running repeated simulations and repeated trial and error, combined with long periods of feeling like getting nowhere. As engineers they often point to the differences between engineered systems and biological ones in these respects. Engineering systems behave better—errors are easier to identify with certainty, and the outcome of modifications more predictable. But the basic method or approach is something akin to engineering work nonetheless. For instance, as a modeler put it in one conversation:

> Modeller: It's not like rocket science, but it is not easy as well. Like, you really need to be very much focused when you're working on it. So, cause you. . .you focus on something and then something else goes wrong and then how you go about. . .umm. . .checking everything and putting things together . . .umm. . .you gotta be very. . .umm. . .like, present minded if that works.
>
> Interviewer: Sure. And are these new skills that you've learned or are they carrying over from engineering in some way?
>
> Modeller: Yeah. I think. . .umm. . .so many engineering people already have it.

As a result graduate modelers tend not to build more than two models during their PhD. They report the need for high levels of concentration and importantly the need to build over time—through repeated simulations—what they call "a feel for the model." This capacity was well described by one modeler in the following words:

> So when you get here, you're like very frustrated. Like, nothing is known to any extent [with emphasis]. Like, and then you. . .you think like can all these like errors here and there and not, like, non exactnesses here and there. Like, I don't think you can get any truth out of the system. That's what you think in the beginning. But then after a while, you know what to expect and you know that kind of thing is not gonna. . .you can reason that. . .umm. . .that kind of thing and that error in there is not gonna effect the whole system like that.

To us what modelers were expressing here was the importance of building a deep understanding of cause-and-effect relations between parameters and variables in a model in order to better localize errors and propose corrections. Doing so allows modelers to anticipate where the causes of problems might be located—by, for instance, ruling out options as the modeler suggests above and also proposes solution. Doing so helps limit the complexity of the search tasks modelers face. However acquiring these capacities requires an intensive work. Modelers spend much time running simulations, experimenting on their models, and, by virtue, building up an understanding of the cause and effect in the models. As such modeling this kind of modeling can be described as intense, requiring a high level of dedication to the task and ability to cope with failure.

4.4 Cognitively Strategic

The fourth aspect here is the cognitively strategic dimensions of modeling, which represents a more theoretical understanding of how modelers were operating. On our analysis of their practices, modelers operated on their complex systems using strategies which met their cognitive constraints. For example, modelers tended to exhibit qualitative reasoning when trying to interpret the cause and effect, rather than more information-rich mathematical description. That is, modelers would try to reduce more complex causal relations into basic assessments of how element X would affect element Y. Consider this quote:

> I find glitches in the model, and why is it that, for example. . .And when you look at it there's no way it can get better because it depends on two things, and those two other things, for example, are increasing. So you can never get it decreasing for a period of time from those two. Maybe something else has a role that I haven't taken into account.

Modelers would typically translate mathematical information into qualitative information using expressions like "increasing" and "decreasing" and "inhibiting" and "promoting" to simplify relationships. Simulation and analysis of the equations themselves help produce these judgments. A complex set of relationships could then be handled with a relative ease cognitively and calculated with by tracking interactions and assessing effects through basic qualitative assessments, e.g., if X is increasing, then Y should also increase. Our modelers were not unique in this regard. People in general when faced with dynamic causal relationships will try to calculate using qualitative assessments rather than quantitative ones (Roschelle and Greeno 1987; de Kleer and Brown 1981). Doing so comes at some risk of error, but it affords a greater capacity to reason over a number of causal steps.

Additionally modelers were skilled at using the affordances of computation to keep the interactions in a network over which they needed to reason to within working memory. As Forrester puts it "The number of variables [people] can in fact properly relate to one another is very limited. The intuitive judgment of even a skilled investigator is quite unreliable in anticipating the dynamic behavior of a simple information system of five or six variables" (Forrester 1994, p. 60). Knowing these constraints modelers will use computation to derive the cause-effect relations between dislocated sets of elements, so that they can reason over large parts of a network, rather than just closely connected parts. Doing so helps build a more sophisticated "feel for a model" which can be used to locate errors which have causes some distance upstream in a network. We saw the relevance of this capacity to reason over large parts of a network in the case above in which the modeler hypothesized a novel unknown element.. In that case the solution had affects at multiple points in the network which were not directly connected. The modeler proposed a novel element which would solve problems operating at two different ends of the network.

4.5 Pragmatic

The last aspect of modeling practice I want to cite here is the essential pragmatism modelers exhibited to us. Modelers were pragmatic in a variety of ways with respect to mathematical and simplification decisions in service of obtaining a result. Indeed the whole process of modeling through trial and error, rather than a more rational directed process, could be considered a pragmatic response, sometimes resulting in rather cobbled together outcomes (MacLeod and Nersessian 2013a, 2013c). But the dimension of pragmatism I want to focus on here relates to the purposes models were put to by modelers, many purposes of which were discovered during the modeling process, rather than initially intended. In general modelers set out to produce models for predictive and control purposes. If a highly reliable model can be produced, then this should afford intervention on a system to produce a desirable outcome. However obtaining that level of reliability or credibility in a model during at least a PhD project is often difficult. Modelers face many challenges which I am sure many readers from the field are aware, including uncertainty regarding data and network structure. Hence they need to spend much time and effort on correcting models. As such the modelers we studied would often pivot their goals toward the production of new information about their networks in lieu of a predictive model. This itself can be considered a pragmatic step, which represents the ability of modelers to find ways to get useful information out otherwise difficult situations and thus find a use for the modeling process.

However modelers can also be quite ingenious in coming up with other alternative uses for their modeling, often as part of a realization that their initial modeling intentions are not capable of being realized. In one interesting case, for instance, the modeler we observed used information on which model structures *could* be consistent with the available data to reach conclusions on system structure without ever producing an exact model. In this case the modeler was faced with three alternative network structures based on available information for a certain system. Her initial intentions were to produce a predictive model of the system. But of course this uncertainty on the structure represented a significant hurdle to her capacity to do this. She also faced a lack of data, meaning that even if she could determine the correct structure, finding a predictive model would be challenging. She did however have information on various biological performance conditions or constraints which could be expected of the network, in addition to certain parameter values. So she decided to test which of the alternative structures could meet the various conditions using a Monte Carlo procedure to sample unknown potential parameter sets. She found only one of the structures was capable of satisfying all conditions across a range of tested parameter sets. With this result she put forward the conclusion that this particular structure was the only possible one given the available information.

These kinds of move are pragmatic on behalf of the modelers involved. They represent creative explorations by modelers of ways to draw information from their models in the face of complexity and the difficulties of producing predictive models. What they do reinforce again is the importance of the competence and familiarity modelers gain of their models in the process of modeling, which inspires creativity in response in how models can be used in these complex contexts.

5 Implications

Above I have identified a set of features which describe the practices my group observed among a certain type of systems biology engaged in bottom-up modeling. These features represent the outcomes of this fine-grained investigation of modeling practices. In the first instance, we think together they provide a rich picture of modeling and in turn of what kinds of skills are involved and expectations required to do it. Together they offer some information to that question raised at the beginning, as to how it is that modelers are able to produce inferences on biological systems while often lacking biological training. Modelers are able to use the affordances of their partially correct models to narrow down on sets of possibilities. Mathematical constraints can then help zoom in on relevant hypotheses. Models provide a large cognitive crux in these regards.

Further some of this information I think can be useful for senior researchers to impart to junior researchers to help craft a better sense among modelers of what their tasks actually involve and what their expectations should be. Most modelers entering these tasks did not necessarily have a good set of realistic expectations about what they could achieve in a modeling project. Educational programs could be fashioned accordingly which concentrate on trial-and-error skills using more uncertain problems (Newstetter et al. 2010). Much emphasis in scientific education has traditionally been placed on showing competence with specific methodologies on exact solvable problems and learning to apply those methods in formulaic ways. But, based on what we have seen above, this is not necessarily what modeling requires.

With those points in mind however, I want to focus here on two other potential implications which have in my view deeper significance. The first is on the role engineers in systems biology. The second are the aims of systems biology and its underlying research strategies. The second contributes to the goal of using an analysis such as this study of practices to help rationalize scientific investigation given contextual constraints we mentioned above.

5.1 The Role of Engineers and the Impact of Engineering

Those observing modern biology often cite the role of engineers in shaping modern investigative strategies. In discussion of synthetic biology, this is a prominent topic of discussion, for instance, and many philosophers and others themselves are interested in the merging of engineering approaches and concepts with biological materials (Green 2022). As such when it comes to understanding the consequences of engineering in biology, there is a focus on what might be called a traditional understanding of what engineering pertains to, namely, the productive manipulation of materials. One of the primary questions philosophers analyze, for instance—one raised by synthetic biologists themselves—is whether processes of building and construction facilitate different kinds of scientific investigation, ones perhaps more suitable to handling complexity. Those in favor of such a view often cite Feynman's dictum: "what I cannot build. I do not understand" (Nordmann 2015).

In systems biology—which does not ostensibly engineer biological materials itself—the import of engineering is rather different. Emphasis is often placed on the relevance of concepts from engineering fields to help understand and describe biological systems, such as control systems (see, e.g., metabolic control theory). And the thought is that analogies from engineered systems could help recognize patterns or motifs within biological systems (Alon 2007).

Undoubtedly engineering concepts are playing a role in systems biology. That said, the observations we draw here point to different capacities of engineers as most relevant to modeling complex systems and which should perhaps play a larger role in how we understand the impact of engineering on biology. The practices we have described above are arguably hallmarks of typical engineering practices in the course, say, of a design project. Although engineers are not building physical structures when modeling, they are of course building models and using heuristics and trial and error in order to do so. As in other engineering cases, modelers put an emphasis on the ability to maintain concentration despite failure and to be pragmatic with respect to outcomes. In this sense what is valued about engineers in systems biology and why perhaps they are readily given roles in the field are not their conceptual knowledge, but rather certain skills and values they bring to a problem-solving process—many of which are honed on physical cases and examples.

One modeler made this kind of point to us:

> Yeah, sometimes method development is mathematical like the smoothing thing…it's mathematical. But some other times, it is like an engineering problem solving sort of approach. It does help that you know a lot of mathematics. Like when I…I have sometimes ideas that might not use in the end, but when I look at this, I think about this state machines and that whatever I learned on Grad logical circuits. I think a little bit like that, maybe because I learned it in the past. But it's more engineering problem solving. You have this <the model>, and that's what it is. What do you want to do with it?

The modeler here is less certain, as many were, about the value of engineering concepts for what they were doing. But they were very sure that engineering values and engineering cognition were essential. Once again it would thus be good for engineers to have a clear sense of why they in particular are asked to work in systems biology. If we generalize from the cases in our study, then the capacity for engineers and engineering to influence and shape productively biological investigation does not just come from the application of engineering concepts and theory; rather it comes from the abilities of engineers to handle complex systems through certain cognitive patterns, behaviors, and values. These are not strictly scientific ones, in the sense of abilities to apply theory or operate experimental systems, but are rather orientated toward effective exploration of difficult problem or design spaces.

5.2 *Rationalizing Systems Biology*

This brings us to the second implication I want to bring up here, which is that the study result above provides the means to situate and strategically frame systems biology. Initial framings of systems biology put a heavy emphasis on the need for high-fidelity and large-scale models in order to effectively predict and control biological systems (Kitano 2002). Biological systems are complex, and as we saw above, only models which can capture their details will in most circumstances effectively predict how those systems will respond to interventions. Until these models are in place, medical treatments will not be optimal (Hood et al. 2004). However despite the original optimism regarding the production of predictive models, it is clear that there is no straightforward path to such models (Voit et al. 2012a, b). Data uncertainty remains a considerable constraint, with the added disadvantage that the more complex the model, the more sensitive it becomes to this data uncertainty. This put sometimes severe constraints on this overall project of prediction using detail-rich complex models. Of course by putting an emphasis on prediction and control as central motivations, systems biology has arguably put itself in a rhetorical corner. And there is perhaps a need to rethink or reframe these goals along the lines which give a better sense of what detail-rich modeling can achieve.

The kinds of practices and their cognitive underpinnings we looked at above suggest ways to do this. Rather than prediction and control, the modeling contributions were more consistently contributions to discovery and understanding. By discovery we mean here the use of models to produce inferences on the structure or parameters of biological systems. Emphasizing these goals as primary is not inconsistent with how practitioners often describe their goals in individual papers (see, for instance, Tran et al. 2008; Kuepfer et al. 2007). We also think it would assist systems biologists in their relations with experimenters. In fact the experimenters we talked to tended to be rather enthusiastic about the use of models to derive hypotheses—which helps direct experimenters toward productive areas of investigation. Experimenters otherwise tended to be negative on assertions that the models they were familiar with or helped with should be relied for prediction and control goals and were in fact skeptical that such goals could be easily obtained. One modeler summed up the attitude of his experimental collaborators as follows:

> They think of [models] as something that's... just hooked up to—to, you know, match figures.... I had somebody who was... data fitting, so, for them, it's just like, you're using your data and then, you know, you're plugging in some numbers to fit the output of your model to that. And then they would not pose a lot of faith in those models or what they predict.

An inferential role of models is a less ground-breaking role, but it is one which seems to meet the capacity of modelers with complex systems and facilitate collaborative relationships.

Turning to understanding, prediction and control goals do not themselves necessitate it, in the sense that computational work and rich data can facilitate the construction or production of models without insight into the underlying systems. However it seems clear from our results that modeling has to proceed through a deep understanding—in the sense that modelers do need to have a rich understanding of how their models (and by virtue their systems) function. The concept of a "feel for a model" embodies this idea. This concept of understanding is not necessarily a "systems-level understanding" which is commonly used in systems biology which often seems to pertain to emergent systems-level properties (see O'Malley and Dupré 2005) but is grounded rather in the mechanics of the models or systems and the capacities to work with those to correct errors.

This concept of understanding dovetails with the notion of a more longer-term "middle-out" or "mesoscopic" strategy, which might be thought of as a broader strategy relevant to the complexity and other constraints modelers face (Noble 2008). On this view models are constructed at a relatively middle scale, but once built they help afford incremental improvements (including these discovery practices). A middle-scale or mesoscopic model is so called because such models are simplified both in an upward and downward sense. In the upward sense, they only try to represent certain system functions (e.g., just dominant behaviors). In the downward sense, models employ abstraction and simplifications which limit their complexity. For instance, interactions between pathway elements are usually represented using some simplified or abstract mathematical form, e.g., power laws per biochemical systems theory (Voit 2000). These often serve to simplify more complex relations in favor of tractability. At the same time, more complex interactions might be represented in terms of single variables ("black-boxing") or elements influencing a system left out. As such middle-out models represent give a broad account which roughly captures some dominant or salient aspects of system behavior. Such models are still nonetheless complex.

Building a more middle-scale model first (rather than trying to encompass the entire complexity at one hit) maintains a cognitive control or at least the ability of modelers over time to gain cognitive control despite the complexity. As such these models can be built out in either direction, by adding opening up interactions within the model, and including a broader range of behaviors. Importantly middle-scale models do not need to be perfectly predictive to perform this role indeed almost by definition they will not be since they seat inaccuracies. As long as they are in the neighborhood of reliable solutions (the structure is not radically wrong), then they can provide a platform for improvement, as such middle-scale models afford longer-term model development.

Ultimately middle-out modeling (and the "feel for a model" concept) fits with cognitive theories on *hierarchical learning* as Voit et al. (2012a, b)—systems biologists themselves—point out in their advocacy for this view on modeling. Humans in general learn to cope with complex systems by working from simplified cases to more complex cases. We generally struggle when faced with complexity upfront. The practice we identified above, to begin from a middle-scale model, learn that model through trial-and-error analysis, and that increase its complexity, is prefaced on the fact that actual complexity of a reliable model is too large at the outset. Certainly computational simulation and engineering approaches and attitudes help ratchet up the ability to learn or come to terms with relatively complex systems at the start. Nonetheless there is a limit, and PIs can help modelers choose appropriate modeling tasks which can be performed at this middle scale (MacLeod and Nersessian 2019).

Middle-out modeling helps situate the modeling strategies we observed in our labs as a part of longer-term strategy built around incremental investigation (MacLeod and Nersessian 2019; Voit 2014). Models require time to assemble, optimize, and earn credibility. The inferential use of models (the discovery use) fits with this strategic framing, as it is a core part of the process of improving models. As such the middle-out modeling concept helps provide a rationalization of these practices, imagining that models produced are not just one off results, but contribute to an ongoing collective enterprise which will accumulate understanding of the systems models represent over the longer term. These sentiments are in conflict however with the notion that systems biology (of this bottom-up type at least) can move quickly or can rely on computational power to render highly accurate and predictive models. The state of knowledge does not seem at present to facilitate such a strategy. On this view systems biology is understood less as a radical break with previous scientific traditions but rather akin to those traditions, which have operated through the production of hypotheses and testing. Models and computational power, when used *mesoscopically*, afford an ability to identify knowledge issues inaccessible to those who do not possess a systems model. Perhaps it is good to sum up at the end then with a certain analogy. Models are something like a magnifying glass in the way they operate in these practices. They bring into focus and allow us to perceive precise errors in our knowledge and at the same perceive the solutions.

6 Conclusion

In this chapter I have described some of the results of a 5-year ethnographic investigation of two systems biology labs. We focused in particular on the ways in which modelers assemble models—in the face of complexity—and what they are able to achieve with them, as we saw much of the value of the models derive from what is learned during the modeling process itself. In turn we think our results can help instruct modelers on the expectations they should have of the modeling process, as well as what skills are required. Further it can also help reframe (or add to) the purposes of systems biology. The practices of our modelers can be rationalized as part of a broader strategy of investigation into biological systems.

References

Alon U (2007) An introduction to systems biology: design principles of biological circuits. Chapman and Hall, Boca Raton

Corbin JM, Strauss A (1990) Grounded theory research: procedures, canons, and evaluative criteria. Qual Sociol 13(1):3–21

de Kleer J, Brown JS (1981) Mental models of physical mechanisms and their acquisition. Cognitive skills and their acquisition, 285–309

Fagan MB (2016) Stem cells and systems models: clashing views of explanation. Synthese 193: 873–907

Forrester J (1994) Policies, decisions, and information sources for modeling. In: Morecroft J, Sterman J (eds) Modeling for learning. Productivity Press, Portland

Green S (2022) Philosophy of systems and synthetic biology. In: Zalta EN (ed) The Stanford encyclopedia of philosophy (Summer 2022 Edition). https://plato.stanford.edu/archives/sum2022/entries/systems-synthetic-biology/

Hood L, Heath JR, Phelps ME, Lin B (2004) Systems biology and new technologies enable predictive and preventative medicine. Science 306(5696):640–643

Kitano H (2002) Looking beyond the details: a rise in system-oriented approaches in genetics and molecular biology. Curr Genet 41(1):1–10

Kuepfer L, Peter M, Sauer U, Stelling J (2007) Ensemble modeling for analysis of cell signaling dynamics. Nat Biotechnol 25(9):1001–1006

MacLeod M (2020) Model-based inferences in modeling of complex systems. Topoi 39(4): 915–925

MacLeod M (2021) The applicability of mathematics in computational systems biology and its experimental relations. Eur J Philos Sci 11(3):84

MacLeod M, Nersessian NJ (2013a) Building simulations from the ground-up: modeling and theory in systems biology. Philos Sci 80:1–24

MacLeod M, Nersessian NJ (2013b) Coupling simulation and experiment: the bimodal strategy in integrative systems biology. Stud Hist Phil Biol Biomed Sci 44:572–584

MacLeod M, Nersessian NJ (2013c) The creative industry of systems biology. Mind Soc 12:35–48

MacLeod M, Nersessian NJ (2018) Modeling complexity: cognitive constraints and computational model-building in integrative systems biology. Hist Philos Life Sci 40:1–28

MacLeod M, Nersessian NJ (2019) Mesoscopic modeling as a cognitive strategy for handling complex biological systems. Stud Hist Phil Biol Biomed Sci 78:101201

Nersessian NJ, MacLeod M (2022) Rethinking ethnography for philosophy of science. Philos Sci 89(4):721–741

Newstetter WC, Behravesh E, Nersessian NJ, Fasse BB (2010) Design principles for problem-driven learning laboratories in biomedical engineering education. Ann Biomed Eng 38:3257–3267

Noble D (2008) The music of life: biology beyond genes. Oxford University Press, New York

Nordmann A (2015) Synthetic biology at the limits of science. In: von Gleich A, Pade C, Wigger H (eds) Synthetic biology. Springer International Publishing, Cham

O'Malley MA, Dupré J (2005) Fundamental issues in systems biology. BioEssays 27(12): 1270–1276

Roschelle JR, Greeno JG (1987) Mental models in expert physics reasoning. ERIC Document ED, Office of Naval Research, Washington, DC (1987), pp. 285–736

Tran LM, Rizk ML, Liao JC (2008) Ensemble modeling of metabolic networks. Biophys J 95(12): 5606–5617

Voit EO (2000) Computational analysis of biochemical systems: a practical guide for biochemists and molecular biologists. Cambridge University Press, Cambridge

Voit EO (2014) Mesoscopic modeling as a starting point for computational analyses of cystic fibrosis as a systemic disease. Biochim Biophys Acta Proteins Proteom 1844(1):258–270

Voit EO, Newstetter WC, Kemp ML (2012a) A feel for systems. Mol Syst Biol 8(1):609

Voit EO, Qi Z, Kikuchi S (2012b) Mesoscopic models of neurotransmission as intermediates between disease simulators and tools for discovering design principles. Pharmacopsychiatry 45(1):22

Emergent Entanglement in Evolution of Biological Networks

Gustavo Caetano-Anollés and Kelsey Caetano-Anollés

Contents

Abstract Biological systems are made up of interacting parts that collectively form cohesive wholes. They are structured and highly dynamic, integrating molecular, cellular, and organismal processes. Systems can be modeled with networks, graph theoretic representations that describe how parts interact with each other at different time scales and levels of biological organization. A number of properties help understand network structure and dynamics, including measures of centrality, community structure, and hierarchy. Morphospaces of structure and hierarchical organization dissect what is possible and impossible in network makeup, locating biological networks in areas of phenotypic space that foster scalefreeness,

G. Caetano-Anollés (✉)
Evolutionary Bioinformatics Laboratory, Department of Crop Sciences, C.R. Woese Institute for Genomic Biology, and Carle Illinois College of Medicine, University of Illinois at Urbana-Champaign, Urbana, IL, USA
e-mail: gca@illinois.edu

K. Caetano-Anollés
Callout Biotech, Albuquerque, NM, USA

J. Barciszewski (ed.), *Systems Biology II*, RNA Technologies 15,
https://doi.org/10.1007/978-3-031-62178-9_5

modularity, and a bow-tie structure but away from randomness. Here we show the structure of time-dependent networks can hold deep evolutionary history. We focus on the oldest network and metabolism and study its origin and evolution. Tracing centrality measures onto a putative ancestral autotrophic metabolic core revealed the importance of purine biosynthesis and supporting metabolites of carbohydrate pathways. Vectors of protein fold abundance of enzymes layered metabolic subnetworks in a multidimensional space but placed nucleotide metabolism far away from the rest. Phylogenomic tree reconstructions revealed metabolism originated explosively in the purine biosynthesis subnetwork by gradual replacement of prebiotic chemistries and later evolved as a patchwork through waves of enzymatic recruitment. More importantly, phylogenomic analyses showed that hierarchical modularity and bow-tie structure were emergent and pervasive properties of network evolution. The rise of a nested hierarchy of modules in bow-tie entanglement substantiates a "linkage" theory of module emergence that explains biphasic evolutionary patterns of growth in various biological and technological systems through the coevolution of accretion and diversification processes. An "entanglement" theory extends the linkage theory by invoking an entropic-like force capable of driving network growth amid an interplay of neutral and adaptive evolutionary drivers. Our exploration of metabolic evolution not only deepens the understanding of life's history but also charts a course for future investigations at the nexus of network science, evolutionary genomics, and structural biology.

Keywords Bow-tie · Dynamics · Entanglement · Graphs · Hierarchy · Metabolism · Modularity · Morphospace · Structure · System · Prebiotic chemistry · Purine metabolism · Temporal networks

Abbreviations

ET	Entanglement theory
Gy	Billions of years
Gya	Billions of years ago
HTH	Helix-turn-helix
KEGG	Kyoto Encyclopedia of Genes and Genomes
MANET	Molecular Ancestry Network
SCOP	Structural Classification of Proteins
WWW	World Wide Web

1 Introduction

An intricate web of networks exists in biology. These networks describe a wide range of phenomena ranging from molecular and genetic interactions to metabolic reactions and ecological relationships. Biological networks are often dynamic in nature, constantly changing and evolving at different time scales and within different environmental contexts. They are also confined by spatial, energetic, information, and functional boundaries. Unraveling the underlying principles governing the emergence of structural patterns and order within these evolving networks is a crucial, challenging, and captivating pursuit in the realm of evolutionary biology (Bogdan et al. 2022).

Capturing temporal and dynamic patterns of evolving networks represents a multifaceted challenge (Strogatz 2001; Newman et al. 2009; Boccaletti et al. 2006). In particular, deciphering how evolution influences the network structure involves untangling an intricate interplay of interactions, recruitments, and adaptive behaviors responsible for the emergence of both network connectivity (Caetano-Anollés et al. 2022) and persistent communication (Caetano-Anollés 2023). The significance of understanding the evolving network structure extends beyond theoretical knowledge, with practical applications in fields such as medicine, where the modular organization of genetic networks significantly influences disease dynamics, and ecology, where the resilience of ecosystems depends on modular structures. The real-world implications of this research underscore its broad relevance and impact.

Here, we discuss the enthralling interplay between the structure and evolution of networks, focusing on network systems that unfold at time scales spanning millions of years of evolutionary history. We first describe how systems are modeled with networks and how morphospaces help dissect network structure and hierarchy. We then focus on modeling network dynamics at short and moderately short time scales, describing how "time-dependent networks" can advance the study of network dynamic behavior. We then focus on deeper time frames, exploring how phylogenomic approaches can advance our understanding of the emergence and deep evolution of the oldest biological network, metabolism. We end by highlighting the centrality of the "emergent entanglement" of network structure in evolution of biological systems, a view strengthened by a biphasic theory of linkage that is supported by the rise and growth of biological innovations.

2 Systems Are Modeled with Networks

Systems consist of individual component parts that form cohesive wholes (Simon 1962). Cohesiveness here refers to dynamical stabilities arising from constraints, restrictions on parts that arise from the integration of forces and flows (Collier 2003). Parts are capable of interacting and/or combining with other parts of the system in an organized and hierarchical manner (Salthe 1985), making up levels of biological organization. Interactions or combinations are often represented using *graphs*, mathematical structures that connect points known as vertices with lines (Wasserman and Faust 1994; Newman 2003). Vertices are the fundamental units of a network as they embody dynamic behavior. They usually represent the individual parts of a system. Conversely, lines are fundamental units of connection in a network. They depict pairwise "relations" between nodes that can be gained and lost. For example, an "interaction" graph illustrating how proteins interact with each other in a biological system (e.g., an organism) can be constructed by defining vertices as proteins and lines as molecular interactions that are known to be established (e.g., a network of receptor kinases in *Arabidopsis*, a plant model system; Smakowska-Luzan et al. 2018). Similarly, a "compositional" graph illustrating the combination of structural domains in multidomain proteins can be constructed by defining vertices as structural domains and lines as the presence of these domains in proteins (e.g., Aziz and Caetano-Anollés 2021). This type of network has information on how domains are recruited into protein makeup in the course of evolution.

Figure 1a describes the anatomy of a toy network containing both undirected (*edges*) and directed (*arcs*) connections, a loop (a line whose end points coincide), multiple lines (a same ordered or unordered pair of vertices assigned to several lines), and an isolated vertex. Note that *edges* are line connections between vertices that represent unordered pairs of vertices and lack a specific direction (they are undirected). Conversely, *arcs* are line connections between vertices that represent ordered pairs of vertices, i.e., each line has an initial and terminal vertex and has a specific direction (it is directed).

A network is made of sets, i.e., collections of well-definable elements or members of the set capable of forming a whole. When explicitly describing sets using the roster (enumeration) notation, the graph of Fig. 1a can be dissected into the following: (1) a set of 11 vertices, $V = \{a, b, c, d, e, f, g, h, i, j, k\}$; (2) a set of 11 edges embodying unordered pairs of elements $E = \{(a, b), (a, f), (a, e), (a, d), (a, g), (b, f), (b, g), (c, e), (d, e), (f, g), (i, j)\}$; and (3) a set of 9 arcs embodying ordered pairs of elements $A = \{(a, c), (b, h), (h, c), (c, b), (c, d), (e, f), (h, i), (i, h), (j, j)\}$. The graph can be defined as $G = \{V,L\} = \{V,A,E\}$, with L representing the set of lines.

Graphs transform into dynamic networks when value functions, generally in the form of properties, weights, or stamps, are assigned to the vertices and lines of the graphs. Consequently, a network N is made of four sets, a set of vertices, a set of lines in the form of undirected (*edges*) and/or directed (*arcs*) connections, and sets of vertex and line value functions that are mapped onto the sets of vertices and lines, respectively. Mathematically, $N = \{V,L,P,W\}$, where N consists of:

1. A *graph* $G = \{V,L\}$, where V is a set of vertices and L is a set of lines in which each element of L is a pair $\{u,v\}$ of elements of V, with $u \neq v$ and $L \subseteq \binom{V}{2}$, $L = E \cup A$, where E is the set of *edges* and A is the set of *arcs*, and $n = card(V)$ and $m = card(L)$.
2. A set P of *vertex value functions*—mainly properties: $p\colon V \rightarrow A$.
3. A set W of *line value functions*—mainly weights or stamps: $w\colon L \rightarrow B$.

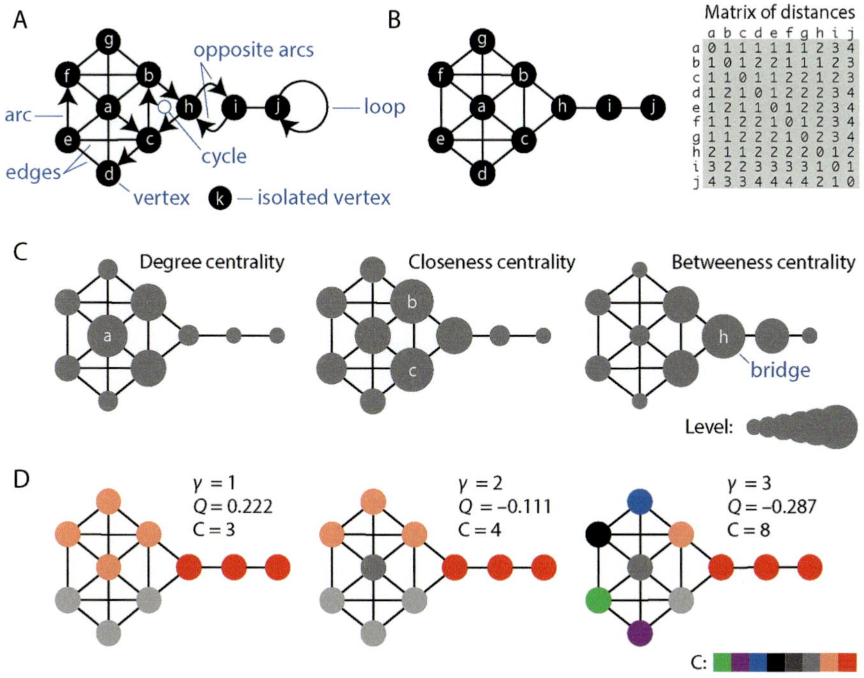

Fig. 1 Dissecting the anatomy and structure of a mixed network. (**a**) The scaffold of a network is a graph holding a set V of vertices (nodes) and a set L of lines (links). Lines connect the pairs of vertices (lines are two-element subsets of V) and can be *edges* or *arcs* if they are undirected or directed, respectively. The network that is shown is a "mixed network" because it contains both types of lines. (**b**) A network can be represented with an adjacency matrix. The directed network described in panel A was simplified by making it undirected (i.e., by uncovering its *underlying graph*) and by eliminating loops and disconnected network components. Such simplification makes its adjacency matrix symmetric. (**c**) Network centralities dissect the influence of nodes in a network. *Degree centrality* estimates how well a node is connected to other nodes. The degree of a node (its connections) provides a local view of network connectivity. *Closeness centrality* estimates how easy is for a node to reach other nodes. Finally, *betweenness centrality* estimates how important is a node in terms of its capacity to connect to other nodes. It offers a global view of connectivity. Other centralities (not shown) offer views of *prestige*, how important is a node in terms of the importance of its neighbors. The size of nodes reflects centrality measures in the networks. (**d**) Louvain modularity (Q) dissects communities in networks at different levels of resolution (γ). Nodes are colored according to communities they belong to. Note that community number (C) increases with γ and that modularity values range [−0.5,1]

Note that vertices and lines are called "nodes" and "links" in computer science, "sites" and "bonds" in physics, and "actors" and "ties" in social science. For simplicity and consistency, we will refer to vertices as *nodes* and lines connecting them as *links*.

3 Networks Are Structured

Natural network abstractions are far from homogeneous. Their makeup is both entangled and structured in ways such that interactions, behavior, and goals of subsets of parts may be different than the rest (Bogdan et al. 2022). Biological networks resemble complex tangles that are large and particularly difficult to model and understand. Their wirings reflect the integration of a multiplicity of biological subsystems. Nodes and links can be diverse. For example, metabolic networks must portray at least two types of agents, protein enzymes and molecular substrates (metabolites), and must embody enzyme-catalyzed reactions exhibiting distinct kinetics, which sometimes makes them reversible or irreversible. Nodes are indexed with properties, weights, or stamps that portray different aspects of a system, including space, time, energy, and information. Links can have different directions and signs and can be part of different link communities, network modules that describe, for example, different classes of biological functions. Nodes and links change over time and must describe non-linear systems and multifractal and long-range memory behaviors. Wiring diagrams unfold in time at different time scales, some involving processes that take billions of years (Gy) to materialize. Finally, each aspect previously mentioned above can influence each other in complex ways generating *meta-complication*. For example, network evolution can affect topology, connectivity, and complexity (e.g., networks describing protein domain organization) and can integrate space (e.g., Rentian scaling in neural networks). Biology involves a daunting diversity of molecular, cellular, organismal, and population processes that must be integrated to portray the functional reality of biological systems.

Some network properties help visualize and study network structure and makeup. For example, a network can be represented by an *adjacency matrix*, A, an $n \times n$ square matrix in which n is the number of nodes. A has elements:

$$A_{ij} = \begin{cases} 1 & \text{if there is an link between nodes } i \text{ and } j \\ 0 & \text{otherwise} \end{cases} \tag{1}$$

When networks are undirected, their adjacency matrix is symmetric: $A_{ij} = A_{ji}$ (Fig. 1b). This means that when there is a link between i and j, then there must be a link between j and i. In sharp contrast, the adjacency matrix of a directed or mixed network is asymmetric. This means that when $A_{ij} = 1$, the existence of a link pointing from j to i is often independent of the existence of a link pointing from i to j. In metabolic networks, for example, many metabolites cannot revert back into reactants, and their irreversible enzymatic reactions are modeled with arcs and asymmetric adjacency matrices.

Adjacency matrices and incidence matrices (not described) help dissect adjacency, incidence, and neighbors in network structure. Two nodes are *adjacent* in the network if they are joined by a link. These nodes are considered *neighbors*, and the links that join them *incident* to them. Mathematically, neighboring nodes N are end points of one and the same element of the set of links. To illustrate, the neighbors of node *c* in Fig. 1a are defined by end points connecting arcs and edges, $N_A(c) = \{b,d\}$ and $N_E(c) = \{e\}$. Since $N = N_A \cup N_E$, then $N(c) = \{b,d,e\}$.

Incidence, adjacency, and neighbors in network structure help measure network *centrality*. Centrality seeks to identify and characterize the most important nodes of a network using real-valued functions on the nodes. These functions provide a ranking of nodes according to an "importance" criterion describing node prominence or structural importance, which can then be mapped onto the nodes of the network (Fig. 1c). While the concept of centrality is intuitive, its elaboration can be mathematically complex (Borgatti and Everett 2006). In fact, estimates of importance are gathered by evaluating how a node influences the connectivity or information flow of the network and how such connectivity or flow makes the network cohesive. This is usually accomplished by traveling (walking) along the links of the network and assessing the "walk structure" of the network with three types of walks: trails, paths, and cycles. A *trail* is a walk with all links being different. A *path* is a walk with all nodes different. A *cycle* is a path in which the initial and terminal nodes of the walk are the same (see cycle in Fig. 1a). One special kind of path, the *geodesic path*, is particularly important for centrality. It is the shortest path between two nodes, the path that requires the least number of links to traverse. Radial centralities count walks that start/end on a given node. Examples include *degree centrality*, which uses walks of length 1; *eigenvector centrality* and its variants (e.g., Google's *PageRank* centrality for ranking web pages), which are bound by walks of infinite length; and *Katz centrality*, which generalizes degree centrality by measuring all nodes that are connected through a path, penalizing distant ones. In turn, medial centralities count walks that pass through a given node. The typical example is *betweeness centrality*, which counts the numbers of the shortest paths. An alternative to these measures is to record the length of walks rather than their numbers. This is the case of *closeness centrality*, which uses geodesic distances from a given node to the rest. We will mathematically describe and illustrate three popular measures of centrality:

(i) *Degree centrality*. The measure uses the *degree* of nodes to provide a "local view" of how well nodes are connected that can be computed as the marginals of the adjacency matrix. The *degree* of node *i*, deg(*i*), is the set of nodes that are adjacent or the number of times links are incident to that node:

$$\deg(i) = \sum_{j=1}^{n} A_{ij} \tag{2}$$

Nodes that have higher degrees are considered more hub-like or central. When dealing with directed networks, the degree of a node can be decomposed into an indegree (indeg) and an outdegree (outdeg) measure of centrality, i.e., links that point into a node or out of a node, respectively. For example, node *c* of the mixed network of Fig. 1a that establishes both directed and undirected links has $\deg(c) = 5$, $\text{indeg}(c) = 4$, and $\text{outdeg}(c) = 3$.

(ii) *Closeness centrality.* This popular measure, which is often highly correlated with degree centrality, offers a "global view" of how easy a node can reach other nodes in the network, measuring "independence" or "efficiency." As previously mentioned, closeness centrality takes advantage of *walks* along the network diagram, w, which represent ordered lists of nodes with lengths $|w|$ corresponding to the number of links the walks contain. The measure calculates the length of a geodesic path between nodes i and j, $d(i,j)$, as the number of links in that path. For example, the length $d(e,b)$ in the network of Fig. 1a is $d(e,b) = |(e,a,b)| = 2$, but note that other paths with a same length are possible. Mathematically, *closeness centrality* of node i is the mean length of geodesic paths (geodesic distances) from node i to every other node:

$$\text{closeness}(i) = \frac{1}{\sum_{i \neq j} d(i,j)} \tag{3}$$

The measure is the reciprocal of "farness" and therefore assigns lower distances to more central nodes. Closeness centrality is often given as a normalized distance, which represents an average length instead of a sum of lengths.

(iii) *Betweenness centrality.* Betweeness centrality is also a popular measure of centrality based on the shortest paths. It provides a "global view" of the importance of a node in terms of connecting to other nodes. Betweenness centrality quantifies the number of times node i acts as a "bridge" along the shortest paths between any two other nodes in the network, providing a measure of how relevant is node i over the flow of communication between nodes. Mathematically,

$$\text{betweenness}(i) = \sum_{u \neq i \neq v} \frac{P_{uv}(i)}{P_{uv}} \tag{4}$$

where P_{uv} is the total number of geodesic paths from node u to v and $P_{uv}(i)$ is the number of those paths that go through node i. Betweenness can be normalized by the number of pairs of vertices that exclude i. Betweenness centrality is widely applied in network theory because it measures how nodes control information flow that pass through a modeled system. It also measures the ability of a node to fragment networks into components when the node is removed, testing network robustness.

Figure 1c shows how these three popular measures of centrality portray different aspects of connectivity in the underlying graph (the undirected network representation) of the mixed network of Fig. 1a. The graph has a subgraph (a subset of nodes and links of the graph) that is highly connected (most pairs of nodes are adjacent) and a spanning subgraph that is minimally connected and is linked to the highly connected subgraph through node h. Degree centrality correctly identified the main hub (node a) and cluster of hubs (nodes a, b, and c) located in the highly connected subnetwork, showing their central overall importance. As expected, closeness centrality identified the same hub region, but its ability to unravel how nodes reach other nodes in the network crucially highlighted node *h*, a necessary connecting step to reaching nodes of the poorly connected subgraph. Finally, betweeness centrality uncovered the capacity of node *h* to act as central bridge between the two subnetworks, with nodes *b* and *c*, constituting internal bridges of connectivity to other nodes of the highly connected subgraph.

4 Network Communities Highlight the Importance of Modularity in Biological Systems

Biological networks harbor *communities*, groups of nodes that are more highly connected with each other than with other nodes of a network. These communities of densely connected nodes make networks modular. Modular structure is a prevalent and essential organizational feature of biological systems and is therefore expected to impact network structure. In biology, modules are defined as sets of integrated parts that cooperate in coordinated fashion to perform a task (Hartwell et al. 1999). They interact more extensively with each other than with other parts and modules and in doing so play pivotal roles in orchestrating biological functions. In fact, the concept of modular organization (*modularity*) underlines the degree to which the components of a system can be separated and rearranged without compromising system functionality (Simon 1962). This property is integral to Herbert Simon's broader idea of "near-decomposability," a concept expounded in his work on systems interacting with their environment (Simon 1997). According to Simon, a nearly decomposable system, which is often hierarchical, exhibits lower-level parts that establish strong internal connections between them, while top-level parts are bound together by relatively weak linkages. This unique arrangement allows "long-term behavior to be studied on an aggregative basis without concern for internal details of the parts, and allows the short-term behavior of each part to be studied independently of the behavior of the other parts" (Simon 1997). This seminal concept is centrally important for network science.

Community structure can be evaluated with a variety of methods. The path-pruning *Newman-Girvan (NG)* algorithm (Newman and Girvan 2004), for example, is a hierarchical method of community structure analysis that relies on the iterative elimination of links that have the highest number of the shortest paths between nodes passing through them. This is accomplished by calculating betweenness centrality for every link, removing the link with the highest value, calculating betweenness centrality for the remaining links, and repeating previous steps until no links are left. Communities can also be identified by measuring *modularity*, the maximization of the difference between the actual number of links in a community and their expected number according to a "configuration" model. Mathematically, modularity Q is given by

$$Q = \frac{1}{2m}\sum_{c}\left(e_c - \gamma\frac{K_c^2}{2m}\right) \tag{5}$$

where e_c denotes the actual number of links in a community c, m is the total number of links in the network, K_c is the sum of degrees of the nodes in community c, and γ represents a resolution parameter. Note that increasing γ leads to a larger number of communities in the network. Because optimizing modularity is NP-hard, a number of heuristic algorithms have been used that, for example, use hierarchical agglomeration by a *Fast-Greedy-Community* (*FGC*) detection algorithm (Clauset et al. 2004), simulated annealing (e.g., Guimerà and Amaral 2005), or spectral algorithmic approaches (Newman 2006). In particular, the Louvain method is a popular optimization method that uncovers communities in networks with millions of nodes thanks to its computational efficiency (Blondel et al. 2008). The method first greedily identifies "small" communities by local optimization of modularity. It then aggregates nodes of the same community into a reduced graph structure. The two steps are applied iteratively until a maximum of modularity is obtained. Iterations provide network partitions of increasing community size in a process that leads to hierarchical decomposition of networks. Figure 1d analyzes the Louvain modularity of the network we used as toy model at three resolution levels, showing how increasing γ leads to dissection of the network into smaller but more numerous communities. Note that while the Louvain method is the fastest and best performing, an alternative recent method, the Leiden algorithm, guarantees communities are not sometimes internally disconnected (Tragg et al. 2019). We expect continuous improvements in the identification of community structure and testing of new methods with increasingly more complex algorithms and networks.

Modularity is often hierarchical. Hierarchy embeds simpler network modules into modules of higher complexity. An "authority relation" subordinates lower-level modules into higher-level modules by either classifying modules into levels of a hierarchy according to some external criterion (e.g., ontology) or internally by criteria that apply to network structure (e.g., clustering). Hierarchical modularity, i.e., the hierarchy associated with modular structure, can be measured with a number of descriptors. The most widely used is the power-law scaling of the *clustering coefficient*, C_i (Ravasz 2009). C_i is a metric of the average probability that two neighbors of a node are also connected (Wasserman and Faust 1994; Watts and Strogatz 1998). The local implementation of C_i is simply the ratio between the number of closed triplets and the total number of triplets connected to a node, with triplets being groups of three nodes connected by either two or three links (open and closed triplets, respectively). Note that the formation of closed triplets ("triangles") follows the adagio "the friend of my friend is also my friend" used to describe network communication in the social sciences and the *small world* idea that all people are $\leq$6 social connections apart ("six degrees of separation"). C_i ranges from 0 (no clustering) to 1 (maximal clustering). Since higher values suggest a small-world structure, scaling of C_i represents a useful description of hierarchical modular structure (Ravasz 2009).

5 Morphospaces Represent Multidimensional Views of Network Structure

Modularity helps systems adapt and innovate by compartmentalizing their structure and function and by allowing the reuse of substructures in different contexts (Simon 1962). From that point of view, modularity elicits *flexibility* in network evolution. Modularity and its hidden links to hierarchy complement two other important properties of networks, randomness, and heterogeneity.

Randomness describes the stochastic addition of nodes and links to a growing network. The property pushes uniform connectivity throughout a network structure providing resilience to removal of nodes and links and shielding against fragmentation of the network from failure or deliberate attack on a system (Newman 2003). This resilience elicits *robustness* in network structure. The distribution of geodesic paths of betweenness centrality follows a power law and allows to classify networks into two groups according to the scaling exponent (Goh et al. 2002). Betweenness centrality also informs how many paths will be extended when removing a node from the network, providing an impact measure of resilience (Newman 2001; Holme et al. 2002).

Heterogeneity describes the "scalefreeness" of a network, a small-world behavior that elicits *economy*, including economy of connections or information flow running through the modeled system. Heterogeneity can be measured with the degree distribution p_k, the probability distribution of the degree of nodes k over the entire network. For some networks, such as metabolism where $p_k \sim k^{-\gamma}$ and γ is a constant parameter that generally ranges 2–3 (Jeong et al. 2000), the distribution can be fitted with a power law and generally results in few nodes being highly connected (e.g., ATP in metabolism). For other networks, scalefreeness may embody heavy-tailed distributions, exponential cut-offs, or even Gaussian distributions. Recent work estimated the best-fitting power-law models for almost a thousand empirical networks, tested their statistical plausibility, and compared them to alternative non-scale-free distributions (Broido and Clauset 2019). While most networks (57%) followed some kind of scale-free behavior, only 5% exhibited strong "scalefreeness" levels, and most networks could be better fitted to log normal than power-law distributions. A number of generative models explain the small-world behavior of these networks, of which the "preferential attachment" model that generates a "rich-get-richer" scenario is the best known (Barabási and Albert 1999). The lack of universality of scale-free networks describing real-world systems and the identification of power-law scaling, both of which remain controversial (Holme 2019), may be explained by not taking into account time dependencies in attachment of growing networks that would describe some form of preferential attachment (Falkenberg et al. 2020). Alternatively, it may simply reflect the well-known fact that the behavior of biological systems is frustrated.

Modularity is in a trade-off relationship with heterogeneity and randomness (Solé and Valverde 2004). This property allows construction of a "morphospace," a phenotypic space that explores the "limits of the possible" in the structure of

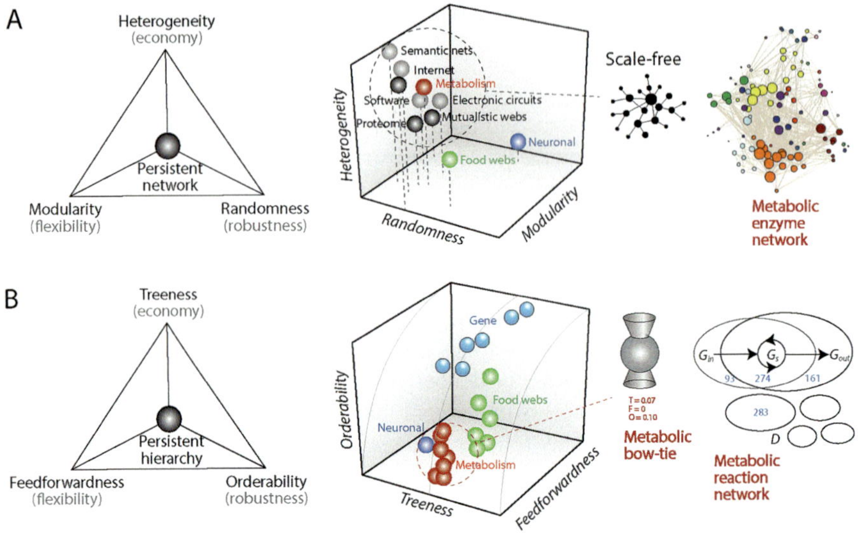

Fig. 2 Morphospaces describing the structure and hierarchy of networks. (**a**) A triangle of network persistence (left) explains trade-off solutions between the heterogeneity, modularity, and randomness of network structure (proxies for economy, flexibility, and robustness, respectively). The sphere at the center represents optimal trade-off solutions in Pareto fronts within a fitness landscape of the networks. The corresponding structural morphospace (right) maps different man-made and natural networks onto the trade-off solution space. Most networks map onto a heterogeneity-randomness projection plane with high heterogeneity making them scale-free. A scale-free network of metabolic enzymes connecting KEGG subnetworks is shown as inset, with nodes made proportional to their degree and colored according to mesonetwork (Caetano-Anollés et al. 2009). (**b**) A triangle of network hierarchy (left) explains trade-off solutions between the treeness, feedforwardness, and orderability of network flow in directed networks (proxies for economy, flexibility, and robustness, respectively). Remarkably, most biological networks are located around an orderability-feedforwardness plane that is centered in the middle of the treeness axis, stressing the importance of the bow-tie structure. An icon of cones and balls in the inset shows the feedforward and cycle bow-tie structure of the metabolic network of *Escherichia coli* (Corominas-Murtra et al. 2013), and a diagram of network components shows a giant strongly connected component (G_s) receiving connection from a an "in" component (G_{in}) and eliciting connections to an "out" component (G_{out}) (Ma and Zheng 2003) (Morphospaces are from Solé and Valverde (2004) and Corominas-Murtra et al. (2013))

networks (e.g., metabolism; Schuetz et al. 2012). This exploration involves only a limited set of variables accounting for the most salient features of a system. Morphospaces can distinguish "what exists, what is possible but does not exist, and what is impossible" by comparing simulated networks satisfying general properties (e.g., multifractality) and real networks constructed from data (Avena-Koenigsberger et al. 2015). Figure 2a describes a "triangle of network persistence" capable of dissecting the scale-free nature of biological networks. In this triangle, optimal trade-off solutions locate in Pareto fronts as networks evolve according to factors that increase the fitness of network structure. This triangle explains trade-off solutions of the network properties of heterogeneity, modularity, and randomness that act as proxies for economy, flexibility, and robustness, respectively. These proxies represent central design principles for biological and engineered systems (Yafremava et al. 2013). Economy imposes a matter-energy budget on a system, maximizing performance while minimizing cost. Flexibility responds to internal and external cues by streamlining structural and functional mechanisms that process information. Robustness channels information to resist structure-altering damage and change. The triangle of network persistence translates into a morphospace describing the heterogeneity, modularity, and randomness of network structure (Solé and Valverde 2004), in which real networks locate in central areas of the multidimensional space fostering small-world behavior (Fig. 2a). For metabolism, the scaling of metabolic networks for almost 30,000 organisms appears universal (Kim et al. 2019). The structure of networks has been proposed to be driven by nature and nurture (Zhou et al. 2023), suggesting that both the environment and evolution are important drivers. This proposal validates the economy-flexibility-robustness landscape of the morphospace as it is driven by both internal and external cues that impact network evolution.

Similarly, hierarchy can be organized into a morphospace describing network flows along wiring diagrams (Corominas-Murtra et al. 2013). This allows to compare network hierarchies generated by simulation and those from real networks. Figure 2b shows a "triangle of network hierarchy" defining trade-off solutions between three metrics of network flow in directed networks used by Corominas-Murtra et al. (2013), treeness, feedforwardness, and orderability. These metrics act as proxies for economy, flexibility, and robustness, respectively. Treeness measures reversibility and pyramidal structure of a hierarchical network, i.e., how economy in chains of hierarchical command is affected by the ambiguity of reversing paths with more than one commander for any commanded. Increases in treeness shift an anti-hierarchical pyramidal structure to a "bow-tie" symmetrical structure and then to a fully hierarchical structure. Feedforwardness measures the impact of cycles (e.g., the non-empty directed trail in Fig. 1a), modular regions that cannot be ordered in a directed network but provide flexibility in establishing hierarchical flows. Higher values penalize modules close to the top of the hierarchy. Finally, orderability measures order by describing how nodes arrange unambiguously without the influence of cycles. Higher orderability decreases stochastic flows but diminishes resistance to structure-altering damage and change, decreasing robustness. Remarkably, most biological networks are located around an orderability-feedforwardness plane centered in the middle of the treeness axis, stressing the importance of the bow-tie structure (Fig. 2b). Metabolic networks, for example, have lower orderability levels because their metabolic reactions include a large number of cycles. This cycle bow-tie structure is illustrated for the metabolic network of *Escherichia coli* with an icon of cones and balls. Remarkably, this bow-tie structure shows a strongly connected cycle component stabilized by a feedforward structure of inputs and outputs (Ma and Zeng 2003) similar to that found in the World Wide Web (WWW) (Broder et al. 2000) but without tendrils or tubes (Fig. 2b).

6 Networks Describe Systems That Are Dynamic and Change Over Time

As intimated by the bow-tie structure of metabolic networks, biological networks describe dynamical systems in which mass, energy, and information flow through their constantly changing wiring diagrams. Networks change over time and do so at different time scales. In network science, time-dependent (temporal) networks refer to networks in which nodes (entities) and links (connections, interactions) between them are active only for restricted periods of time (Holme and Saramäki 2012). Unlike static networks, where nodes and connections between them remain constant over time, time-dependent networks capture the dynamic nature of connectivity, which plays a crucial role in delimiting the structure, behavior, and evolution of a network. When a system evolves its network, representation adds/subtracts nodes (e.g., system's components) and/or links (e.g., interaction between components) according to processes that, for example, convey function or growth. Metabolomic networks, for example, have nodes describing metabolites and links describing a correlated change in metabolite levels, linking the transcriptome and proteome with matter-energy dynamics of metabolic reaction chemistries (Amara et al. 2022). When processes of matter-energy flow in metabolomic networks are perturbed by non-lethal stress (e.g., cold, heat, diauxic stress), networks rewire quickly (in minutes) in biphasic patterns to form modules that fulfill structural, functional, and cellular responses to induced environmental change (Aziz et al. 2012). In sharp contrast, models of metabolic growth resulting in evolving scale-free (Jeong et al. 2000) and hierarchical modular (Ravasz et al. 2002; Guimerà and Amaral 2005) network structures (enabling economy-driven connectivity and functional versatility, respectively) must by definition unfold at time scales of billions of years of evolution. Clearly, the time scale of processes affecting nodes, links, and their value functions (e.g., weights) becomes important when modeling time-dependent networks (Holme and Saramäki 2012). Three general scenarios are possible: (1) If time scales of network evolution are order-of-magnitude larger than the time scales of processes that drive network dynamics, then network structure evolves slowly compared to its dynamics. This allows to map dynamics on static network representations by, for example, following reaction kinetics in metabolic networks or building chronologies of metabolic evolution. (2) If time scales of network evolution and dynamics are comparable, then their interplay becomes important. Such is the case of temporal networks that model disease transmission or e-mail or mobile phone communication. (3) If network evolution is faster than network dynamics, then processes can be approximated using a time-averaged version of the network. This is the case of the phenomenon of behavioral contagion that is modeled in social networks. Note that the analytic simplification endowed by scenario (1) is complicated by the fact that usually these networks require some sort of retrodiction (traveling back in time), which in itself can be a daunting task. Scenarios (2) and (3) require real-time tracking of changes in the system being modeled, but the highly

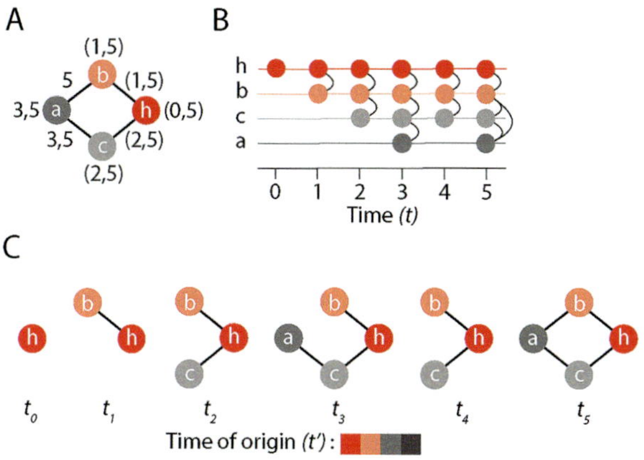

Fig. 3 Tracing the evolution of time-dependent networks. (**a**) The toy network of Fig. 1 was reduced by collapsing communities defined at $\gamma = 2$ into single nodes. Community history was then mapped by labeling nodes and links with times or intervals of existence. (**b**) A chronology of nodes and links is made explicit by transferring temporal information of panel A into a "contact sequence" diagram that unfolds how time impacts the existence of nodes and links in the evolving network along a time series. (**c**) The same chronology is now seen as a series of "time events," static network snapshots that illustrate the growth and/or dynamics of network components. In the figure, node colors describe the time of origin of nodes. A similar coloring could be applied to links

dynamic network structure necessitates addressing methodological complexity and large data sets.

The evolution of time-dependent networks can be visualized with a number of different representations. To illustrate, we reduced the toy network of Fig. 1 by collapsing communities into single nodes (Fig. 3a). We then traced its history in time using two types of chronologies, contact sequences (Fig. 3b), and time events (snapshots) (Fig. 3c). When using *contact sequences* (sensu Holme and Saramäki 2012), network chronologies explicitly describe the existence of nodes and links at particular time points or intervals of a time series. This usually corresponds to sets of pairs of nodes and times (t_i) or intervals (t_i, t_j) when these node sets exist and connections are active. This "local" analysis approach dissects the entire evolving network into its evolving node and link components, keeping track of their existence in the time series. The local representation also allows to capture recurring temporal patterns (temporal motifs) (Kovanen et al. 2011). These motifs represent topologically equivalent subnetworks unfolding in time. In sharp contrast, the *time event* representation takes snapshots of the entire network as the network evolves (Fig. 3c). This global analytical view, inspired by computer-based discrete event simulation (DES) tools (Delaney and Vaccari 1989), allows to visualize the totality of the network at each time event of the temporal series while losing detail of changes that occur to individual nodes and links. A focus on specific events or critical time points helps understand how they influence network structure and dynamics. Both representations require *time stamps*, i.e., labeling times or intervals of node and link existence onto a network representation that incorporates all reported nodes and links of the network (Fig. 3a). Note that the chronology of the toy network of Fig. 3 exhibits two distinct behaviors. At first, during time interval (t_0, t_3), the network grows without loss of nodes or dissolution of links in an evolutionary pattern consistent with scenario (1) discussed above. Later on, during time interval (t_4, t_5), however, the temporal behavior of node *a* and its associated links follows a more dynamic sequence that would be compatible with scenarios (2) or (3). These contrasting *deep* and *shallow* evolutionary behaviors highlight the dynamic complexity offered by a single chronology when modeling network evolution. Deep network behavior is particularly relevant when networks evolve at time scales associated with deep evolutionary processes of biology, such as the rise and evolutionary growth of metabolic networks we will later describe.

A number of properties characterize time-dependent networks. Here we will mention four of them, temporal paths, latency, reachability, and burstiness. *Temporal paths* (also known as "journeys") are sequences of links between nodes that unfold sequentially over time (Kempe et al. 2002). Links in these series are each indexed with a specific time stamp, capturing the chronological order of connectivity of an evolving system. Analyzing these temporal paths can reveal patterns, trends, and the flow of matter, energy, and information (including influence) that is associated with network dynamics. Unlike static paths in traditional networks, temporal paths take into account the temporal order of connections and are mathematically non-transitive, providing a more nuanced understanding of how relationships between parts of a system evolve. They help understand processes such as information diffusion, disease spread, or collaboration dynamics. *Latency*, also known as temporal distance, is the fastest temporal path between two nodes, i.e., the path exhibiting the shortest duration (time taken to follow the path) (Pan and Saramäki 2011). Latency is a temporal concept analogous to distance in static networks. However, it depends on the start time of the path. *Reachability* is a property of nodes that measures how fast and comprehensively temporal flows spread through node pairs of a network (Holme 2005). Because there is not always a temporal path from one node to another, the *reachability ratio* estimates the fraction of node pairs that have a temporal path between them. Networks with higher fraction of nodes in temporal paths exhibit high reachability. Finally, *burstiness* measures departures from a Poisson distribution of times in contact sequences (known as inter-contact times or inter-event times) (reviewed in Holme and Saramäki 2012). Such inhomogeneities are often displayed as heavy-tailed distributions of scaled inter-contact times that describe unpredictable timings. Intermittency and burstiness are common in highly dynamic networks and can slow dynamic flows. However, tracing spreading dynamics in social networks of honey bees has shown that burstiness and fast spreading can occur (Gernat et al. 2018).

In addition to the dynamic properties of links, nodes in time-dependent networks may also display *dynamic attributes* that change over time such as appearance or dissolution, changing characteristics, activity levels, states, or roles of nodes in the network. Analyzing dynamic node properties help understand how individual nodes contribute to network evolution, including identifying influential nodes at specific time points and uncovering patterns in the evolution of node characteristics. When studying biological evolution, time-dependent networks reveal evolutionary transformations and changes that occur in the topology, connectivity, community structure, and other properties of networks over different time scales. Trends, patterns, and dynamic behaviors help uncover how networks adapt and reconfigure over time, providing insights into drivers of network evolution, emergence and dissolution of connections, the formation of communities, and the overall temporal dynamics of the system under study. We end by noting that time-dependent networks have been referred to as "temporal," "time-varying," "time-aggregated," "time-stamped," "dynamic," and "evolving" networks or graphs. When referring to biological networks that are the subject of evolutionary change, the term "evolving networks" is most appropriate.

7 Network Structure Can Hold Deep Evolutionary History

The structure of biological networks is expected to hold evolutionary information. Here, we illustrate how network structure can help make inferences about the origin and early evolution of metabolism. Metabolism is made up of life-sustaining chemical reactions that transform small molecules (metabolites) with the help of enzymatic protein catalysts (enzymes) and organic and inorganic cofactors that assist chemical transformations. Metabolic pathways are a series of successive reactions that often combine metabolites to form larger molecules in "anabolic" reactions or degrade them to smaller molecules in "catabolic" reactions. These pathways make up intricate metabolic networks that describe how enzymes transform a multiplicity of chemical compounds to produce, degrade, and store matter, energy, and biological information. There are thousands of enzymes and metabolites in the pathways of metabolism, all of which have been carefully curated in numerous databases, including the Kyoto Encyclopedia of Genes and Genomes (KEGG) (Kanehisa et al. 2023) and MetaCyc (Caspi et al. 2020). For example, the KEGG database is a manually curated repository of published experimental data that is used to create models of biological networks in thousands of organisms, including those used to describe diseases and drugs. Release 108.1 (November 2023) contains 569 pathway maps (metabolic "subnetworks") with chemical information about 11,971 biochemical reactions grouped into 3194 classes, 8077 enzymes, 19,270 metabolites, and other compounds and 11,222 glycans. Subnetworks are unified into what we call "mesonetworks," upper-level categorization of pathways with similar functional capabilities. Similarly, MetaCyc is a database of experimentally elucidated metabolic pathways in organisms belonging to all domains of life (Caspi et al. 2020). MetaCyc version 27.1 (released in 2023) catalogs 3128 pathways (the classification units of the database) together with their 18,818 enzymatic and 961 transport reactions and their associated 14,320 enzymes, 679 transporters, 4851 protein complexes, and 19,001 compounds that include metabolites and cofactors. Pathways involve the activity of 15,377 genes with their 76,877 indexed Gene Ontology (GO) terms describing biological functions.

Metabolism is an excellent model system to study network evolution as it represents the oldest and best studied biological network. Our understanding of metabolic evolution benefits from recent advances in a wide range of fields of inquiry, including top-down explorations in evolutionary genomics and structural and systems biology (Caetano-Anollés et al. 2021) and bottom-up explorations in prebiotic systems chemistry, synthetic biology, and cheminformatics (Ruiz-Mirazo et al. 2014). Computer-assisted organic chemistry in particular now provides a vast network of plausible molecules that can be synthesized from prebiotic feedstocks (Shen et al. 2021). This accumulating knowledge can be used to test a number of previously described models of metabolic origin and evolution as well as phylogenetic reconstruction efforts (Caetano-Anollés et al. 2009). While the "origin of life" field is controversial, there is an agreement that a number of chemistries and prebiotic chemical pathways developed on primitive Earth prior to the emergence of proteinaceous enzymes, replicating units, and primitive cells (e.g., Maden 1995). These prebiotic chemical reactions, most of which are nonexistent today, acted as seeds for more complex chemistries, leading eventually to the production of the main building blocks of life, amino acids, ribonucleotide triphosphates, and their transforms. In fact, a forward-synthesis algorithm (Allchemy; https://allchemy.net) produced from seven generations of synthesis a full and robust network of prebiotic reactions from a limited set of chemical substrates considered starting points in a number of OL studies (H_2O, N_2, HCN, NH_3, CH_4, and H_2S) (Wołos et al. 2020). The modeled network included 82 biotic and 36,603 abiotic molecules, contained experimentally reported and unidentified routes to biotic targets, and described chemical emergence in the form of catalysis, self-regenerating cycles, and production of surfactants for biological compartmentalization. Biotic molecules included high-yielding amino acid and peptides, nucleobases and nucleosides, carbohydrates, and other metabolites needed to jumpstart primordial enzymes and modern metabolic networks (Fig. 4). Remarkably, the 614 reaction rules involving C, O, N, S, and P elements employed by the Allchemy algorithm produced synthetic network connectivity that vastly matched knowledge generated from prebiotic chemistry experimentation but also described emergence of new self-regenerating cycles such as that of iminodiacetic acid and (2-cyanoethyl)glycine, as well as biomimetic routes to prebiotic fatty acid and peptide surfactants necessary for the formation of membrane structures.

One appealing model of metabolic origins in a prebiotic world is metabolic expansion through sequential addition of "shells" of chemical reactions (Morowitz 1999) facilitated by "gateways" of prebiotic chemistries and, later on, enzymatic catalysts (Morowitz et al. 2000). In this chemical accretion model, an autocatalytic "energy amphiphile" core (shell A) was first established that acted as starting point for other metabolic pathways and did so without the assistance of enzymes. The core was organized around a *reductive citric acid cycle*, which functioned as a simple autotrophic synthetic system. It transformed simple precursor molecules such as CO_2, acetate, and formic acid into larger molecular structures by condensation followed by dehydration-reduction reactions to form increasingly longer carbon backbones (involving perhaps the aldol and Claisen condensations of modern

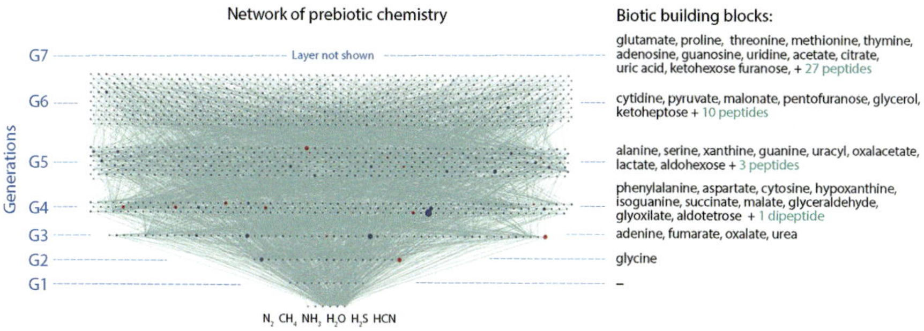

Fig. 4 Synthetic connectivity in a network of prebiotic chemistry. The layered synthetic network was produced in seven generations (G1 to G7) from six primordial substrates, N2, CH4, NH3, H2O, H2S, and HCN in *Allchemy*. The seventh layer is not shown because of its complexity. Red and blue nodes in the network are biotic and abiotic molecules with mass <300 g/mol, respectively. The node size is proportional to the node degree. Each layer produced new biotic building blocks. A total of 82 biotic and 36,603 abiotic molecules were generated, of which 41 simple biotic compounds are listed in the right plus an additional 41 peptides that were generated between generations G4 and G7, including 20 dipeptides, 18 tripeptides, and 3 tetrapeptides (Gly_4, Ser-Gly_3, and Ala-Gly_3), all combinations of Gly, Asp, Ala, Ser, and Phe (the amino acids produced in G2, G3, and G4). Note that the synthesis of amino acids His, Arg, Val, and Leu materialized in additional generations (G8–G16) (Data from Wołos et al. (2020))

metabolism). Because acetate is linked to pyruvate in the glycolytic and pentose phosphate pathways, the core also included reactions that are now central to these pathways. Finally, the core developed reactions that allowed the synthesis of long-chain hydrocarbon precursors of amphiphiles present in the modern pathways of fatty acid biosynthesis. These reactions facilitated the formation of molecular envelopes needed for primordial compartmentalization of the prebiotic world, a prerequisite to the formation of modern cells. The energy needed for all reactions of the central shell was assumed to come from non-nitrogenous compounds such as polyphosphates, formic acid, H_2S, or CH_3SH. In the second shell (shell B), reactions that convert keto acids into amino acids provided gateways to incorporation of nitrogen into emerging metabolic reactions and the synthesis of the nitrogenous building blocks of proteins. The third shell (shell C) incorporates sulfur into amino acids cysteine and methionine. The final shell (shell D) involved ring closure and synthesis of nitrogen and dinitrogen-containing heterocycles, most notably purines-pyrimidines and their derivatives and a large number of cofactors, including water-soluble vitamins. The last shells developed chemistries that are considered entry points for a primordial world of protein enzymes.

The citric acid cycle seated in the middle of the amphiphile core and working in reverse as an autocatalytic carbon fixation system in the absence of enzymes was in line with the previous proposal of Wächtershäuser (1990) of an autocatalytic and cycle-concatenated reductive citric acid cycle in which thioesters are replaced by thioacids, and the reducing power is obtained from oxidative pyrite formation. These cycles have been criticized for lacking solid empirical support (Ruiz-Mirazo et al. 2014); side reactions in the network could act as drains of intermediate molecules, and pathways could progressively become more restricted in their ability to redistribute energy and facilitate kinetics (e.g., Weber 2002). However, three lines of evidence provide a significant boost to the centrality of the amphiphile core. First, a combination of genomic, metabolomic, and enzymatic analysis of metabolic networks of the deep-branching chemolithoautotrophic *Thermosulfidibacter takaii* uncovered a previously unknown reversible citric acid cycle controlled by available carbon sources (Nunoura et al. 2018). These results support previous phylometabolic analyses of metabolic routes in *Aquifex aeolicus* that showed reconstructed citric acid and branched-chain amino acid biosynthesis being more ancestral than in other microbes (Braakman and Smith 2014). Second, combinatorial modeling of chemical transformations allowed to build a chemical supernetwork consistent with reaction types of the reverse citric acid cycle, showing it represents an optimal autocatalytic cycle (Zubarev et al. 2014). Third, recent studies showed the reaction of glyoxylate with pyruvate under mild aqueous conditions produced a series of α-keto acid analogues of the reductive citric acid cycle without metals or enzyme catalysts, providing a route for the biosynthesis of amino acids by transamination with glycine (Stubbs et al. 2020). These findings obviate the search for metal-catalyzed abiotic reductive aminations, reinforcing instead extant chemical mechanisms present in modern metabolism. Findings support the glyoxylate pathway being an evolutionary precursor of the reductive citric acid cycle that bypasses decarboxylation steps (Zubay 2003).

Note that the "shell" accretion model somehow resembles the layering of chemical synthesis produced by the Allchemy algorithm (Wołos et al. 2020) that is illustrated in Fig. 4. In fact, the generated biotic molecules of the prebiotic network already included many elements of the citric acid cycle at the center of the amphiphile core (acetate, oxaloacetate, citrate, pyruvate, glyoxylate, malate, and fumarate), many of the amino acids of shell B and C (Gly, Asp, Phe, Ala, Ser, Thr, Met, Pro, Glu), and many nucleotide building blocks of shell D (Fig. 4), all needed to jumpstart life and connected by numerous cycles of the 47,000 of length ≤ 10 that were present in the prebiotic network. Clearly, compounds of shell D that require ring closure and synthesis of nitrogen and dinitrogen heterocycles seem to be already generated in G3, suggesting their facile prebiotic synthesis in the absence of enzyme catalysts. This challenges the validity of shell D being a late prebiotic newcomer but boosts the idea of it being facilitated by primordial peptides and enzymes that could have been already generated at that time.

To further the biogenetic model of the shell hypothesis, Srinivasan and Morowitz (2009a, b) constructed a canonical network of autotrophic intermediary metabolism derived from information in the KEGG database. The network was modeled based on metabolic pathways from chemoautotrophic organisms, including *Aquifex aeolicus*, *Hydrogenobacter thermophilus*, *Thiomicrospira denitrificans*, and *Chlorobaculum tepidum*. It represented a minimal metabolic chart of 286 reactions involving 287 metabolites. From this metabolic set, they then built a metabolic core of 125 metabolites capable of producing the basic building blocks of life and withstanding some empirical generalizations: (1) C, H, O, N, P, and S are universal metabolic atomic constituents; (2) all autotrophic pathways are anabolic; (3) no reactants are left behind, ensuring conservation of matter, energy, and information; (4) all anabolic pathways have starting points in five *nodal molecules*, acetate (acetyl-CoA), pyruvate, phosphoenol pyruvate, 2-oxoglutarate, and oxaloacetate, which appear universal in autotrophs employing the reductive, oxidative, or incomplete (horseshoe) citric acid cycle; and (5) the network is both "brittle" (sparse, non-redundant, and sensitive to perturbation) and "robust" (resilient to 3.8 billion years of evolution). Figure 5 describes this core metabolic network. The nodes colored green (amino acids) and red (ribonucleotide triphosphates) represent the basic building blocks of life and those colored yellow metabolite intermediates introduced prior to the integration of nitrogen in their makeup.

To test if the structure of the network carried evolutionary information that would match the shell hypothesis, we made nodes proportional to a number of measures of centrality, two of which are shown in the figure. Remarkably, degree centrality, a local view of the connection of a metabolite to other metabolites, highlighted three of the five nodal molecules proposed to be autotrophic starting points: phosphoenolpyruvate, pyruvate, and acetate (Fig. 5a). These starting points led to all amino acids and ribonucleotide triphosphates. Degree centrality also highlighted the central connectivity of Glu and Asp in the network, which clustered with Gln, Asn, Leu, Ile, Phe, and Tyr, and the more distant connectivity of Ser, which clustered with Gly. In turn, betweeness centrality, a global view describing the capacity of nodes to mediate connections to other nodes of the network, made evident the importance of purine biosynthesis metabolic pathways and carbohydrate metabolites linked to nucleotide biosynthesis, including metabolites at the intersection of the pentose phosphate, glycolysis/gluconeogenesis, and starch and sucrose pathways (Fig. 5b), which are believed to represent ancestral starting points of metabolism (*see* below). Note that betweenness centrality highlights network resilience (Newman 2003), showing these ancestral pathways are evolutionarily stable. These results are compatible with the rise of biotic compound hubs in the network of prebiotic reactions that we illustrate in Fig. 4.

8 Evolutionary Genomics Help Untangle the Origin and Evolution of Modern Metabolism

We tested the shell accretion hypothesis with a top-down approach that mined protein structural data at the genomic level (reviewed in Caetano-Anollés et al. 2009). We found in more than two decades of research that the world of protein structure can be best explored with *structural domains*, the structural, functional, and evolutionary units of proteins (Caetano-Anollés et al. 2021). Currently, about 1300 known protein folds represent the structural diversity of the protein world. These folds have been hierarchically cataloged in domain databases such as the Structural Classification of Proteins (SCOP) (Murzin et al. 1995) and SCOPe (Fox et al. 2014). Increasing database entries indicate very few new folds are expected to be found,

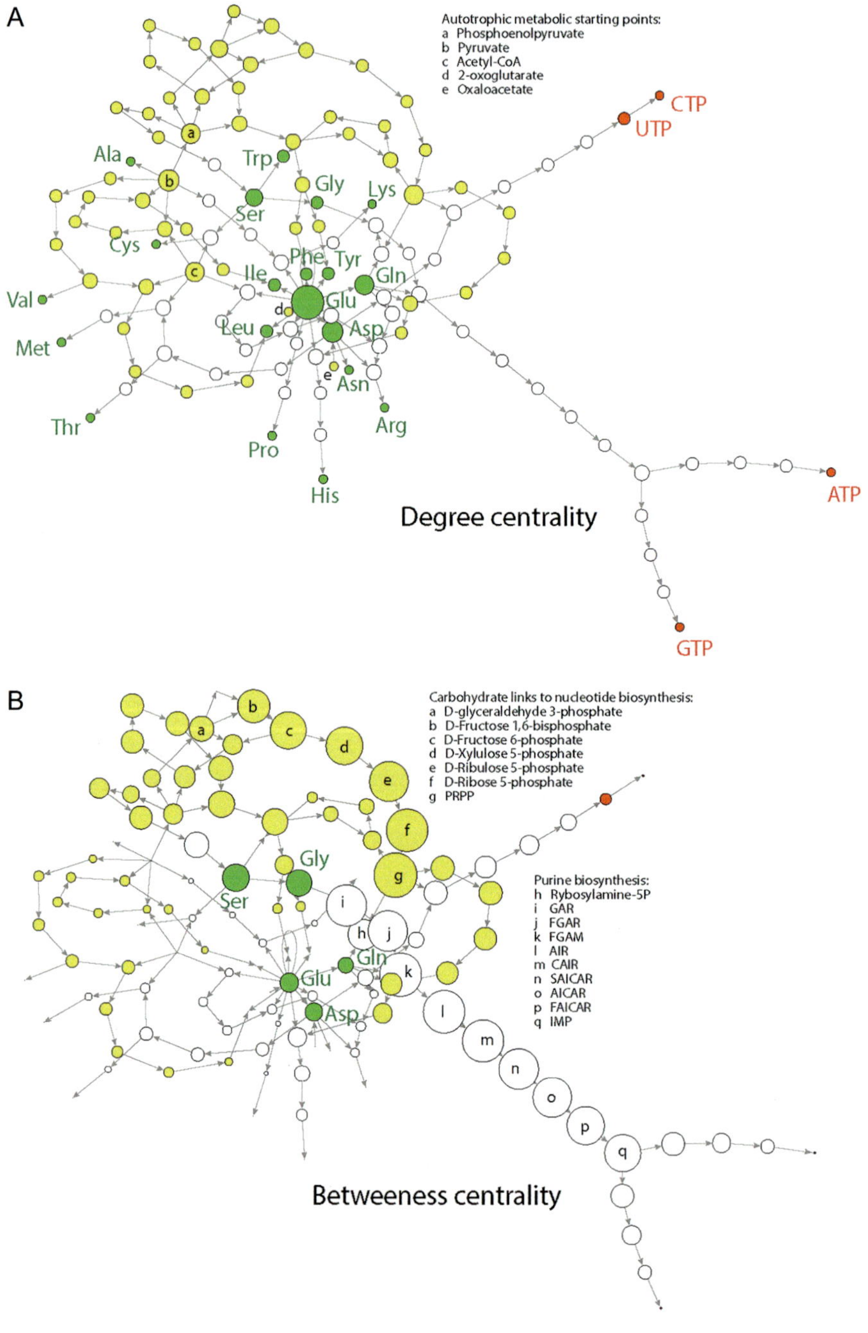

Fig. 5 An autotrophic metabolic network believed to represent the ancient core of intermediary metabolism. Nodes colored green and red describe the basic building blocks of life, i.e., the 20 amino acids and the 4 ribonucleotide triphosphates necessary to build proteins and nucleic acids. Nodes colored yellow represent metabolites introduced prior to the entry and integration of

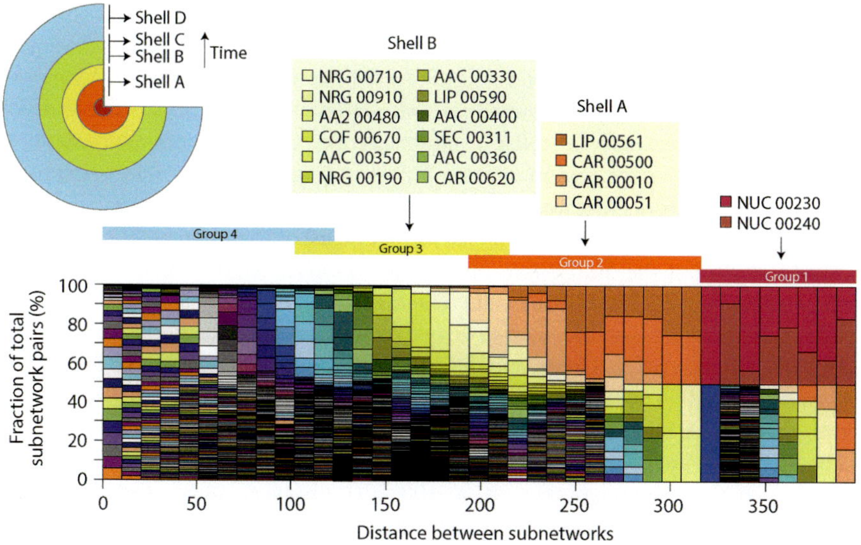

Fig. 6 The shell structure of modern metabolism. Metabolic subnetworks were represented as vectors of protein fold abundances of their respective enzymes in an 877-dimensional space. Pairwise comparison of subnetworks in a Euclidean distance space allowed to place subnetworks participating in comparisons in a continuum of distances, which was then divided in bins. The number of pairs a given subnetwork participates in for every distance bin was displayed as a fraction of the total number of pairs in that bin. Colors of stacked bar plots correspond to individual subnetworks, with the most abundant listed above using KEGG nomenclature. The inset in the left describes the metabolic shell theory of Morowitz (1999), which for the most part matches the metabolic groupings identified in the study (The figure was modified from Caetano-Anollés et al. (2009))

suggesting that the world of protein domain structure is essentially complete at this high level of structural classification. In a comparative genomic study, we compared metabolic subnetworks of KEGG in terms of the folds that their enzymes used (occurrence) and reused (abundance), searching for similarities that would convey evolutionary and functional relatedness (Caetano-Anollés et al. 2009). For that purpose, subnetworks were described as vectors of SCOP fold abundance in an 877-dimensional space. This allowed to calculate Euclidean distances between subnetworks and to use those distances in pairwise analyses (Fig. 6). Our expectation was that subnetworks that were similar in fold abundance levels would cluster (with small Euclidean distances between them) close to an origin, while those that had higher and idiosyncratic abundances would be spread far apart from each other and

Fig. 5 (continued) nitrogen into compounds with C, H, O, P, and S. The node size of the 125 metabolites in the directed network core is proportional to *degree centrality* (**a**), a local view of the connection of a node to other nodes, or *betweeness centrality* (**b**), a global view describing the capacity of nodes to mediate connections to other nodes of the network. Networks were visualized using the Kamada-Kawai free energy representation

away from that origin. Indeed, we found that while most subnetworks were clustered at small distances (about 0–50 distance range), two subnetworks were extremely distant from each other and the rest (>300). These subnetworks were the *purine biosynthesis* (NUC 0230) and *pyrimidine biosynthesis* (NUC 0240) subnetworks of the *nucleotide metabolism* (NUC) mesonetwork. Their remote placement in multidimensional space (group 1) was followed sequentially by three additional groups of subnetworks that were increasingly less remote (Fig. 6). Group 2 included the *glycerolipid metabolism* (LIP 00561) subnetwork of the *lipid metabolism* (LIP) mesonetwork and the *starch and sucrose metabolism* (CAR 00500), *glycolysis/ gluconeogenesis* (CAR 00010), and *fructose and mannose metabolism* (CAR 00051) subnetworks of the *carbohydrate metabolism* (CAR) mesonetwork at a 200–300 distance range. Group 3 included half of the six subnetworks of the *energy metabolism* (NRG) mesonetwork, *oxidative phosphorylation* (NRG 00190), *carbon fixation* (NRG 00710), and *nitrogen metabolism* (NRG 00910) and several important subnetworks of the *amino acid metabolism* (AAC) and *metabolism of cofactors and vitamins* (COF) mesonetworks at a 100–220 distance range. Finally, group 4 included the rest of subnetworks, including those involved in the synthesis of complex amino acids and cofactors, spanning distances that overlap within the 0–120 range. Remarkably, the groupings in fold abundance space support Morowitz's shell hypothesis, both in content and sequence, with the important exception of the boost of nucleotide metabolism that placed it in a separate basal remote group. The initial energy amphiphile core (shell A) represented by groups 2 and 3, with group 2 most likely embodying a "gateway" of prebiotic reactions for energy and lipid amphiphile construction, was followed by conversion of keto acids into amino acids (shells B and C) in the subnetworks of metabolism of amino acids and cofactors that appeared in groups 3 and 4. Group 4 also included many subnetworks involved in the synthesis of amino acids and of nitrogen and dinitrogen-containing heterocycles necessary for production of nucleobases, cofactors, and vitamins (shell D). A consequence of these results is that the folds of metabolic enzymes appear to recapitulate the possible chemistries of a prebiotic world. Therefore, modern metabolism either acts as a palimpsest of ancient metabolism or ancient and modern metabolism originated at some point in parallel and later on replaced each other. We note that the biotic molecules of the prebiotic world generated by the Allchemy algorithm (Fig. 4) included nucleotide building blocks and reactions for prebiotic nucleotide biosynthesis (adenine in generation G3, cytosine in G4, guanine in G5, cytidine in G6, and adenosine, guanosine, uracil and uridine in G7), supporting the basal evolutionary placement of chemical reactions of purine and pyrimidine biosynthesis (group 1). In this respect, prebiotic chemical experiments have already shown that prebiotic pathway to purine (Becker et al. 2016) and pyrimidine (Powner et al. 2009) nucleosides occur in stepwise reaction sequence with excellent regioselectivity and yields. Thus, the primordial appearance of nucleotide metabolism appears highly likely.

We recognized early on that in order to make statements of origin and evolution of biological networks, we needed to embark on a quest for retrodiction (reviewed in Caetano-Anollés et al. 2021). With that in mind, we have been reconstructing phylogenomic trees by optimizing the fit of genomic data along branches of all trees that are possible according to an optimality criterion and an evolutionary model describing how phylogenetic characters change. Trees are branching diagrams of the history of biological systems. Useful phylogenetic characters are those that are "shared and derived" genomic features that are phylogenetically conserved, such as the structure of molecules. Figure 7 shows one of such phylogenies. The tree, which describes the evolution of SCOP protein folds, was reconstructed from a structural domain census in 185 fully sequenced genomes from organisms in superkingdoms Archaea, Bacteria, and Eukarya (Wang et al. 2006). The tree is different from standard trees used in systematic biology. The leaves of the tree are protein folds (not organisms), branch lengths are temporal changes in fold abundance (not changes in molecular sequence), and branching (cladogenic) events are instances of fold diversification (not diversification of sequences representing organisms). Because trees are rooted and are highly unbalanced, their comb-like appearance accommodates stepwise generative models of innovation (Tal et al. 2016). This property justifies counting the number of branching events along the tree and calculating a "node distance" (*nd*) that measures the time of origin (ancestry) of each fold structure. This allows to generate chronologies describing the stepwise appearance of structural novelty in the protein world, mapping, for example, fold ancestries onto the structures of enzymes in metabolic networks and publishing illustrations in the Molecular Ancestry Network (MANET) database (Kim et al. 2006; Mughal and Caetano-Anollés, 2019; https://manet.illinois.edu). The exercise essentially colors times of origin onto the nodes of time-dependent networks (Fig. 3).

Our phylogenomic approach provided unanticipated knowledge on the origin and evolution of modern metabolism. First, tracing enzyme history in the colored metabolic subnetworks of MANET revealed pervasive recruitment of enzymes occurring within and between subnetworks (Kim et al. 2006; Caetano-Anolles et al. 2007; Caetano-Anollés et al. 2009; Kim et al. 2013; Caetano-Anollés and Caetano-Anollés 2013; Mughal and Caetano-Anollés 2019). This exercise eliminated a number of hypotheses as major scenarios of metabolic evolution, including the retrograde and forward-evolution hypothesis that posit that the most downstream or upstream enzymes will be the first to appear in evolving pathways, respectively. Instead, it supported the "patchwork" evolution model, in which early recruitment of enzymes with broad substrate specificities enhanced acquisition of novel chemistries and later fostered enzyme specialization (Ycas 1974; Jensen 1976). Second, tracking recruitment events along the chronology of folds revealed that only the first 9 folds (Fig. 7), and to a lesser degree the first 24 folds, were responsible for most enzymatic activities (Caetano-Anolles et al. 2007). This "big bang" of metabolic innovation occurred during the first 100–300 million years of metabolic evolution and seeded the pathways of about 80% of metabolic subnetworks. Third, phylogenomic trees describing the evolution of subnetworks and mesonetworks from abundance of the nine most ancient folds (Caetano-Anolles et al. 2007) revealed that the two subnetworks of nucleotide metabolism (NUC 00230 and NUC 00240) were the oldest metabolic pathways of metabolism (Fig. 7). Thus, the betweenness centrality mappings onto the putative ancient core of intermediary metabolism that highlighted

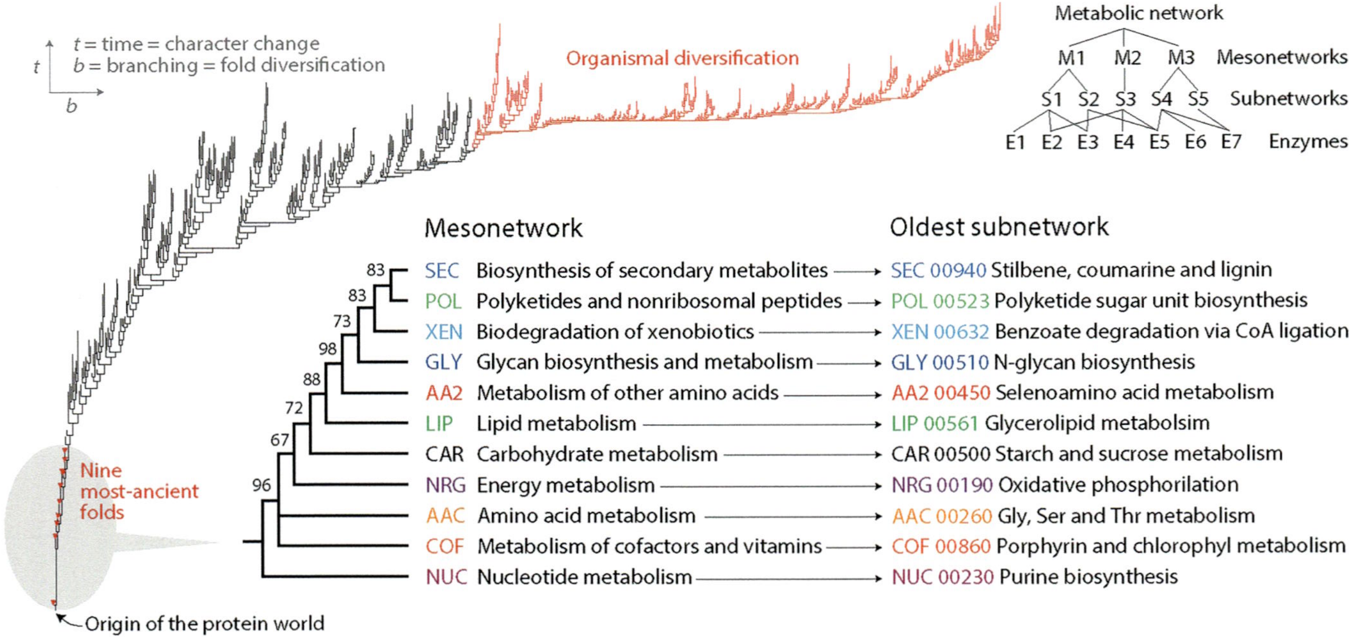

Fig. 7 Evolution of metabolic mesonetworks and subnetworks inferred from phylogenomic analysis of protein domain structure. The phylogenomic tree, which describes the evolution of 766 protein folds defined according to SCOP protein taxonomy, was built using data from a structural domain census in 185 fully sequenced genomes from organisms in the three superkingdoms of life (Wang et al. 2006). Bullets identify the nine most ancient folds shared by all genomes analyzed (c.37, a.4, c.1, c.2, d.58. c.23, c.55, b.40, and c.66, from ancestral to derived). All other terminal leaves of the tree are unlabeled because they would not be legible. Evolution of many folds in the organismal diversification epoch (red branches) is specific to a group of organisms. The cladogram depicted below represents phylogenetic relationships of mesonetworks inferred from abundance of the nine most ancient folds in mesonetworks. The oldest subnetworks in mesonetworks were derived from separate phylogenetic reconstructions. The inset illustrates levels of organization in metabolic networks of KEGG. Branches with bootstrap support (BS) values >50% are shown above the nodes (Data from Caetano-Anollés et al. (2007) and MANET 1.0 (Kim et al. 2006))

purine biosynthesis and carbohydrate links to nucleotide metabolism (Fig. 5), and the remote appearance of nucleotide metabolism in the multidimensional space of fold abundance (Fig. 6) reflected evolutionary history. The COF and AAC mesonetworks appeared later in evolution, with the subnetworks of *porphyrin and chlorophyll metabolism* (COF 00860) and *Gly, Ser, and Thr metabolism* (AAC 00260) being the oldest of those mesonetworks. Again, the distant clustering of degree centrality of Gly and Ser in the ancient metabolic core reflects their ancient origin (Fig. 5). Finally, *metabolic wheels* revealed that the most ancient waves of enzymatic recruitment originated in pathways of nucleotide interconversion of the purine biosynthetic subnetwork, distribution of chemical energy in acid-anhydride bonds of nucleotides, and terminal production of nucleotides and cofactors (Caetano-Anolles et al. 2007).

Figure 8 shows metabolic wheels for the two most ancient fold structures, the P-loop-containing nucleoside triphosphate hydrolase (c.37) and the DNA/RNA-binding three-helical bundle (a.4) folds. Each wheel consists of nodes representing subnetworks arranged in a cycle and colored according to their time of origin. Subnetworks are connected by weighted links representing levels of sharing of enzymatic activities, with activities defined at different levels of Enzyme Commission (EC) classification. Wheels establish which subnetworks were "donors" and which were "acceptors" and a directionality of recruitment of enzyme activities. The metabolic wheel of the c.37 fold included 29 subnetworks, three of which were the most ancient and harbored the highest fold abundance levels: subnetworks of NUC (NUC 00230 and NUC 00240) and the oldest COF subnetwork (COF 00860). Purine biosynthesis (NUC 00230) constituted the central donating vertex of the wheel. Phylogenetically it was the most ancient; it harbored the largest number of enzymes with the c.37 fold and acted as a sharing hub of enzyme activities delivering fold functionalities (mostly transferases) to 19 subnetworks, including the other two most ancient subnetworks. Applying this rationale to each and every link of the wheel explicitly unfolded the c.37 recruitment wave as a temporal blueprint of network evolution. The metabolic wheel of the a.4 fold included 19 subnetworks, revealing that NUC 00230 donated enzyme activities exclusively to NUC 00240 and that COF 00860 was the most ancient recruitment hub, donating activities to LIP and AAC subnetworks.

The recruitment waves uncovered by the two oldest metabolic wheels (Caetano-Anolles et al. 2007) found corroboration in chronologies of protein loops (Aziz et al. 2016, 2023) and chronologies of structural domain organization (Aziz and Caetano-Anollés 2021). Protein loops are prior molecular states that combine with each other in evolution to form structural domains with novel molecular functions (Goncearenco and Berezovsky 2015). The Rossmanoid $\alpha/\beta/\alpha$-layered structure of the c.37 fold is the most ancient and popular in the chronology of structural domain history (reviewed in Caetano-Anollés et al. 2021). The structure embeds a P-loop prototype (and its associated Walker A sequence motif), which crucially enables the nucleotide triphosphate-binding functions of the fold. The DNA/RNA-binding three-helical bundle structure of the a.4 fold, which is flanked by a four-strand β-sheet, exposes crucial elbows between an helix-turn-helix (HTH) loop and

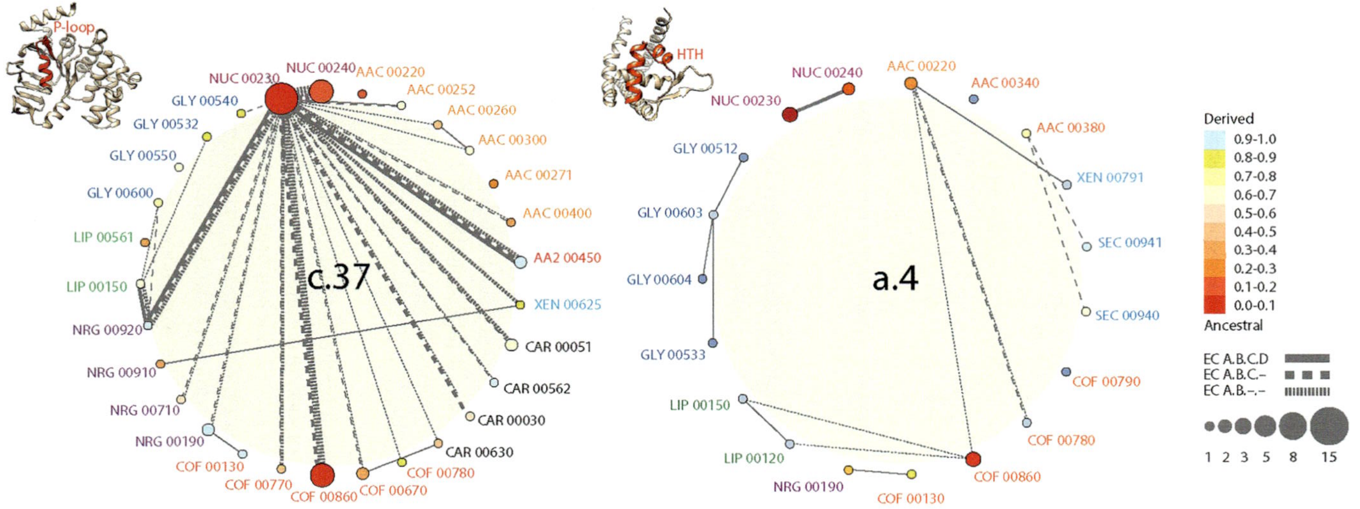

Fig. 8 Metabolic subnetwork wheels for folds involved in the first two waves of enzymatic recruitment, the P-loop hydrolase (c.37), and the DNA/RNA-binding three-helical bundle (a.4) folds. The c.37 fold structure is organized around a central β-sheet flanked by α-helices that originated in a nucleotide-binding P-loop that is Gly-rich. The a4 fold has a core of three helices, a partly opened or closed bundle with a right-handed twist that originated in a helix-turn-helix (HTH) loop. The ancient loops are colored red in the insets. The metabolic wheel graphs show nodes representing metabolic subnetworks containing enzymes with either the c.37 or the a.4 folds and links representing the sharing of Enzyme Commission (EC) number at different levels of classification. The node area is proportional to fold abundance, and the line width is proportional to sharing of enzymatic activities. Subnetwork ancestries derived from phylogenomic trees of subnetworks given as a function of distance in nodes from a hypothetical ancestral subnetwork (*nd*) are color coded and used to paint the nodes of each subnetwork wheel (Data from Caetano-Anollés et al. (2007) and MANET 1.0 (Kim et al. 2006))

sequence motifs that harbor the specificity of protein-protein and protein-RNA interactions typical of these enzymes. The "winged" domain structure of the a.4 fold plays central roles in transcription, providing flexibility and nucleic acid clamping capacity to RNA polymerases and crucial surfaces for domain-domain recognition in complexes such as ubiquitin gases and condensins. Evolving networks of "elementary functional loops" (EFLs) (bioinformatically derived from sequence profiles) that combine to form active sites (including cofactor-binding sites) and exert molecular functions (Aziz et al. 2016) and loop prototypes defined by a density search-clustering algorithm that explores the structure, conformation, and geometry of protein loops (Aziz et al. 2023) revealed two initial loop recruitment waves involving the P-loop and the HTH loop of the c.37 and a.4 folds. The waves involved cysteine-rick loop prototypes and strong recruitment pathways involving the NAD(P)-binding Rossmann-fold domain (c.2.1), the S-adenosyl-L-methionine-dependent methyltransferase domain (c.66.1), and the OB-fold of the nucleic acid-binding protein domain (b.40.4), all necessary to develop translation. At higher levels of protein organization, structural domains in proteins combine with other domains to form multidomain proteins, many of which are metabolic enzymes. A time series of networks describing the gradual combination of structural domains in proteins revealed again the two primordial waves (Aziz and Caetano-Anollés 2021). However, most of the connectivity of these major recruitment gateways was established relatively recently in evolution, during the organismal diversification epoch less than 1.5 billion years ago (Gya) (nd $\geq$ 0.6), showing percolation of evolutionary constraints from lower to higher levels of organization manifest as evolutionary delay, a tendency we often observe at the molecular level.

It is remarkable that the first two waves of enzymatic recruitment that were congruently identified in numerous studies at different levels of protein organization involved the same primordial α/β/α-layered sandwich, β-barrels, and helical bundle structures we identified as part of the first 54 structural domain families that appeared in evolution (Caetano-Anollés et al. 2012). Retrieval of congruent evolutionary history with different data sets provides strong support to our statements of metabolic evolution.

9 An Emergent Entanglement Generates Network Structure, Modules, and Communication

Since the modern purine biosynthesis subnetwork (NUC 00230) was the very first subnetwork to appear in metabolism, we followed its history by constructing snapshots of its early evolution (Caetano-Anollés and Caetano-Anollés 2013). Figure 9 shows a time series of growing networks. The time series reveals the subnetwork assembled as a patchwork through processes of both enzyme recruitment and enzymatic replacement of prebiotic chemistries. These processes likely operated at the planetary scale. Note how red nodes and links representing prebiotic

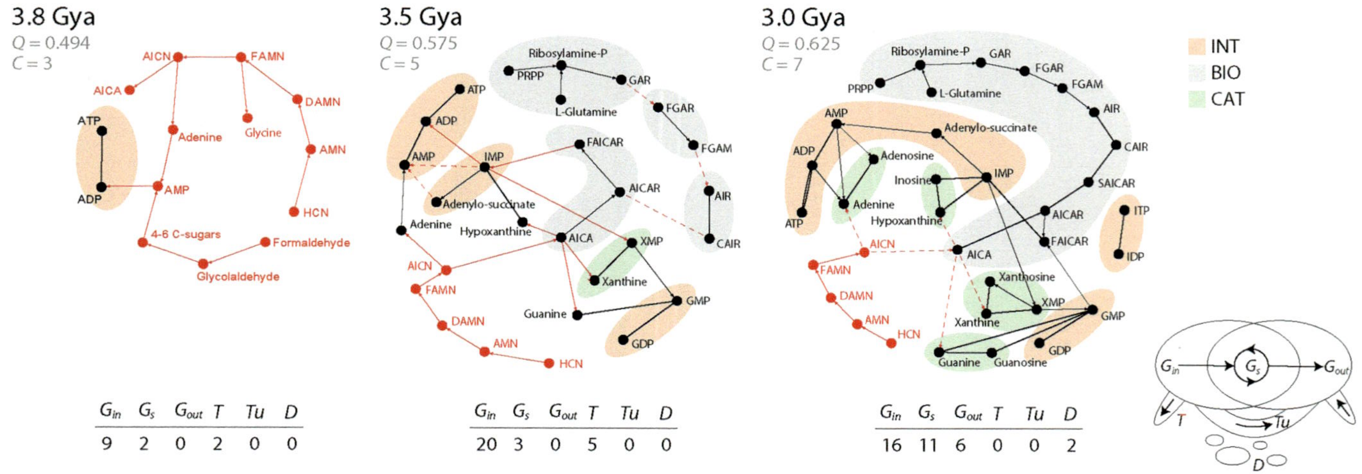

Fig. 9 The evolutionary rise of the purine biosynthesis subnetwork and its bow-tie structure. The reconstruction of subnetworks that were present 3.8, 3.5, and 3 billion years ago (Gya) reveals the piecemeal recruitment of functional modules for the nucleotide interconversion (INT), catabolism and salvage (CAT), and biosynthetic (BIO) pathways of the purine biosynthesis subnetwork (NUC 00230). Plausible metabolites and prebiotic chemical reactions that are needed to support the activity of emergent enzymes are depicted with red nodes and links. Unknown reaction candidates or withering prebiotic pathways are indicated with dashed lines. These ancient chemistries are gradually replaced by modern pathways and are unified from separate components into a cohesive network of INT, CAT, and BIO modules making up a bow-tie structure. The network was rendered using the energy spring embedders and the Fruchterman-Reingold algorithm. Note how modularity (Q) measures of community structure and numbers of communities (C) increase in evolution. A diagram of network components (described on the right) for each network snapshot shows bow-tie structure emerges 3 Gya as a strongly connected component (G_s) increases its numbers, receives connections from an "in" component (G_{in}), and then elicits connections to an "out" component (G_{out}). In the process, "tendrils" (T) disappear in exchange of "disconnected components" (D), while "tubes" (Tu) remain inexistent (Full metabolite names can be found in the work by Caetano-Anollés and Caetano-Anollés (2013))

metabolites and associated chemical reactions were increasingly replaced by black nodes and links representing modern metabolites and their enzyme-mediated transformations. The chronology shows the initial prebiotic subnetwork that was present 3.8 Gya at the start of the protein world was composed of three small communities (identified with the Louvain method) that were part of a larger prebiotic reaction network of the type illustrated in Fig. 4. This subnetwork replaced one of its chemical reactions with an enzyme-driven transformation mediated by primordial ATP phosphohydrolase enzymes (EC 3.6.1.3 and EC 3.6.4.1) harboring the P-loop hydrolase fold structure. This transformation enabled ADP-ATP-mediated energy interconversion reactions and involved the first and second structural domain families of the protein world, the ABC transporter ATPase domain-like (c.37.1.12) and the AAA-ATPase domain (c.37.1.20), both of which harbored a primordial bundle and were likely associated with mechanoenzyme-linked stability of primordial membranes and protocell persistence (Caetano-Anollés and Caetano-Anollés 2013). We have shown that the c.37.1.12 structure was given birth by either modular (DS.EH.6.1.1; Aziz et al. 2023) or non-modular (DS.EH.6.17.1 and DS.EH.7.6.1; Mughal and Caetano-Anollés 2023) P-loop prototypes. These prototypes attained more complex structure by quickly recruiting other loop prototypes for rapid convergence into the core c.37 fold structure, as evidenced by AlphaFold2 structural modeling (Aziz et al. 2023). The subnetwork that existed 3.5 Gya benefited from a number of enzymatic replacements of prebiotic chemistries, which gradually seeded the nucleotide interconversion (INT), metabolism/salvage (CAT), and biosynthetic (BIO) pathways that make up the NUC 00230 subnetwork. While several recruitments of sixfold family structures (c.23.16.1, c.30.1.1, d.142.1.2, c.6.1.1.1, c.37.1.1, and a.127.1.1, in order of recruitment) expanded the subnetwork (Caetano-Anollés and Caetano-Anollés 2013), pathways were not completely enzymatic, and as previously suggested, several crucial prebiotic chemistries had to hold them together (Becerra and Lazcano 1998). This resulted in a network with five Louvain communities. Finally, and remarkably, enzymatic pathways became fully functional 3 Gya, at a time when a functional ribosome was already active with a peptidyl transferase center (PTC) and processive ribosomal regions responsible for ribosomal dynamics (Harish and Caetano-Anollés 2012). The pathway primordium had seven Louvain communities but appeared fully integrated into the prebiotic network, suggesting the appearance of a functional ribosome occurred when metabolism was still a semi-enzymatic process. The 11 enzymatic activities of the BIO pathway provided a steady supply of IMP for emerging RNA by sequentially building the purine ring on a phosphoribosyl scaffold.

The Louvain maximization method revealed that modularity and community number increased in network evolution, suggesting hierarchical modularity increases with time (Fig. 9). Similarly, a bow-tie analysis of network components of the type used to dissect the WWW (Broder et al. 2000), metabolism (Ma and Zeng 2003), and a neural network of *Caenorhabditis elegans* that combines both chemical and electrical synapses (Timár et al. 2017) revealed that the bow-tie network structure was initially absent but emerged 3 Gya with the fully functional semi-enzymatic pathway. Note the early existence of "tendrils" and absence of an "out" component (G_{out}), both indicating the full integration of the initial enzymatic steps onto a larger prebiotic network. The pathway established 3 Gya, however, had all components present in modern metabolism (G_{in}, G_s, G_{in}, and D) (Ma and Zeng 2003). Analysis of the WWW reveals that this man-made evolving network is generating multiple layers of tendrils (T) and a multiplicity of tubes (Tu) that connect them, suggesting networks have the potential to develop hierarchy at this level of network structure (Timár et al. 2017). The absence of such structure in metabolism may simply indicate significant constraints on network growth operating at this low level of biological organization, especially because the *C. elegans* neural network develops tendrils upon network damage suggesting trade-offs between robustness (minimizing damage) and flexibility (establishing redundant connections through tendrils and tubes) (Timár et al. 2017).

The rise of hierarchical modularity appears to be a general phenomenon in evolution of biological networks (Caetano-Anollés et al. 2019). A number of phylogenomic studies and their associated chronologies reveal that hierarchical modularity is an emergent property of networks that unfolds at different time scales. Hierarchical modularity emerges in networks that describe the nanosecond dynamics of protein loops studied with molecular dynamics (MD) simulations (Mughal et al. 2017; Caetano-Anollés et al. 2019), the rewiring of metabolomic networks occurring at second-to-hour timeframes (Aziz et al. 2012), tripartite metabolic networks describing minute-to-hour changes in gene expression and metabolic pathway activities (Koç et al. 2018), bipartite networks evolving at timeframes of billions of years (Gy) describing the evolution of elementary functionomes (Aziz et al. 2016), emergence of loop and domain makeup (Aziz et al. 2023) and protein domain organization (Aziz and Caetano-Anollés 2021), and bipartite metabolic networks describing the evolution of recruitment of enzyme activities (Mughal and Caetano-Anollés 2019). For example, chronologies of the entire metabolic network showed increases in clustering coefficients, modularity, and community structure confirming the emergence of modularity and hierarchy in metabolic evolution (Mughal and Caetano-Anollés 2019). Figure 10 illustrates selected aspects of a chronology of bipartite networks and their subnetwork and enzyme projections showing the rise of hierarchical modularity.

The rise of hierarchical modularity in networks supports a "linkage" theory (Mittenthal et al. 2012) that explains the existence of temporal biphasic patterns in biological systems (Caetano-Anollés et al. 2019). The biphasic (bow-tie) pattern that emerges in the purine metabolic subnetwork (Fig. 9) is actually widely present in biological and man-made technological systems (Mittenthal et al. 2012; Sabrin and Dovrolis 2017), including structural biology, metabolism, gene expression, biological signaling, immunology, development, vision, and cognition but also in large-scale integrated circuits and the WWW (reviewed in Caetano-Anollés 2021). The "linkage" theory explains biphasic patterns as the result of two interlinked evolutionary processes, one of unification and the other of diversification (Mittenthal et al. 2012). The theory also explains the rise of modules in biology. Figure 11 uses one of several paradigms to describe the theory: a time series of evolutionary change. In the first "construction" phase (Phase 1), a series of events of accretion increases the

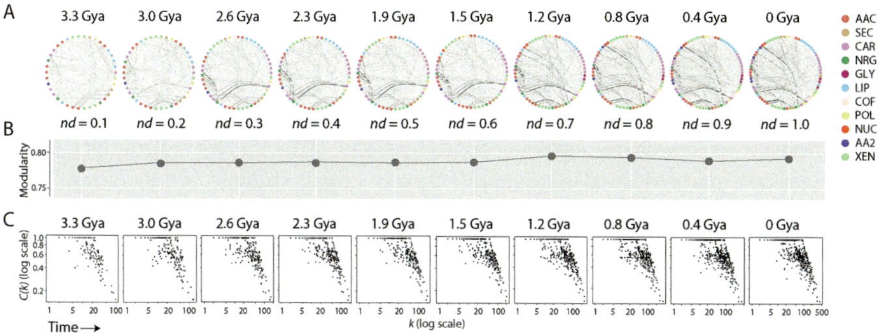

Fig. 10 Evolution of a bipartite metabolic network of subnetworks and enzymes. (**a**) Evolutionary snapshots of a reduced subnetwork projection of the bipartite network. The evolving networks describe how subnetworks and the enzymes they share grow in time, with time expressed as a "node distance" that describes time of origin *(nd)* or as geological time (Gya). The projection shows major subnetworks (nodes) colored according to their mesonetworks connected to each other through shared enzymes (links). Grayscale values of links indicate the number of enzymes shared among the subnetworks. (**b**) Maximum modularity measured using the Fast-Greedy-Community (*FGC*) algorithm. (**c**) Log-log plots of the clustering coefficient *C(k)* as a function of the number of links *k* for evolving enzyme projections of the bipartite network. The scaling, which is the hallmark of hierarchical modularity, increases in evolution (Data from Mughal and Caetano-Anollés (2019))

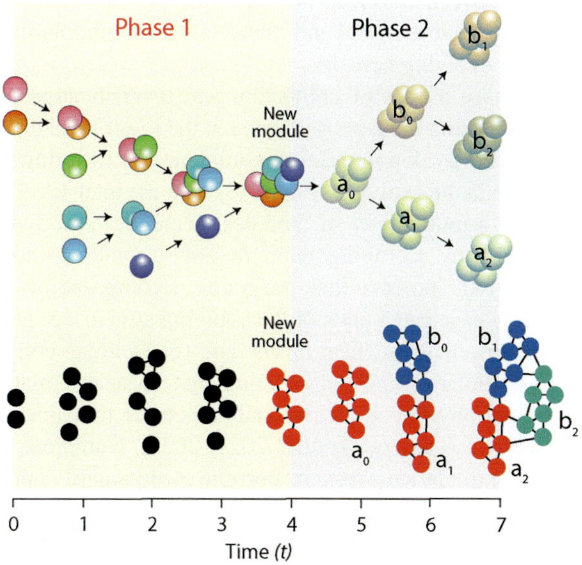

Fig. 11 The biphasic (bow-tie) paradigm. The biphasic model is explained with a time series describing accumulating events of accretion and diversification of the parts of a biological system (top) or snapshots of a network in which nodes describe parts and links describe their interactions (bottom). (**a**) Individual parts (spheres) come together to form a new module (Phase 1), which then multiplies and diversifies in a sequential (modules a_0, a_1, and a_2) or branching (modules b_0, b_1, and b_2) manner in the absence of other emerging modules (Phase 2). Note that the network representation makes explicit the interaction between the evolving modules

structure of the evolving system by gradually merging individual parts (spheres in the figure) into a module. The dynamics of this phase changes in time. At first, it is driven by establishing weak links (connections) between parts as they associate with each other in various configurations. With time, links compete with each other for performance in search for optimal configurations. This causes constraints in the association of parts, decreasing dynamic turn-over and structuring the system. This search for optimality in performance space finally results in increasingly stronger links, leading to self-organization and the rise of modules and modular structure. In the second "diversification" phase (Phase 2), the module diversifies in a sequential (modules a_0, a_1, and a_2) or branching (modules b_0, b_1, and b_2) manner in the absence or presence of other evolving modules. This process generates variants of the module that can become new parts for a new generative cycle of a higher level of organization. It also generates redundancy, which furthers system's persistence.

The linkage theory can also be described with a paradigm of evolving networks (Fig. 11). In Phase 1, temporal paths are at first short, with emerging networks exhibiting intermittency and burstiness. This highly dynamic behavior that foster spreading dynamics is gradually slowed down by the establishment of stronger links, shorter paths, and the rise of small communities that gradually merge into modules. Here, short-range interactions that stabilize emerging communities are followed by establishment of long-range interactions that foster inter-modular connections. In Phase 2, modules coalesce into new communities generating higher-level network structures that act as network units for a new generative cycle of network structuring. The two phases are tightly interlinked and cause the rise of modularity and hierarchy in the structure of the evolving network.

The two evolutionary phases of unification and diversification are as follows: (1) create novelty by reusing the parts generated at the lower level of organization in novel combinations, (2) develop intra-level redundancy by fostering abundance and module diversification in the evolving system, (3) increase inter-level redundancy by embedding parts into a multiplicity of modules (accretion) and diversifying those modules, for example, by recruiting them in novel combinations (modularity), (4) foster a "nucleation" process that generates decomposability and cohesion (dynamical stabilities) in selected parts of the evolving system, (5) generate increasing "granularity" at lower hierarchical levels, and (6) facilitate communication by helping develop vocabularies and computation that dissipates matter-energy and information in an entropy-like directive that structures dynamical systems and extends their persistence (Caetano-Anollés 2021, 2023). Numerous agents of evolutionary change can explain why systems become evolutionarily entrenched. Computer simulations have shown that non-adaptive (neutral) drivers such as reuse, connection costs, or information transmittal and adaptive drivers such as competitive optimization or natural selection can further the structure and persistence of a system and its network representation. An "entanglement" theory (ET) addresses the reason why nodes that represent parts of a system get entangled (Caetano-Anollés 2021; Caetano-Anolles et al. 2022). Since macroscopic (entropic) forces originate in systems holding numerous degrees of freedom that tend to increase entropy by dissipation (Caetano-Anollés et al. 2010), ET postulates that network growth results from an entropic-like attractive force driven by short-distance and long-distance entanglements of neighboring degrees of freedom. We view this putative entropic force as an emergent phenomenon analogous to that operating in the collapse of polymers, polymer elasticity, the hydrophobic effect, and Brownian motion. The emerging phenomenon satisfies a tendency to increase entropy of the system, with entropy being broadly defined, projecting entanglements into a holographic-like horizon. The short-range entanglement process generates modules, and the long-range entanglement generates hierarchy, pushing growth via exploration of informational spaces. Thus, entanglement unfolds during the first phase of the linkage paradigm in Fig. 11 in sequential but integrated manner, with short-range entanglements that occur early (time events 0–2) constraining possible long-range entanglements (time events 2–4) that lead to the formation of modules.

10 Conclusion

Our exploration of metabolic evolution through a phylogenomic lens provides a captivating narrative of emergence and entanglement in network evolution. The genomic-driven reconstruction of phylogenies of protein structure was able to uncover the dynamic history of metabolic networks, revealing a pivotal "big bang" moment of metabolic innovation that seeded the pathways of the majority of metabolic subnetworks. The departure from traditional trees used by systematic biology allowed to trace steps of enzymatic recruitment in the metabolic wiring diagrams, challenging prevailing hypotheses and supporting a "patchwork" evolution model. In our phylogenomic quest, metabolic wheels vividly portrayed the chronological waves of enzymatic recruitment that originated in pathways crucial for nucleotide interconversion and energy distribution in the purine biosynthesis subnetwork. The integration of structural data, protein loops, and chronologies of loop and structural domain organization corroborated our findings, reinforcing the validity of our approach across different levels of molecular organization. The two primordial waves of enzymatic recruitment that we identified at the origin of the protein world align consistently with the first structural domain families that appeared in evolution, further solidifying the congruence of our metabolic evolution narrative.

We also uncovered emergent properties of biological network structure. The analysis of chronologies across diverse time scales, from nanosecond dynamics of protein loops to evolutionary processes unfolding at Gy timeframes, consistently supported the rise of hierarchical modularity as a general evolutionary phenomenon of network evolution. This not only reaffirms the universality of the phenomenon but also underscores its fundamental role in shaping the complexity and robustness of biological systems. The rise of hierarchical modularity, illustrated through bipartite network chronologies, echoes the biphasic patterns observed in various biological systems and technological networks. Emergent patterns of network structure also support a "linkage" theory that provides a conceptual framework to elegantly explain the coexistence of accretion and diversification processes, shedding light on the intricate balance of evolutionary forces shaping network evolution. The linkage theory not only elucidates the generation of novelty and redundancy but also provides insights into the increasing granularity of structure that exists at lower levels of hierarchical organization. This granularity, akin to the development of vocabularies and computation, becomes a conduit for effective communication that dissipates matter-energy and information in an entropy-like directive. In essence, the theory captures the essence of evolutionary forces shaping the dynamical systems, extending their persistence and resilience across the evolutionary spectrum.

Our findings underscore the entangled nature of biological systems. The entropic-like forces driving network growth in ET add a layer of understanding to the dynamics of evolutionary change. The proposed emergent entanglement force underscores the intricate interplay of short-distance and long-distance interactions that is common in the processes of molecular biology, including the collapse of RNA molecules into helical regions of structure or the vestibular co-translation process of protein folding in the ribosomal exit tunnel. By acknowledging the interplay of neutral and adaptive drivers, we recognize the myriad factors contributing to the structure and persistence of biological systems. Our journey through the evolutionary tapestry of metabolism not only enriches understanding of life's history but also offers a roadmap for future explorations at the intersection of network science, evolutionary genomics, and structural biology.

Acknowledgments We dedicate this manuscript to the memory of Harold J. Morowitz, vigorous proponent of a metabolic-first thermodynamic view of the origin of life, and Russell F. Doolittle, pioneer of protein evolution. We much appreciated their friendship, inspiration, and encouragement. Research on metabolic evolution was supported by grants from the National Science Foundation (MCB-0749836 and OISE-1132791) and the United States Department of Agriculture (ILLU-802-909 and ILLU-483-625) to GCA.

References

Amara A, Frainay C, Jourdan F et al (2022) Networks and graphs discovery in metabolomics data analysis and interpretation. Front Microbiol 9:841373

Avena-Koenigsberger A, Goñi J, Solé R et al (2015) Network morphospace. JR Soc Interface 12: 20140881

Aziz MF, Caetano-Anollés G (2021) Evolution of networks of protein domain organization. Sci Rep 11:12075

Aziz MF, Chan P, Osorio JH et al (2012) Stress induces biphasic-rewiring and modularization patterns in metabolomic networks of *Escherichia coli*. IEEE Intl Conf Bioinf Biomed 2012: 593–597

Aziz MF, Caetano-Anollés K, Caetano-Anollés G (2016) The early history and emergence of molecular functions and modular scale-free network behavior. Sci Rep 6:25058

Aziz MF, Mughal F, Caetano-Anollés G (2023) Tracing the birth of structural domains from loops during protein evolution. Sci Rep 13:14688

Barabási A-L, Albert R (1999) Emergence of scaling in random networks. Science 286:509–512

Becker S, Thoma I, Deutsch A et al (2016) A high-yielding, strictly regioselective prebiotic purine nucleoside formation pathway. Science 552:833–846
Bercerra A, Lazcano A (1998) The role of gene duplication in the evolution of purine nucleotide salvage pathways. Orig Life Evol Biosphere 28:539–553
Blondel VD, Guillaume J-L, Lambiotte R et al (2008) Fast unfolding of communities in large networks. J Stat Mech Theory Exp P10008
Boccaletti S, Latora V, Moreno Y et al (2006) Complex networks: structure and dynamics. Phys Rep 424:175–308
Bogdan P, Caetano-Anollés G, Jolles A et al (2022) Biological networks across scales—the theoretical and empirical foundations for time-varying complex networks that connect structure and function across levels of biological organization. Integr Comp Biol 61:1991–2010
Borgatti SP, Everett MG (2006) A graph-theoretic perspective on centrality. Soc Networks 28:466–484
Braakman R, Smith E (2014) Metabolic evolution of a deep-branching hyperthemophilic chemoautotrophic bacterium. PLoS One 9:e87950
Broder A, Kumar R, Maghoul F et al (2000) Graph structure in the web. Comp Networks 33:309–320
Broido AD, Clauset A (2019) Scale-free networks are rare. Nat Commun 10:1017
Caetano-Anollés G (2021) Untangling molecular biodiversity. World Scientific, Singapore
Caetano-Anollés G (2023) Agency in evolution of biomolecular communication. Ann N Y Acad Sci 1525:88–103
Caetano-Anollés K, Caetano-Anollés G (2013) Structural phylogenomics reveals gradual evolutionary replacement of abiotic chemistries by protein enzymes in purine metabolism. PLoS One 8:e59300
Caetano-Anollés G, Kim HS, Mittenthal JE (2007) The origin of modern metabolic networks inferred from phylogenomic analysis of protein architecture. Proc Natl Acad Sci USA 104: 9358–9363
Caetano-Anollés G, Yafremava LS, Gee H et al (2009) The origin and evolution of modern metabolism. Int J Biochem Cell Biol 41:285–297
Caetano-Anollés G, Yafremava L, Mittenthal JE (2010) Modularity and dissipation in evolution of macromolecular structures, functions, and networks. In: Caetano-Anollés G (ed) Evolutionary bioinformatics and systems biology. Wiley-Blackwell, New York, pp 443–449
Caetano-Anollés G, Kim KM, Caetano-Anollés D (2012) The phylogenomic roots of modern biochemistry: origin of proteins, cofactors and protein biosynthesis. J Mol Evol 74:1–34
Caetano-Anollés G, Aziz MF, Mughal F et al (2019) Emergence of hierarchical modularity in evolving networks uncovered by phylogenomic analysis. Evol Bioinforma 15: 1176934319872980
Caetano-Anollés G, Aziz MF, Mughal F et al (2021) Tracing protein and proteome history with chronologies and networks: folding recapitulates evolution. Exp Rev Proteomics 18:863–880
Caetano-Anollés G, Aziz MF, Mughal F et al (2022) Recruitment: a problem of entangled temporal parts. Front Biosci (Landmark Ed) 27:128
Caspi R, Billington R, Keseler IM et al (2020) The MetaCyc database of metabolic pathways and enzymes - a 2019 update. Nucleic Acids Res 48:D445–D453
Clauset A, Newman MEJ, Moore C (2004) Finding community structure in very large networks. Phys Rev E 70:66111
Collier J (2003) Hierarchical dynamical information systems with a focus on biology. Entropy 5: 100–124
Corominas-Murtra B, Goñi J, Solé RV et al (2013) On the origins of hierarchy in complex networks. Proc Natl Acad Sci USA 10:13316–13321
Delaney W, Vaccari E (1989) Dynamic models and discrete event simulation. Marcel Dekker Inc, New York
Falkenberg M, Lee J-H, Amano S et al (2020) Identifying time dependence in network growth. Phys Rev Res 2:023352

Fox NK, Brenner SE, Chandonia JM (2014) SCOPe: structural classification of proteins—extended, integrating SCOP and ASTRAL data and classification of new structures. Nucleic Acids Res 42:D304–D309

Gernat T, Rao VD, Middendorf M et al (2018) Automated monitoring of behavior reveals bursty interaction patterns and rapid spreading dynamics in honeybee social networks. Proc Natl Acad Sci USA 115:1433–1438

Goh K-I, Oh E, Jeong H et al (2002) Classification of scale-free networks. Proc Natl Acad Sci USA 99:12583–12588

Goncearenco A, Berezovsky IN (2015) Protein function from its emergence to diversity in contemporary proteins. Phys Biol 12:45002

Guimerà R, Amaral LAN (2005) Functional cartography of complex metabolic networks. Nature 433:895–900

Harish A, Caetano-Anollés G (2012) Ribosomal history reveals origins of modern protein synthesis. PLoS One 7:e32776

Hartwell LH, Hopfield JJ, Leibler S et al (1999) From molecular to modular cell biology. Nature 401:c47–c52

Holme P (2005) Network reachability of real-world contact sequences. Phys Rev E 71:046119

Holme P (2019) Rare and everywhere: perspectives on scale-free networks. Nat Commun 10:1016

Holme P, Saramäki J (2012) Temporal networks. Phys Rep 519:97–125

Holme P, Kim BJ, Yoon CN et al (2002) Attack vulnerability of complex networks. Phys Rev E 65: 056109

Jensen RA (1976) Enzyme recruitment in evolution of new function. Ann Rev Microbiol 30:409–425

Jeong H, Tombor B, Albert R et al (2000) The large-scale organization of metabolic networks. Nature 407:651–654

Kanehisa M, Furumichi M, Sato Y et al (2023) KEGG for taxonomy-based analysis of pathways and genomes. Nucleic Acids Res 49:D545–D551

Kempe D, Kleinberg J, Kumar A (2002) Connectivity and inference problems for temporal networks. J Compu Syst Sci 64:820

Kim HS, Mittenthal JE, Caetano-Anollés G (2006) MANET: tracing evolution of protein architecture in metabolic networks. BMC Bioinformatics 7:351

Kim HS, Mittenthal JE, Caetano-Anollés G (2013) Widespread recruitment of ancient domain structures in modern enzymes during metabolic evolution. J Integr Bioinform 10:214

Kim H, Smith HB, Mathis C et al (2019) Universal scaling across biochemical networks on earth. Sci Adv 5:eaau0149

Koç I, Yuksel I, Caetano-Anollés G (2018) Metabolite-centric reporter pathway and tripartite network analysis of Arabidopsis under cold stress. Front Bioeng Biotechnol 6:121

Kovanen L, Karsai M, Kaski K et al (2011) Temporal motifs in time-dependent networks. J Stat Mech P11005

Ma H-W, Zeng A-P (2003) The connectivity structure, giant strong component and centrality of metabolic networks. Bioinformatics 19:1423–1430

Maden BEH (1995) No soup for starters? Autotrophy and the origins of metabolism. Trends Biochem Sci 20:337–341

Mittenthal JE, Caetano-Anollés D, Caetano-Anollés G (2012) Biphasic patterns of diversification and the emergence of modules. Front Genet 3:147

Morowitz H (1999) A theory of biochemical organization, metabolic pathways, and evolution. Complexity 4:39–53

Morowitz HJ, Kostelnik JD, Yang J et al (2000) The origin of intermediary metabolism. Proc Natl Acad Sci USA 97:7704–7708

Mughal F, Caetano-Anollés G (2019) MANET 3.0: hierarchy and modularity in evolving metabolic networks. PLoS One 14:e0224201

Mughal F, Caetano-Anollés G (2023) Evolution of intrinsic disorder in protein loops. Life 13:2055

Mughal F, Gräter F, Caetano-Anollés G (2017) How function shapes dynamics in protein evolution. In: Szuch S, Watkins C (eds) Blue waters annual report. National Center for Supercomputer Applications, Champaign, pp 198–199
Murzin AG, Brenner SE, Hubbard T et al (1995) SCOP: a structural classification of proteins database for the investigation of sequences and structures. J Mol Biol 247:536–540
Newman MEJ (2001) Scientific collaboration networks: II. Shortest paths, weighted networks, and centrality. Phys Rev E 64:016132
Newman MEJ (2003) The structure and function of complex networks. SIAM Rev 45:167–256
Newman MEJ (2006) Finding community structure in networks using the eigenvectors of matrices. Phys Rev E 74:036104
Newman MEJ, Girvan M (2004) Finding and evaluating community structure in networks. Phys Rev E 69:26113
Newman MEJ, Barabási AL, Watts DJ (2009) The structure and dynamics of networks. Princeton Univ Press, Princeton
Nunoura T, Chikaraishi Y, Ikazi R et al (2018) A primordial and reversible TCA cycle in a facultatively chemolithoautotrophic thermophile. Science 359:559–563
Pan RK, Saramäki (2011) Path lengths, correlations, and centrality in temporal networks. Phys Rev E 84:016105
Powner MW, Gerland B, Sutherland JD (2009) Synthesis of activated pyrimidine ribonucleotides in prebiotically plausible conditions. Nature 459:239–242
Ravasz E (2009) Detecting hierarchical modularity in biological networks. Meth Mol Biol 541: 145–160
Ravasz E, Somera AL, Mongru DA et al (2002) Hierarchical organization of modularity in metabolic networks. Science 297:1551–1555
Ruiz-Mirazo K, Briones C, de la Escosura A (2014) Prebiotic systems chemistry: new perspectives for the origin of life. Chem Rev 114:285–366
Sabrin KM, Dovrolis C (2017) The hourglass effect in hierarchical dependency networks. Network Sci 5:490–528
Salthe SN (1985) Evolving hierarchical systems: their structure and representation. Columbia University Press, New York
Schuetz R, Zamboni N, Zampieri M et al (2012) Multidimensional optimality of microbial metabolism. Science 336:601–604
Shen Y, Borowski JE, Hardy MA et al (2021) Automation and computer-assisted planning for chemical synthesis. Nat Rev Meth Primers 1:23
Simon HA (1962) The architecture of complexity. Proc Am Phil Soc 106:467–482
Simon HA (1997) Models of bounded rationality. MIT Press, Cambridge, Massachusetts
Smakowska-Luzan E, Mott AG, Parys K et al (2018) An extracellular network of *Arabidopsis* leucine-rich repeat receptor kinases. Nature 553:342–346
Solé RV, Valverde S (2004) Information theory of complex networks: on evolution and architectural constraints. Lecture Notes Phys 650:189–207
Srinivasan V, Morowitz HJ (2009a) The canonical network of autotrophic intermediary metabolism: minimal metabolome of a reductive autotroph. Biol Bull 216:126–130
Srinivasan V, Morowitz HJ (2009b) Analysis of the intermediary metabolism of a reductive chemoautotroph. Biol Bull 217:222–232
Strogatz SH (2001) Exploring complex networks. Nature 410:268–276
Stubbs RT, Yadav M, Krishnamurthy R et al (2020) A plausible metal-free analogue of the Krebs cycle composed entirely of α-ketoacids. Nat Chem 12:1016–1022
Tal G, Boca SM, Mittenthal J et al (2016) A dynamic model for the evolution of protein structure. J Mol Evol 82:230–243
Timár G, Goltsev AV, Dorogovtsev SN et al (2017) Mapping the structure of directed networks: beyond the bow-tie diagram. Phys Rev Let 118:078301
Tragg VA, Waltman L, van Eck NJ (2019) From Louvain to Leiden: guaranteeing well-connected communities. Sci Rep 9:5233

Wächtershäuser G (1990) Evolution of first metabolic cycles. Proc Natl Acad Sci USA 87:200–204
Wang M, Boca SM, Kalelkar R et al (2006) A phylogenomic reconstruction of the protein world based on a genomic census of protein fold architecture. Complexity 12:27–40
Wasserman S, Faust K (1994) Social network analysis: methods and applications. Cambridge Univ Press, New York
Watts DJ, Strogatz SH (1998) Collective dynamics of 'small-world' networks. Nature 393:440–442
Weber AL (2002) Carbon group transformations under mild aqueous conditions. Orig Life Evol Biosph 32:333–357
Wołos A, Roszak R, Zadło-Dobrowolska A et al (2020) Synthetic connectivity, emergence, and self-regeneration in the network of prebiotic chemistry. Science 369:eaaw1955
Yafremava LS, Wielgos M, Thomas S et al (2013) A general framework of persistence strategies for biological systems helps explain domains of life. Front Genet 4:16
Ycas M (1974) On earlier states of the biochemical system. J Theor Biol 44:145–160
Zhou B, Holme P, Gong Z et al (2023) The nature and nurture of network evolution. Nat Comm 14: 7031
Zubarev DY, Pappoport D, Aspuru-Guzik A (2014) Uncertainty of prebiotic scenarios: the case of the non-enzymatic reverse tricarboxylic acid cycle. Sci Rep 5:8009
Zubay G (2003) The glyoxylate cycle, a possible evolutionary precursor of the TCA cycle. Chemtracts 16:783–788

Engineering Microbial Evolution for Biotechnological Applications

Kangsan Kim and Byung-Kwan Cho

Contents

K. Kim
Department of Biological Sciences, Korea Advanced Institute of Science and Technology, Daejeon, Republic of Korea

KAIST Institute for the BioCentury, Korea Advanced Institute of Science and Technology, Daejeon, Republic of Korea

B.-K. Cho (✉)
Department of Biological Sciences, Korea Advanced Institute of Science and Technology, Daejeon, Republic of Korea

KAIST Institute for the BioCentury, Korea Advanced Institute of Science and Technology, Daejeon, Republic of Korea

Graduate School of Engineering Biology, Korea Advanced Institute of Science and Technology, Daejeon, Republic of Korea
e-mail: bcho@kaist.ac.kr

J. Barciszewski (ed.), *Systems Biology II*, RNA Technologies 15,
https://doi.org/10.1007/978-3-031-62178-9_6

Abstract Spontaneous mutations in replicating genomes enable permutations, allowing organisms to evolve genotypes favorable for survival. This defining biological trait has long been exploited by microbes to explore evolutionary dynamics. With the aid of systems biology techniques, it is now possible to decipher the basis of adaptation at the molecular level. These findings provide a basis for effective strain engineering designs for biotechnological applications. Additionally, advances in synthetic biology have permitted the effective coupling of non-growth-associated traits to adaptive laboratory evolution (ALE), thereby expanding the scope of ALE as a highly versatile engineering toolkit. In this chapter, we outline the growing scope of ALE and novel experimental designs proposed based on systems and synthetic biology principles.

Keywords Adaptive laboratory evolution · Systems biology · Synthetic biology · Bioproduction · Genome-streamlined bacteria

1 Introduction

Mutations are an intrinsic feature of genome replication that contributes to genetic diversity and drives evolution. The natural process of evolution was recapitulated in a seminal long-term evolution experiment (LTEE), which elucidated the tempo and mode of evolutionary dynamics since its inception in 1988 (Good et al. 2017). In principle, iterative propagation of bacterial cultures under certain selection pressures is enriched for specific mutations, among many spontaneously arising mutant progenies, with phenotypic traits best suited for survival. This artificial selection process, referred to as adaptive laboratory evolution (ALE), has been proven to be a robust engineering tool in biotechnology, as exemplified by the pioneering work of Professor Frances Arnold on the directed evolution of protein function (You and Arnold 1996). Laboratory evolution is also relevant in the directed evolution of strain phenotypes, which is the main scope of this chapter, particularly when confounding biological factors render rational engineering design unpredictable (Kim et al. 2023). Therefore, ALE is widely utilized in biotechnological and synthetic biology applications, including industrial bioproduction and synthetic microbial construction (Kim et al. 2023; Sandberg et al. 2019; Wu et al. 2022).

The central design principle of ALE is the coupling of the desired phenotype with selection pressure. In parallel with rapid advances in synthetic biology, strategies for coupling synthetic traits with adaptive fitness are becoming increasingly sophisticated. The integration of various features of synthetic biology, such as metabolite-specific biosensors, ribozyme cleavage systems, and phage-assisted evolution platforms, has streamlined the evolution of production strains by coupling the production of target compounds with strain fitness (Johnston et al. 2020; Seok et al. 2021; Yang et al. 2013). The ad hoc nature of synthetic biology-driven ALE designs expands their scope and application to the construction and testing of novel biological systems.

Since the advent of high-throughput sequencing technologies, an immense volume of datasets pertinent to ALE have flooded the scientific community. To streamline the systematic analysis, whole-genome resequencing (WGS) data from independent studies were collectively tabulated into informative matrices (databases), enabling a comprehensive assessment of mutations and their implications in adaptive evolution (Phaneuf et al. 2019). Similarly, a host of machine learning algorithms provide fine-grained descriptions of remodeled transcription regulatory networks and offer inferences on the underlying topology of transcription regulatory networks that contribute to evolved phenotypes (Maeda et al. 2020; Sastry et al. 2019). Furthermore, in silico simulation frameworks that reconcile multi-omics datasets help predict multiscale changes in cellular physiology, facilitating the interpretation of adaptive changes in a highly accurate manner (Chen et al. 2017). Thus, systems biology provides a comprehensive understanding of the biological processes underlying the emergence of complex phenotypes during evolution.

This chapter discusses the significance and utility of ALE in the context of biotechnology, focusing mainly on industrial and synthetic biology. We introduce readers to variations in ALE experimental designs that complement the limitations of rational engineering approaches for constructing optimized cell systems and genetically intractable microbes with potential industrial relevance. We also describe mainstream systems biology approaches that help decipher the mechanistic basis of evolution in both natural and synthetically engineered microbes. Finally, we outline the perspectives on ALE-guided strain engineering in the context of emerging technologies.

2 Principles of Adaptive Laboratory Evolution for Bioengineering Applications

Microbial ALE is based on the principles of Darwinian evolution, in which clones that acquire mutations beneficial for fitness in a selective environment are selected and enriched over time (Barrick et al. 2009). Owing to the rapid propagation of progenies that generate several hundreds of generations within a matter of weeks, microorganisms have served as model biological systems to elucidate various facets of evolution. These include the following: (1) the genetic basis of adaptive evolution (Barrick et al. 2009) and the magnitude and context dependency of mutations (Conrad et al. 2009; Herring et al. 2006); (2) epistatic effects that confound the evolvability of bacteria (Woods et al. 2011) impart order dependency of evolution (Bajic et al. 2018) and metabolic innovation (Szappanos et al. 2016); (3) negative epistasis that slow adaptation (Arjan et al. 1999) or act antagonistically on non-adaptive environments (Leiby and Marx 2014); (4) ramification of genome mutability on the pace of adaptive evolution (Kang et al. 2019), strain survival

(Giraud et al. 2001a), and metabolic readiness in non-adaptive niches (Sprouffske et al. 2018); and (5) evolvability as the function of sexual recombination (Peabody et al. 2017) and population size (Blank et al. 2014), alongside other important principles underpinning natural evolution (Carpenter et al. 2023).

Studies have also shown that adaptive fitness gains in LTEE populations that have evolved over three decades resemble a power-law model, suggesting that adaptation may continue indefinitely but at a significantly decelerated pace (Chou et al. 2011; Good et al. 2017). The rate of fitness gain is most pronounced in the first few thousand generations or less (Hua et al. 2007; LaCroix et al. 2015; Lenski et al. 1991), effectively reaching an "endpoint" of evolution, a technical term for describing evolved lineages that have acquired a core set of beneficial mutations within a relatively short timescale (LaCroix et al. 2015; Sandberg et al. 2019). Leveraging this phenomenon, ALE as a tool of biotechnology explores the immediate "genotypic optimality" within the vast sequence space of genomic mutations, which is akin to hijacking the natural process of evolution to construct microbial strains with the desired phenotypes (Kim et al. 2023). In its simplest form, ALE entails the iterative pass aging of bacterial populations under user-defined culture conditions to enrich mutant clones with improved adaptive phenotypes. The adaptive changes, such as increased doubling time between passages, can be monitored throughout evolution to identify the point at which populations reach their "endpoint" of evolution (Fig. 1) (LaCroix et al. 2015). The propagating population size of 10^8–10^{10} colony-forming units, with a basal genomic mutability of 2.0×10^{-11}–3.5×10^{-9} nucleotides per generation (Jee et al. 2016), provides genetic diversity sufficient to reconfigure target microbes toward the desired objective functions within a timescale of several weeks to months (Gresham and Dunham 2014). It is important to note that the "objective functions" described herein refer to the phenotypic attributes for which ALE aims to maximize survival under specific selective conditions. Objective functions are highly subjective and vary in accordance with the ALE experimental settings, such as the maximization of biomass yield (Hua et al. 2007), non-native metabolite production (Kim et al. 2022; Seok et al. 2021), and environmental resistance (Fait et al. 2022; Overbeck et al. 2017; Zhu et al. 2014), all of which have the shared goal of maximizing strain survival.

Evolution-driven genetic manipulations of this kind complement the inherent shortcomings of bioengineering, particularly when knowledge-driven engineering strategies result in unpredictable outcomes, deviations from optimal performance, and even process failures (Antonovsky et al. 2016). This is largely attributable to an incomplete understanding of the complex biology that interferes with rationally implemented genetic manipulations (Sandberg et al. 2019). The fact that ALE does not require a priori knowledge of the biological changes (i.e., bottleneck reactions or regulations) required for the desired phenotypic changes to manifest; it is routinely adopted in strain construction pipelines to reconfigure and further refine both naturally occurring and artificially implemented cellular functions. In this section, we present the key concepts and mechanisms of ALE that are relevant to biotechnological applications. The details of the selected studies have been elaborated on to aid comprehension.

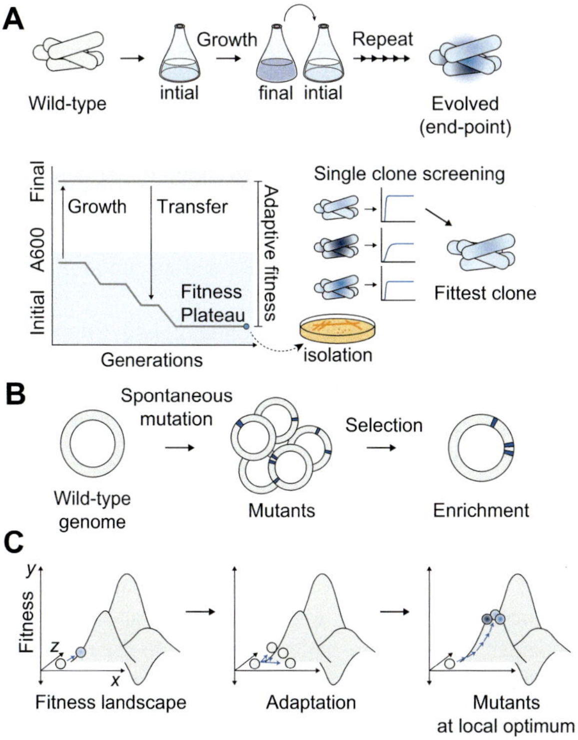

Fig. 1 Schematic illustration of a batch-culture ALE workflow. (**a**) Serial transfer of wild-type bacteria under specific selection pressures. Constant monitoring of changes in adaptive fitness allows characterization of fitness plateau, which is indicative of an "endpoint" of ALE. To account for population heterogeneity, individual clones randomly isolated from the "endpoint" populations are screened based on their adaptive phenotypes. (**b**) Among spontaneously arising mutations, those that confer a competitive edge under the selection pressure become enriched by the end of ALE. (**c**) Illustration of a fitness landscape in an arbitrary environment. The horizontal planes (x- and z-axis) represent a sequence space of a microbe. The vertical plane (y-axis) represents fitness levels corresponding to a sequence space at a set coordinate. Multiple optimal fitness peaks exist in a rugged fitness landscape. In the course of ALE, evolving populations are recurrently enriched at a genotype optimum in the closest proximity to the starting (wild-type) genotype

2.1 A Brief Overview of the Mechanistic Basis of Fitness Optimization

Strain fitness optimization is arguably the most popular avenue for ALE applications. The LTEE experiments have showcased the seemingly limitless potential of *Escherichia coli* strains to evolve, developing even more refined "palates" toward even their most favored carbon source glucose over the 30 years of evolution (Good et al. 2017). Since then, the evolvability of microbes' indifferent cellular and environmental contexts has been tested, revealing that adaptive changes in functional phenotypes are usually communicated in two distinct ways: structural and expression changes in afflicted proteins (Blank et al. 2014). Adaptive evolution produces causal mutations that are reasonably predictable and closely aligned with the existing knowledge. However, it is often difficult to predict and design the genetic alterations required to satisfy an objective function using the existing modeling techniques. The advent of high-throughput genome sequencing tools enables the complete cycle of "Design-Build-Test-Learn" to systematically evaluate genotype-phenotype relationships in the adaptively evolved cell systems, facilitating in-depth comprehension of biology and expanding upon a novel hypothesis (Shendure et al. 2017).

Using classical ALE experiments as an example, *E. coli* endpoint strains that have adaptively evolved under carbon-limiting conditions are repeatedly enriched with deletion mutations in the *rph-pyrE* operon (encoding RNase PH and orotate phosphoribosyltransferase, respectively) (Conrad et al. 2009; Kang et al. 2019; Knoppel et al. 2018; LaCroix et al. 2015; Morgenthaler et al. 2019; Sandberg et al. 2017). The basis of this selection is the upregulation of PyrE, which mediates pyrimidine biosynthesis in *E. coli*. The change in transcriptional and translational regulation is mediated by a deletion mutation which repositions the stop codon of *rph* closer to the 5′upstream region of *pyrE* where an attenuator loop is present, resulting in an upregulation of *pyrE* expression and translation (Bonekamp et al. 1984). In this manner, the deletion mutations alleviate the translational defects in pyrimidine biosynthesis caused by low *pyrE* expression intrinsic to *E. coli* MG1655 (Jensen 1993). The reconstituted single-deletion mutants account for as much as 55% of the increase in endpoint strain fitness in nutrient-limited conditions, irrespective of supplemented carbon sources, which reflects the genomic context of *E. coli* MG1655, which strongly selects for pyrimidine biosynthetic capacity (Conrad et al. 2009; Knoppel et al. 2018; LaCroix et al. 2015; Morgenthaler et al. 2019). The regulatory mutation in *rph-pyrE* is one of the few instances of an informed adaptive outcome with an existing knowledge base and is within the scope of rational engineering design.

Structural mutations and their accumulation within the genome often translate into global changes in metabolic and regulatory networks in a manner that is difficult to understand a priori. This is partly because of the prevalence of epistatic interactions between genes and their mutant derivatives, which have been shown to play a deterministic role in shaping evolution and confounding the predictability of evolutionary paths (Bajic et al. 2018; Woods et al. 2011). The fitness landscape metaphor is used to explain the effect of genetic epistasis on the evolutionary trajectory and the corresponding fitness in a given environment (Wright 1932). The fitness landscape consists of a horizontal plane representing an array of possible genotype spaces (mutations), with fitness levels corresponding to each genotype on the vertical axis. Taking ALE as an example, bacterial populations under certain selection pressure progress from lower basins into higher peaks through mutation fixation events (Nahum et al. 2015). Under the assumption that genetic epistasis is absent, the topography of the bacterial fitness landscape follows a "smooth" landscape characterized by a single fitness optimum accessible upon fixation of all beneficial mutations regardless of mutational backgrounds and that each mutation imparts a fixed effect on fitness. In contrast, the fitness landscape under the influence of genetic epistasis is heavily dependent on background genotypes and the order in which mutations are acquired, thereby displaying a rough terrain with multiple peaks and lower basins that reflect fitness attributes of various genotypes under epistasis (Nahum et al. 2015; Van Cleve and Weissman 2015). Although it is heavily debated whether the topography of the bacterial fitness landscape is smooth or rugged (Kryazhimskiy et al. 2012), a preponderance of empirical and computational data now supports the rugged fitness landscape scheme. Implying that on many occasions, genome evolution is under the heavy influence of epistatic interactions (Bajic et al. 2018; Kvitek and Sherlock 2011; Nahum et al. 2015; Van Cleve and Weissman 2015).

The ALE-Test-Learn cycle facilitates the systematic analysis of genotype-phenotype relationships and the mechanistic basis, such as hidden epistatic interactions. For instance, endpoint strains of *E. coli* serially propagated on glycerol minimal medium in independent experiments were found to harbor mutations in identical coding sequences of glycerol kinase (*glpK*) and RNA polymerase subunits (*rpoS*, *rpoB*, and *rpoC*) (Herring et al. 2006; Kang et al. 2019; Kim et al. 2022; Knoppel et al. 2018; Peabody et al. 2017). Glycerol kinase catalyzes the first step in the phosphorylation of intracellular glycerol into glycerol-3-phosphate, which is allosterically inhibited by fructose 1,6-bisphosphate (Zwaig and Lin 1966) and Enzyme IIA^{Glc} (EIIA^{GLC}) in its free form (Novotny et al. 1985). GlpK mutants (GlpK^{m}) that contribute to increased glycerol uptake rates in the ALE endpoint strains exhibited nearly diminished sensitivities toward fructose 1,6-bisphosphate and unphosphorylated EIIA^{GLC} (Applebee et al. 2011; Kim et al. 2022). Thus, GlpK^{m} contributes to cellular fitness via increased glycerol uptake rates, primarily through the evasion of regulatory constraints, which also results in the onset of carbon overflow, such as non-fermentative acetate secretion (Basan et al. 2015). The apparent "wastage" of biomass by the overflow mechanism in GlpK^{m} can be complemented by the RNAP mutants, such as RpoC^{m}, which reconfigures the global metabolic network to redirect metabolic flux into biomass formation (Cheng et al. 2014). The reconstitution of GlpK^{m} and RpoC^{m} double mutants displayed a positive magnitude epistasis, where the combinatorial effect on strain fitness was greater than the sum of the individual contributions (Cheng et al. 2014; Herring et al. 2006). This example demonstrates that the challenge of maximizing strain fitness can be resolved through structural mutations in two seemingly unrelated proteins that are nonetheless positively epistatic.

2.2 *Adaptive Evolution of Novel Metabolic Phenotypes*

Another interesting development in our understanding of epistatic interactions is the idea that future mutational events are contingent on background mutations acquired from past events. Such historical evolutionary contingencies were showcased in the emergence of aerobically active citrate assimilation in 1 of the 12 lineages of LTEE (Blount et al. 2008). Citrate was supplemented as an iron chelator in LTEE culture medium and did not support aerobic cell growth in wild-type *E. coli.* However, a duplication mutation in *citT* (citrate/succinate antiporter) and subsequent fixation of a mutation in *dctA* (C4 dicarboxylate/H^+ symporter) led to a strong citrate-uptake phenotype under aerobic conditions. Moreover, mutant reconstitution experiments revealed the highly order-dependent nature of the *citT-dctA* mutant pair, in which $DctA^m$ displayed a sign epistasis, characterized by a net negative impact on strain fitness in the absence of $CitT^m$ (Bajic et al. 2018). This is because $CitT^m$ facilitates extracellular citrate import and export of intercellular C4 dicarboxylates from the citric acid cycle, which are recovered back into the cell by the constitutively upregulated $DctA^m$. The sole upregulation of $DctA^m$ in the absence of active C4-dicarboxylate export is futile and may result in a net fitness loss (Quandt et al. 2014). This example of sign epistasis illustrates that the future mutational trajectory scan depend on the ancestral mutations that precede them. Furthermore, it sheds light on possible evolutionary mechanisms by which novel metabolic phenotypes arise (Blount et al. 2008). Such epistatic interactions may have important implications for strain engineering to harness non-permissive nutrients. However, considering that it took approximately 20 years of ALE (~31,500 generations) for *E. coli* to aerobically assimilate citrate, conventional ALE methods may be impractical for non-native substrate utilization.

Instead, a more practical approach to harness the evolution of metabolic innovation entails stepwise ALE on non-utilizable substrates that share similar metabolic pathways at a heightened mutation rate (Szappanos et al. 2016). First, Szappanos and colleagues opted for genome-scale metabolic modeling, an in silico algorithm that simulates flux changes through a virtual metabolic network (discussed in Sect. 3.2), to evaluate the feasibility of novel metabolic innovations among the list of non-growth-supporting substrates with similar metabolic properties. Subsequently, to enable growth on propylene glycol (PG) and ethylene glycol (EG), an *E. coli* hypermutator with 1000-fold elevated genome mutability was evolved on each carbon. While growth on PG (PG^+) was possible after a single-step ALE, the EG^+ phenotype necessitated an extra intermediary evolution step on PG in addition to the hypermutator phenotype. The "enabling mutation" that supported the PG^+ phenotype *fucO** (a mutant derivative of L-1,2-propanediol oxidoreductase) also served as an adaptive route to bridge the EG catabolic pathway to the central metabolism. The second mutation that enabled EG^+ phenotype, *aldA** (a mutant derivative of aldehyde dehydrogenase A), supported the EG^+ phenotype but required the presence of $FucO^*$ (Szappanos et al. 2016). This underscores that the interplay of epistasis between two similar but mutually exclusive fitness landscapes can potentially give rise to novel metabolic capacities.

2.3 *Generality of Mutations Across Multiple Fitness Landscapes*

The effects of mutations on fitness depend not only on the genetic background but also on the environmental context. Evidently, mutant derivatives of *glpK*^{m} recurrently arise in glycerol ALE but are rarely observed in the absence of glycerol (Sandberg et al. 2017). In this regard, GlpK^{m} belongs to "specialist" mutations whose functional benefits manifest and are selected for in a relatively confined niche width (Kassen 2002). In contrast, mutations that remain beneficial across a broader niche width are called "generalist" mutations. Mutations in the *rph-pyrE* operon and RNA polymerases in previous examples represent generalists that are repeatedly enriched in minimal media, regardless of the carbon source (Conrad et al. 2009, 2010; Kang et al. 2019; Knoppel et al. 2018; LaCroix et al. 2015; Morgenthaler et al. 2019). Contrary to the logical prediction that generalists would maintain a competitive edge in maelstrom of fluctuating environments, specialists repeatedly prevailed in a certain pattern and scale of multi-substrate ALE (Jasmin and Kassen 2007), suggesting the presence of an unknown layer of regulations and constraints that favors the emergence of population heterogeneity. Understanding the fundamentals of evolvability in multi-substrate environments is of increasing importance for industrial-grade feedstock utilization, which often contains a mixture of sugar substrates with varying degrees of preference, such as in heterotrophic and mixotrophic culture settings (Bae et al. 2022) and the fermentation of complex lignocellulose hydrolysates and other agro-industrial wastes (Jayakody et al. 2022).

The evolvability of microbes on multiple substrates or periodically shifting single-substrate environments is influenced by a host of intrinsic and extrinsic factors, such as programmed metabolic conditioning that invokes anticipatory regulation (Mitchell et al. 2009); stochastic diversification of phenotypic states within an isoclonal population; termed bet-hedging (Beaumont et al. 2009; Solopova et al. 2014), and the pattern and scale of nutrient shifts (Jasmin and Kassen 2007). One of the most decisive factors that underlies evolution on multiple substrates is the difference in the topology of the metabolic landscape—the divergence in the distribution of proteome, transcriptome, and metabolic flux states as a function of input carbon substrates (Sandberg et al. 2017). The "metabolic distance," an arbitrary unit used to infer divergence in metabolic topologies between carbon substrates, can be computed using a genome-scale model of metabolism and macromolecular expression (ME models), which allows predictive modeling of the transcriptome and proteome composition in an optimally growing cell (Lloyd et al. 2018). Moreover, ME models compute the proteomic composition in an optimally growing cell, in addition to the metabolic flux distributions of conventional metabolic models (discussed in Sect. 3.2). Including proteome biosynthetic costs in the modeling framework has been shown to promote highly accurate prediction of metabolic flux distributions for given experimental inputs, such as carbon uptake rates, metabolite production rates, and gene expression profiles (Lloyd et al. 2018).

Sandberg et al. showed that the propensity of populations to diverge into distinct lineages in multi-substrate environments can be inferred from the metabolic distance between the two substrates. For instance, populations that alternated between glucose and xylose or glucose and glycerol evolved into isogenic generalists, whereas those that evolved in metabolically more distant glucose and acetate environments developed divergent sub-lineages (Sandberg et al. 2017; Saxer et al. 2010). It is presumed that, for certain combinations of carbon substrates, the fitness cost associated with a metabolic shift (i.e., diauxie) in a single generalist outweighs that of having more than one subpopulation specializing in each ecological niche (Friesen et al. 2004). This represents an interesting development in evolutionary biology, allowing us to quantitatively predict strain evolvability in a multi-substrate environment, albeit at a limited scale.

2.4 Evolvability as the Function of Genome Mutation Rates

Inherent errors in the genome replication machinery serve as the engine of the genetic variability that drives natural evolution. Although genetic elements associated with metabolic and regulatory functions are frequent targets of selection, genome mutability is also influenced by selection pressure in the form of second-order selection (Giraud et al. 2001b). Inactivation of DNA repair machinery, such as the methyl-mismatch repair system (MutHSL), increases genome mutability by 50- to 1000-folds, rendering the inflicted mutants highly prone to replication errors (LeClerc et al. 1996). In ALE settings, hypermutators tend to expedite adaptive evolution because of the higher per capita probability of acquiring beneficial mutations, effectively outcompeting wild-type contenders in competitive settings, irrespective of environmental conditions and the mode of pass aging (Gibson et al. 1970; Giraud et al. 2001a; Kang et al. 2019; Loh et al. 2010). However, increasing mutation rates do not always guarantee a proportional increase in the rate of adaptation (Arjan et al. 1999). One possible cause of the reduced adaptability in highly mutagenic genomes, other than the higher chance of the emergence of deleterious mutations, is presumed to be due to multiple clones harboring different beneficial mutations arising simultaneously and competing within an evolving population. The competition between the contending clones may also negate the fitness contribution of otherwise beneficial mutations in the absence of such "clonal interference," decelerating the pace at which beneficial mutations are fixated and slowing down adaptation (Cooper 2007). This may partly explain why not all hypermutators, especially those with an excessively high mutagenic potential, are negatively selected for ALE (Loh et al. 2010; Sprouffske et al. 2018). After rigorous experiments aimed at determining the optimal mutagenic potential of ALE, it was concluded that in *E. coli*, moderate hypermutators with a 10- to 100-fold increase in genomic mutability facilitate adaptation at the most rapid pace (Loh et al. 2010; Sprouffske et al. 2018).

However, early success in adaptive specialization in a single fitness landscape comes at the cost of reduced fitness in other non-adaptive environments, undermining the potential evolvability of endpoint hypermutators in alternative fitness landscapes. Fixation of a plethora of non-beneficial mutations over the long term can lead to debilitating changes in enzymes that are metabolically dormant in one environment but are nonetheless essential in another, leading to metabolic erosion (Leiby and Marx 2014). Sexual recombination can be employed to dissipate the potentially lethal effects of deleterious and neutral mutations, where the hypermutator *E. coli* equipped with the F-plasmid conjugation machinery outperformed all sexual, asexual, and non-mutator counterparts (Peabody et al. 2017). Considering that sexual recombination effectively alleviates clonal interference (Cooper 2007), it is expected to minimize the accumulation of metabolic erosion through constant allelic shuffling. In addition, recent evidence suggests that the transient evolution of moderate hypermutators for fewer than 3000 generations does not undermine strain vigor in alternative fitness landscapes but facilitates the expansion of the metabolic niche (Kang et al. 2019; Sprouffske et al. 2018). Although there are no set standards or quantitative references to predict the adaptive outcome of the hypermutator ALE, this line of evidence highlights the potential utility of "moderate" hypermutators to accelerate adaptation and metabolic niche expansion. The utility of the transient inactivation of DNA mismatch systems as an "add-in" tool in ALE has been demonstrated in the innovation of complex metabolic phenotypes (the PG^+ phenotype in Sect. 2.2), which proved to be challenging in the wild-type strain with basal mutability (Szappanos et al. 2016), fitness optimization in synthetic biology (Choe et al. 2019; Wannier et al. 2018), and improvement of bioproduction capacity and stress tolerance in industrial strains (Overbeck et al. 2017; Zhu et al. 2014).

3 ALE in Biotechnological Applications

The metabolic versatility that enables microbes to convert input feedstock into value-added commodities, coupled with their capacity for large-scale production, is an appealing trait for biotechnological applications (Kamionka 2011; Lee et al. 2019b). The natural capacity of wild-type strains for bioproduction can be further augmented by using ALE in tandem with rational engineering strategies. In addition, synthetic biology synergizes with ad hoc ALE experimental designs, wherein the exhaustive catalogue of genetic devices in synthetic biology enables the development of novel and more effective growth-coupling strategies. In addition, we discuss systems biology approaches to understand the implications of genome streamlining in adaptive evolution. The ALE approach used in this study is summarized in Table 1.

Table 1 Characteristics and limitations of ALE approaches outlined in this study

Approach	Characteristics	Limitations
Growth optimization ALE	—Most straightforward form of ALE designed to select for mutant progenies with superior fitness in a given culture condition —Applicable to a wide range of non-model microbes	—Maximization of strain fitness as the objective function may adversely affect the bioproduction capacity of compound of interests that are not growth-coupled, especially when the production process incurs fitness costs
Genetic perturbation ALE	—Conventional genome engineering to couple production of compound of interest with strain fitness, mostly through replacement of endogenous essential functions —Elimination of competing pathways with shared substrate usage may further augment metabolic flux into compound of interests —May consult flux balance analysis (FBA) of genome-scale metabolic models to probe genome-wide genetic engineering targets that maximize compound-of-interest production	—Requires a priori knowledge on the metabolic network, which may limit the scope of engineering to a few well-known targets —FBA analysis not applicable for microbes lacking a high-quality GEM
Synthetic biology-aided ALE	—Incorporates genetic circuits with a pair of biosensor and a cognate promoter. Biosensors orthogonally sense target compound and relay the signal to downstream actuator, expression of which is coupled with strain fitness —Synthetic libraries may be implemented to expand solution space and expedite evolution —High-throughput screening of heterogeneous populations is possible when appropriate reporter systems are implemented	—Non-target "escapee" mutations can render the entire genetic circuit-based selection system obsolete —Orthogonal biosensors may not exist in nature —Limited to genetically tractable microbes

3.1 Evolutionary Engineering to Improve Growth-Coupled Productions

Owing to the growing demand for biosustainability and carbon neutrality, an increasing number of studies are focusing on the microbe-based bioproduction of commodity products using various industrial, agricultural, and municipal by-products and wastes (hereafter, alternative carbon sources) (Asopa et al. 2022; Bae et al. 2022; Jayakody et al. 2022). However, the non-preferential status of alternative carbon sources owing to metabolic imbalance, compounded by the cellular toxicity inherent to several compounds, undermines the overall bioproduction efficiency. ALE has traditionally been implemented to troubleshoot the impaired strain fitness under various operational conditions, particularly when the mechanistic basis for growth inhibition remains unclear. This straightforward strategy has been proven effective in simultaneously enhancing the production of target products that are growth-coupled or derived from synthetically introduced orthogonal pathways.

Growth optimization ALE is the method of choice for producing metabolites that are growth-coupled or orthogonal (Fig. 2a). For example, industrial bacteria such as *E. coli* exhibit suboptimal fitness when glycerol is used as the carbon source (Ibarra et al. 2002), which is due to regulated glycerol uptake with a presumed role in preventing excessive accumulation of the toxic intermediate methylglyoxal (Freedberg et al. 1971). The suboptimal fitness and productivity of the desired compound could be markedly enhanced by as much as 1.4- and 2.6-folds, respectively, in a 1,000-generation ALE experiment (Kim et al. 2022). Such growth optimization strategies are routinely adopted to enhance the bioproduction capacity of industrially relevant microbes, including lactic acid production in *Lactobacilli* through adaptive evolution at low pH (Overbeck et al. 2017) or crude sugarcane molasses hydrolysate (Liang et al. 2020), utilization of non-preferential sugars for valencene production in *Saccharomyces cerevisiae* (Zhu et al. 2022), ethanol production in *Kluyveromyces marxianus* (Du et al. 2022), and optimization of biomass generation in *Corynebacterium* (Radek et al. 2017).

By leveraging the native DNA replication machinery of host organisms, microbes that are considered biotechnologically challenging to manipulate can be evolutionarily engineered to improve target product synthesis. Notable examples include the synthesis of partially oxidized compounds in *Gluconobacter oxydans* (Jin et al. 2019), a range of value-added chemicals bioconverted from toxic gaseous substrates (CO, CH_4) by acetogenic bacteria and methanotrophs (Awasthi et al. 2022; Jin et al. 2022; Kang et al. 2020; Lee et al. 2019a), microalgae for enhancing lipid and biomass generation (Yu et al. 2013), and antimicrobial activity (as a proxy for antibiotic production) in *Streptomyces* strains (Harwani et al. 2022). For an exhaustive list of growth rate optimization ALE experiments on (non)-model microbes, readers may refer to the latest review articles (Kim et al. 2023; Wang et al. 2023; Wu et al. 2022).

Taken together, the growth optimization strategy is well-suited for enhancing the productivity of growth-coupled compounds of interest in a range of microbial hosts. By harnessing the native DNA replication machinery of the host cell, ALE expands the ability to engineer industrially relevant microbes with a limited understanding of genome functions and the lack of genetic engineering toolkits. However, the primary objective of this method is aimed at maximizing growth rate instead of target compound production, which may render target production pathways with significant metabolic costs non-functional (Rugbjerg et al. 2018a). To circumvent the selection pressures that act against the desired objective functions, genome modification strategies based on rational designs can be implemented to harness evolution for the benefit of bioproduction.

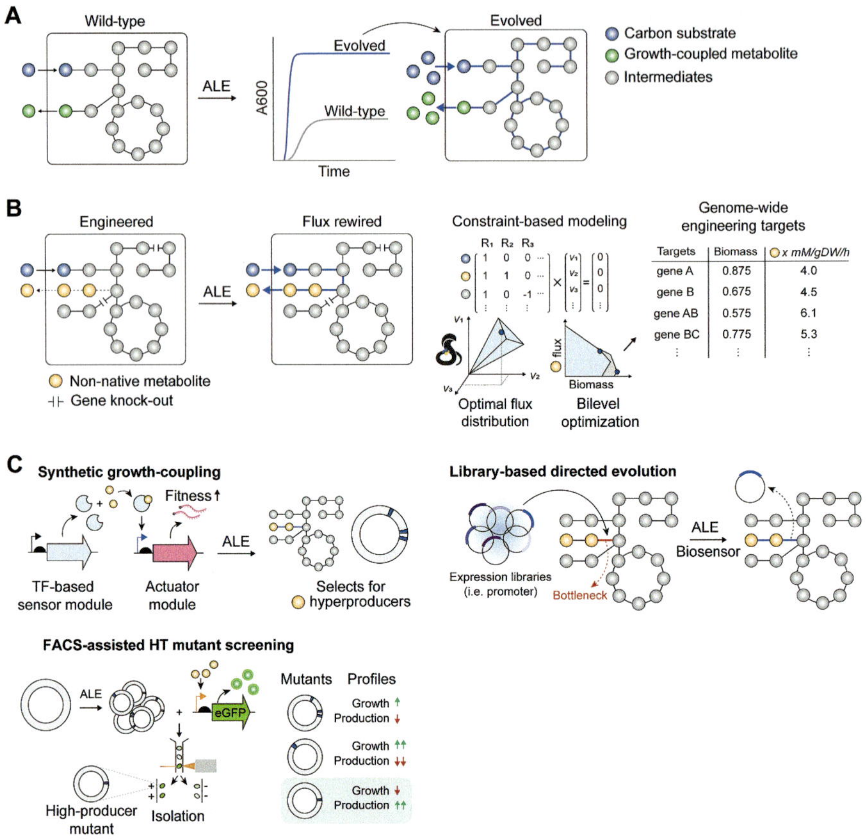

Fig. 2 Growth-coupling strategies in ALE. (**a**) Growth optimization ALE with the primary objective of optimizing strain fitness may translate into an improved production of a growth-coupled compound of interest. (**b**) Growth coupling through genetic perturbation couples target metabolite production with fitness, thereby setting the production of the target compound as the objective function. GEM framework can be implemented to identify and quantitatively assess potential genome-wide engineering targets. (**c**) Examples of synthetic biology-aided ALE. First, synthetic growth coupling by genetic circuits leverages the ability of transcription factors to orthogonally sense target molecules and activate the expression of fitness-associated genes through a cognate promoter. Second, the strains of interest can be transformed with synthetic libraries, such as those of gene expression with random promoters, and enriched with cycles of ALE. Used in tandem with biosensors orthogonal to the compound of interest, synthetic libraries enable the discovery of novel optimal solutions and expedite the pace of directed evolution. Genetic circuits that incorporate fluorescence reporter readout as the proxy for producing compounds of interest enable high-throughput screening and isolation of hyper-producer mutants within the heterogeneous pool of evolving (or endpoint) populations

3.2 *Classic Genetic Perturbation to Growth-Coupled Product Formation*

Targeted deletion of non-essential metabolic pathways that compete for shared resources was strategically employed to further augment the production of the compounds of interest (Fig. 2b). For example, the use of ALE in tandem with rationally designed metabolic engineering—the deletion of non-essential "metabolic sink" pathways (Δ*ackA*Δ*mgsA*Δ*gabT*) in the evolved strain—led to a 3.9- and 4.3-fold increase in titer and productivity of gamma-aminobutyric acid, from glycerol (Kim et al. 2022). The authors showed that implementing evolutionary and metabolic engineering pipelines leads to a comparable or superior improvement in gamma-aminobutyric acid production compared with other strategies (Le Vo et al. 2013; Pham et al. 2015). Similarly, ALE of *E. coli* lacking genes associated with mixed-acid fermentation pathways (Δ*frdA* Δ*ldhA* Δ*pta* Δ*adhE*) on acetate resulted in an endpoint strain with a significantly reduced lag-phase growth (Seong et al. 2020). Inactivation of the Pta-AckA node made the ATP-intensive ACS pathway the sole pathway for acetate influx, selecting for mutations with enhanced ATP and NADH biosynthesis. The elevated energy biosynthesis in the evolved strain was leveraged for the production of n-butanol and mevalonate from acetate, which were 7.5- and 14.3-fold higher in maximum titer (with additional engineering efforts), respectively, than the parental strain (Seong et al. 2020).

Rational genetic perturbations help establish a driving force toward the desired pathways. However, perturbation of native metabolic networks may adversely affect strain fitness, resulting in disparities between expectations and reality. For example, in an attempt to couple the production of non-native 1-butanol with growth in *E. coli*, native NAD^+-generating pathways (Δ*adhE*Δ*ldhA*Δ*frdBC*) were inactivated to establish the 1-butanol biosynthetic pathway as the sole NAD^+-replenishing pathway (Pontrelli et al. 2018). However, the perturbed metabolic network failed to utilize the non-native 1-butanol pathway as an electron sink and failed to thrive in minimal medium. To activate the latent 1-butanol pathway, the authors implemented a "weaning ALE" strategy characterized by a gradual decrease in the supplementation of extra nutrients in a hypermutator background to facilitate the reconfiguration of redox metabolism. This strategy effectively activated and coupled the latent pathway to cellular growth in minimal medium, generating 600 mg/L 1-butanol after 30 serial transfers (Pontrelli et al. 2018). The "metabolic ALE" approach remains as one of the highly efficient approaches to improve the production of various value-added metabolites (Wu et al. 2022). Furthermore, the unintended failure of system design can be complemented by the aforementioned "weaning ALE" approach, which has been recently adopted in yeast reconstructed for lipogenesis (Yu et al. 2018) and widely adopted in synthetic cells that suffer from imbalance in metabolic networks (Choe et al. 2019) or unexpected latency of artificially introduced pathways (Antonovsky et al. 2016; Gleizer et al. 2019).

However, native metabolic networks can be intentionally disrupted to allow the desired phenotypes to be selected through evolution. This approach, referred to as metabolic engineering to guide evolution (MGE), was implemented in *Corynebacterium glutamicum* to improve L-valine biosynthesis, in which the ALE of anaplerotic pathway-impaired mutant (Δ*ppc* Δ*pyc*) led to the activation of glyoxylate shunt and pyruvate dehydrogenase complex, resulting in rerouted flux through L-valine biosynthesis (Schwentne et al. 2018). Overexpression of the L-valine biosynthetic gene in the mutant-reconstituted background resulted in an L-valine titer of 8.9 g/L. Similarly, in an attempt to select for L-serine-overproducing strain, *E. coli*-lacking L-serine degradation pathways—serine deamination to pyruvate (Δ*sdaA*Δ*sdaB*Δ*tdcG*) and conversion to glycine (Δ*glyA*)—were challenged with increasing concentrations of L-serine (from 3 to 100 g/L). While low concentrations of L-serine interfere with cell growth, the evolved strain was robust at an otherwise lethal L-serine concentration (100 g/L) and achieved the highest recorded L-serine titer of 37 g/L (Mundhada et al. 2017). Taken together, traditional metabolic engineering regimes and growth coupling of target pathways by rational genetic perturbation and MGE can be used in tandem or incorporated into ALE design to aid the reconfiguration of host microbes for bioproduction applications.

3.3 ALE Design Informed by In Silico Modeling

Flux balance analysis (FBA) is widely adopted in genome-scale metabolic models to infer flux distribution within an in silico metabolic network reconstructed using the gene annotation information of an organism (Orth et al. 2010). Assuming a steady-state mass balance and other constraints, FBA calculates the optimal flux distributions that maximize user-defined objective functions (i.e., biomass generation) within possible combinations of flux distributions. Such constraint-based mathematical predictions have provided good approximations of optimally growing cell phenotypes, such as those of ALE endpoint strains (Fong et al. 2003; Ibarra et al. 2002).

The utility of in silico modeling in ALE applications overlaps with the approaches described in Sect. 3.2. The main difference is that while the strain engineering schemes in previous approaches may be confined to the scope of informed knowledge, in silico algorithms assist in exploring the most effective engineering strategies by interrogating genome-wide metabolic networks (Fig. 2b). Remarkably, the in silico-guided approach also captures the non-intuitive aspects of network properties that may otherwise remain unnoticed under human inspection (Burgard et al. 2003). For example, to growth-coupled non-native ethylene biosynthesis in *E. coli*, FBA simulation suggested that the deletion of *proB* (glutamate-5-semialdehyde dehydrogenase) renders ethylene biosynthetic pathway as the sole route for L-proline synthesis, thereby coupling ethylene production to cell growth (Vaud et al. 2021). As predicted, the ALE of the Δ*proB* mutant under mutagenic stress without L-proline resulted in an endpoint strain with a 49-fold increase in ethylene productivity.

Metabolic models also enable the quantitative prediction of objective functions, such as the yield of metabolites of interest in response to the deletion of competing pathways (Fong et al. 2005; Tokuyama et al. 2018), substrate-specific metabolic flexibility for biochemical production (Zhu et al. 2022), and validation of the feasibility of growth-coupled designs (Luo et al. 2019). For example, a gene deletion strategy designed by FBA to maximize succinate production was predicted to yield 0.45 C-mol/C-mol of succinate using glycerol (Tokuyama et al. 2018). However, the succinate yield in the designed deletion mutant (Δ*adhE* Δ*pykAF* Δ*gldA* Δ*pflB*) deviated from the prediction (0.08 C-mol/C-mol), and a round of ALE restored the yield to the predicted 0.45 C-mol/C-mol. Downstream analysis of the high-yield mutant revealed that a mutation in *ppc* desensitized the allosteric inhibition toward L-aspartate and increased metabolic flux through Ppc down to the citric acid cycle (Tokuyama et al. 2018). While FBA assumes optimal instances of metabolic states, it does not consider biological regulation (Orth et al. 2010), which is the cause of the disparity between predicted and empirical outcomes. In this regard, ALE offers experimental optimization solutions to reconfigure the metabolic states predicted a priori by FBA (Ibarra et al. 2002).

3.4 Novel Growth-Coupling Strategies Driven by Enabling DNA Technologies

Genetic circuit design powered by elaborate bioparts in synthetic biology catalogues offers novel engineering designs that enable the growth coupling of specific classes of metabolites, which is challenging using conventional methods. The key components of genetic circuit-guided ALE are sensory proteins that sense the presence of specific ligands and transduce signals in the form of transcriptional or translational activation of downstream genes in a concentration-dependent manner (Bentley et al. 2020; Raman et al. 2014; Seok et al. 2021; Yang et al. 2013). The output signal of the genetic circuit can be coupled to strain fitness, such that mutations that produce a higher amount of activator ligands or compounds of interest can be selectively enriched by ALE (Fig. 2c).

Recently, in an attempt to improve 3-hydroxypropionic acid (3-HP) bioproduction, Seok et al. built a genetic circuit constituting a transcription regulation system responsive to 3-HP, the transcription activator LysR, and its cognate promoter P_{C4M}, which was designed to control the expression of the tetracycline resistance gene *tetA* (Seok et al. 2021). By challenging *E. coli* containing the synthetic biosensor with a stepwise increase in tetracycline concentrations in a 180-hour-long ALE experiment, the authors isolated endpoint strains with significantly enhanced 3-HP production. Reconstitution of casual mutants *cyaA*^{m} and *crp*^{m}, which alter intracellular cAMP levels and regulon expression, respectively, resulted in a near-maximum 3-HP yield reported to date (0.91 g/g glycerol), thus providing novel rational engineering targets for 3-HP production in *E. coli* (Seok et al. 2021).

Similarly, the non-coding *cis*-regulatory elements in the 5′-untranslated region of mRNA, called riboswitches, are harnessed for metabolic flux control for L-lysine production (Yang et al. 2013). Based on the a priori knowledge that L-lysine production is affected by the flux through Ppc (phosphoenolpyruvate carboxylase), a promoter library of *ppc* was introduced in *E. coli* containing the "Lysine Riboselector" comprised of the lysine-responsive *lysC* riboswitch with a *tetA* selection marker (Fig. 2c). Promoter variants that allow optimal distribution between glycolytic and an aplerotic fluxes for L-lysine biosynthesis were isolated through several enrichment cycles in the presence of the selection marker $NiCl_2$, resulting in L-lysine accumulation from approximately 0 to 0.6 g/L (Yang et al. 2013). Moreover, synthetic libraries broaden the range of the solution space for a given optimization problem, enabling the discovery of optimal solutions previously unattainable in non-synthetic cellular contexts. Furthermore, introducing ready-made sequences for selection expedites evolution and significantly reduces the number of effective cell transfers required to attain the desired phenotypes.

As demonstrated in the examples above, genetic circuits can relay the production of target compounds to the expression of genes essential for survival, enabling effective coupling between the compounds of interest and strain fitness. However, using antibiotics or toxic analogs that inhibit cell growth may undermine the bioproduction performance and favor the emergence of inactivating mutations, potentially leading to the enrichment of non-target phenotypes (Rugbjerg et al. 2018b). Other genetic circuit-based approaches have been devised to effectively screen for high performers while minimizing false positives. Genetic circuit designs that use fluorescence reporter readouts as a proxy for bioproduction capacity enable screening the desired mutants within an evolving population, which could be as large as 10^{10} colony-forming units (Bentley et al. 2020; Gresham and Dunham 2014). The availability of high-throughput screening systems, such as fluorescence-activated cell sorting, promotes the simultaneous screening and sorting of producers from non-producing mutants and isolates of mutants with superior compound production within a heterogeneous population (Fig. 2c) (Bentley et al. 2020). In addition, the synthetic regulation of native essential genes by the desired compounds (Rugbjerg et al. 2018b), the introduction of redundancies in selection designs through dual biosensor systems and multi-copy selection markers (Umeyama et al. 2013), and toggle-switch-like selection designs to select against disabling mutations (Raman et al. 2014) present viable options for selecting mutants with the desired properties while minimizing escaping mutations.

In addition to genetic circuit-driven growth-coupling strategies, many other customized ALE schematics have been published (Wu et al. 2022). Some examples include the implementation of phage-assisted continuous evolution technology in tandem with a target-compound-orthogonal biosensor for optimization of antibiotic biosynthetic pathway (Johnston et al. 2020), ALE using rare codon-recoded selection markers and reporter proteins to select for amino acid overproducers (Zheng et al. 2018), and in vivo mutagenesis driven by plasmid-bore mutagens or CRISPR/Cas9 to expedite evolution and expand novel solution space for a given optimization problem (Badran and Liu 2015; Jakociunas et al. 2018). Strikingly, the induced expression of six mutator genes via mutagenesis plasmids resulted in a staggering 322,000-fold increase in mutation rates, far surpassing the existing in vivo mutagenesis methods and enabling the directed evolution of antibiotic resistance traits and T7 polymerase variants in less than 24 h (Badran and Liu 2015). More recently, genome-recorded bacterial strains have adaptively evolved for bioengineering applications. In one study, a β-lactamase selection marker structurally altered to harbor a diselenide bond was introduced in genome-recoded *E. coli* to couple selenocysteine dependence to growth in a continuous evolution experiment. Using reporter proteins designed to incorporate selenocysteine in addition to the recorded selection marker, the authors isolated recoded mutants with up to a 20-fold increase in selenoprotein yield (Thyer et al. 2018). Similarly, two of the six "unused" serine codons (TCR) in the genome-recoded *E. coli* synthetic genome, Syn61Δ3 (Robertson et al. 2021), were reassigned into leucine to confer resistance against viral infections (Nyerges et al. 2023). To evolve serine-codon suppressors in *E. coli leuU* tRNA-carrying TCR, a kanamycin resistance marker ($aph3Ia^{29\times Leu\text{-}to\text{-}TCR}$), in which all leucine codons were swapped with TCR, was introduced to select tRNA variants that efficiently translated $\mathrm{Aph3Ia}^{29\times \mathrm{Leu\text{-}to\text{-}TCR}}$ under high kanamycin selection pressures. The resulting tRNA variants exhibit prolonged immunity against viral resistance and horizontal gene transfer (Antonovsky et al. 2016). As such, the ever-expanding toolbox of synthetic biology is bringing forth increasingly robust ALE designs with wide-ranging implications across different sectors of biotechnology.

4 Systems Understanding of Synthetic Microbes

The synthetic traits empowered by engineering biology—synthetic trophic states (Antonovsky et al. 2016; Gleizer et al. 2019), codon compression (Robertson et al. 2021; Wannier et al. 2018), and genome minimization (Choe et al. 2019; Moger-Reischer et al. 2023)—present new-to-nature traits such as viral resistance (Nyerges et al. 2023), non-natural peptide synthesis (Spinck et al. 2023; Thyer et al. 2018), and biochemical production (Choe et al. 2022). However, synthetic microbes often exhibit unexpected phenotypic defects such as severe growth retardation in genome-streamlined organisms (Hutchison et al. 2016; Park et al. 2014) and latent synthetic metabolic pathways (Antonovsky et al. 2016; Gleizer et al. 2019), which require additional reconfiguration steps to fine-tune non-natural phenotypes. In particular, systems biology-driven analytical pipelines remain the most effective approach for debugging and refining perturbed functions in synthetically engineered biological systems (Fig. 3a). In this section, we present a systems biology approach to decipher the molecular and mechanistic bases of adaptation in synthetic strains, with a specific focus on genome-streamlined organisms in which large subsets of non-essential genes are removed from their genomic makeup.

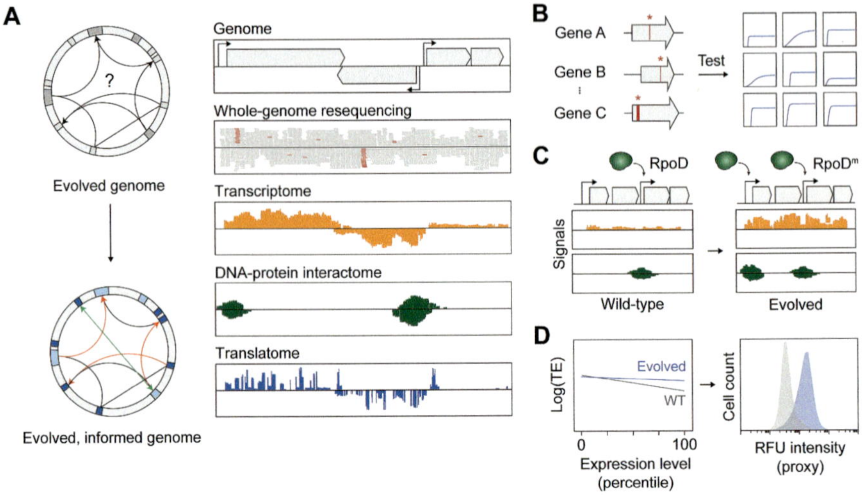

Fig. 3 Systems biology approach to analyze molecular-level changes in adaptively evolved genomes. (**a**) An example of integrative multi-omics methodologies encompassing WGS, RNA-Seq, ChIP-Seq, and ribosome profiling implemented to elucidate the molecular basis of adaptation in eMS57. (**b**) The causality of mutations identified in WGS can be analyzed through the fitness assay of mutants reconstituted in a wild-type background. (**c**) The combinatorial analysis of ChIP-Seq and RNA-Seq datasets helps elucidate rewired regulatory networks. (**d**) Ribosome profiling of the evolved genome (eMS57) revealed enhanced TE in highly transcribed genes compared with MG1655. FACS assay revealed a higher fluorescence intensity of a fluorescence reporter protein, which sets the proxy for enhanced translation in eMS57

4.1 Multi-omics Approach to Analyze Synthetic Optimality

High-throughput sequencing techniques are highly effective in characterizing genome-wide mutation profiles (Barrick et al. 2009), global changes in the transcriptome (Wang et al. 2009), translatome profiles (Ingolia et al. 2012), and alterations in DNA-protein regulatory interactions (Johnson et al. 2007), among many others. Integrative analysis of multi-omics datasets can help uncover hidden layers of regulation that govern genotype-phenotype relationships (Fig. 3a). Discerning causal mutations from those of neutral or deleterious hitchhikers in WGS data is an important step toward characterizing genotype-phenotype relationships (Fig. 3b). The following approaches can be considered. Mutations shared among replicate lineages reflect a strong causality (Conrad et al. 2009). In cases where replicates are not available, manual inspection of gene functions, pathway-mapping analysis, and the public ALE database help infer the functional role of mutations (Kang et al. 2019; Phaneuf et al. 2019).

Taking eMS57 as an example, which is an ALE derivative of one of the most extensively genome-streamlined *E. coli* strains, MS56 (Choe et al. 2019), WGS analysis identified 118 mutations in the eMS57 genome. Based on prior understanding, the authors focused on a large deletion mutation encompassing 21 genes, including *mutS* and *rpoS*. MutS is a component of the MutHSL DNA repair system, the absence of which has been shown to expedite evolution (Kang et al. 2019; Overbeck et al. 2017; Wannier et al. 2018). RpoS is a σ^{38} that governs global gene transcription in response to stress and the stationary phase growth (Cho et al. 2014). Reconstitution of the *rpoS* deletion in the parental MS56 genome increased its growth rate to 80% that of eMS57, indicating that the global reprogramming of stress responses is causal to the fitness of genome-perturbed microbes. In addition to RpoS, the authors identified a subset of mutations associated with environmental (*cspP*) nutrient stress (*ilvN*) that contributed to increased fitness in the wild-type background. Hence, it was concluded that global rewiring of stress-response mechanisms facilitated by the MutS hypermutator phenotype led to fitness restoration in MS56, compared to that of parental MG1655 (Choe et al. 2019).

In addition to *rpoS*, mutations in other RNAP subunits *rpoD* and *rpoC* identified in eMS57 imply large-scale reprogramming of transcription regulation dynamics (Conrad et al. 2010). Variations in high-throughput sequencing techniques, such as chromatin immunoprecipitation followed by sequencing (ChIP-Seq), allow segments of transcription factor-bound DNA to be selectively sequenced (Johnson et al. 2007). Integrative analysis of transcription factor (TF)-binding profiles in tandem with genome-wide gene expression profiling (i.e., transcriptome sequencing) has been established as an analytical pipeline for the identification of TF-DNA interactions and corresponding changes in the genome-wide expression of cognate regulons (Fig. 3c) (Cho et al. 2014; Kim et al. 2018). Integrative analysis of the transcriptional regulation of mutant RpoD revealed distinct promoter-binding patterns that appear to reshape the expression profiles of metabolic genes within eMS57. These included a preference toward the NADPH-generating Entner-Doudoroff pathway for glucose utilization and upregulation of the dNTP salvage pathway, elucidating a distinctive adaptive response that seems to reflect optimal physiology in a genome-streamlined organism (Choe et al. 2019).

Transcription and translation represent two discrete layers of regulatory processes in which an increase in gene expression does not necessarily correspond to a concurrent increase in protein abundance (Jeong et al. 2016; McManus et al. 2014). In this regard, ribosome profiling offers a quantitative outlook on post-translational regulation that affects translatome profiles (Fig. 3d). This technique uses an arbitrary metric called translational efficiency (TE), which is the ratio between the relative abundance of ribosome-protected mRNA fragments and the corresponding total mRNA transcript levels, to quantitatively analyze genome-wide translational changes (McManus et al. 2014). Normally, ribosomal occupancy was less variable than mRNA abundance in bacteria, resulting in highly varied TE levels. However, eMS57 exhibited constant TE levels. This suggests diminished translational buffering in eMS57 validated through a stronger reporter protein activity in eMS57 than in MG1655 (Choe et al. 2019).

As such, an integrative multi-omics approach enables a comprehensive examination of adaptive changes in the venture point of various omics regulations. Additional phenotypic layers not covered in this chapter, such as high-throughput metabolomics, can be incorporated ad hoc to refine the genotype-to-phenotype regulatory logic (Chen et al. 2017). In particular, owing to their reduced genomic redundancy and underlying complexities, genome-streamlined organisms are expected to facilitate genotype-phenotype mapping for a basic understanding of biology. A whole-cell kinetic model was made possible for *M. mycoides* minimal genome JCVI-syn3A. The whole-cell model simulates the three-dimensional spatiotemporal dynamics of principal cellular processes, allowing inspection of the implicit details of cellular processes (Thornburg et al. 2022). In this regard, recent progress in the ALE of a minimal genome, a 2000-generation evolution experiment of the smallest minimal genome, JCVI-syn3B (Moger-Reischer et al. 2023), represents an interesting development in the field of evolutionary biology and warrants system-level investigations to gain insight into the principles of life in its simplest form.

4.2 Other Emerging Technologies for ALE Applications

Automated multiplexing technologies, poised to offer substantial improvements in experimental throughput and quality of outcomes, are gaining traction in biological sciences (Gurdo et al. 2022). Concerning ALE, automated multiplexing platforms promise a precise control over complex high-throughput culture settings and expand the capacity of evolution experiments. Automated ALE platforms in several leading laboratories generate fine-grained datasets for experimental evolution in a multiplexed manner (Sandberg et al. 2016; Wong et al. 2018). This could greatly expand the width at which multiple design elements are evaluated in a single experimental cycle. In parallel with multiplexed experimental capacity, various machine learning toolkits have been showcased to effectively process and analyze the ever-expanding volume of high-throughput data, particularly various -omics datasets. For instance, Sastry et al. (2019) demonstrated that large-scale RNA-Seq datasets could be processed using independent component analysis to statistically delineate groups of genes co-regulated by specific transcriptional or environmental factors (Sastry et al. 2019). This unsupervised machine learning algorithm, called iModulon, proved to be robust in inferring changes in TF activities and their functional attributes in adaptive phenotypes and in identifying groups of genes independently modulated by specific environmental variables (Anand et al. 2022; Kavvas et al. 2022). Integrative analysis of the iModulon in tandem with ME model-guided flux distribution could offer a tremendous insight into the hidden regulatory logic between specific TFs and metabolism in evolved genomes (Anand et al. 2020; Anand et al. 2022). Furthermore, the machine learning-assisted protein structure prediction platform AlphaFold (Jumper et al. 2021) helps predict the mechanistic underpinnings of fitness improvement induced by structural modifications in a mutant enzyme (Jin et al. 2022). Taken together, an array of machine learning-based analytic and simulation platforms have streamlined the processing and interpretation of complex biological phenomena in unprecedented detail.

5 Conclusions

By leveraging the Darwinian natural selection process, ALE is an important workflow in strain development. Experimental setups imposing "adequate" and "sufficient" selection pressures select for mutants at optimal fitness peaks in a given fitness landscape. Selection pressures can be customized using genetic perturbations, synthetic circuits, and other intrinsic and extrinsic factors to effectively couple the desired phenotypes to strain fitness. This paves the way for novel routes to optimize various features of biotechnologically relevant traits, including substrate utilization, tolerance to toxic chemicals, latent pathway activation, and the overproduction of compounds of interest in a timely manner. The contributions of various enabling technologies to strain engineering design are worth emphasizing. The capacity to couple non-natural characteristics, even those that potentially incur fitness costs, widens the scope of ALE applications, such as high-throughput screening of hyper-producing mutants (Bentley et al. 2020), non-native peptide synthesis (Thyer et al. 2018), and viral resistance (Nyerges et al. 2023).

NGS technologies have allowed for systematically assessing the genetic basis of fitness improvement and the molecular mechanisms underlying adaptive changes. Genome-streamlined organisms, with their "simplified" life processes, offer conducive platforms for testing and examining fundamental gene functions and associated phenotypes. Rapid, low-cost, and almost ubiquitous sequencing platforms and services have made enormous volumes of high-throughput datasets available. Novel public databases, in conjunction with machine learning-based analytical platforms, facilitate knowledge mining and discovering novel relationships between different phenotypes that have remained elusive. These advances are expected to continue to drive innovations in biotechnology and improve our understanding of fundamental biological processes.

Acknowledgments This work was supported by the Korea Bio Grand Challenge (2018M3A9H3024759 to B-KC) through the National Research Foundation of Korea (NRF), funded by the Ministry of Science and ICT.

Declaration of Competing Interest The authors declare no conflict of interest.

References

Anand A, Chen K, Catoiu E et al (2020) OxyR is a convergent target for mutations acquired during adaptation to oxidative stress-prone metabolic states. Mol Biol Evol 37:660–667

Anand A, Patel A, Chen K et al (2022) Laboratory evolution of synthetic electron transport system variants reveals a larger metabolic respiratory system and its plasticity. Nat Commun 13:3682

Antonovsky N, Gleizer S, Noo E et al (2016) Sugar synthesis from CO2 in Escherichia coli. Cell 166:115–125

Applebee MK, Joyce AR, Conrad TM et al (2011) Functional and metabolic effects of adaptive glycerol kinase (GLPK) mutants in Escherichia coli. J Biol Chem 286:23150–23159

Arjan JA, Visser M, Zeyl CW et al (1999) Diminishing returns from mutation supply rate in asexual populations. Science 283:404–406

Asopa RP, Bhoi R, Saharan VK (2022) Valorization of glycerol into value-added products: a comprehensive review on biochemical route. Bioresourc Technol Rep 20:101290

Awasthi D, Tang YH, Amer B et al (2022) Adaptive evolution of Methylotuvimicrobium alcaliphilum to grow in the presence of rhamnolipids improves fatty acid and rhamnolipid production from CH4. J Ind Microbiol Biotechnol 49

Badran AH, Liu DR (2015) Development of potent in vivo mutagenesis plasmids with broad mutational spectra. Nat Commun 6:8425

Bae J, Song Y, Lee H et al (2022) Valorization of C1 gases to value-added chemicals using acetogenic biocatalysts. Chem Engin J 428:131325

Bajic D, Vila JCC, Blount ZD et al (2018) On the deformability of an empirical fitness landscape by microbial evolution. Proc Natl Acad Sci USA 115:11286–11291

Barrick JE, Yu DS, Yoon SH et al (2009) Genome evolution and adaptation in a long-term experiment with Escherichia coli. Nature 461:1243–1247

Basan M, Hui S, Okano H et al (2015) Overflow metabolism in Escherichia coli results from efficient proteome allocation. Nature 528:99–104

Beaumont HJ, Gallie J, Kost C et al (2009) Experimental evolution of bet hedging. Nature 462:90–93

Bentley GJ, Narayanan N, Jha RK et al (2020) Engineering glucose metabolism for enhanced muconic acid production in Pseudomonas putida KT2440. Metab Eng 59:64–75

Blank D, Wolf L, Ackermann M et al (2014) The predictability of molecular evolution during functional innovation. Proc Natl Acad Sci USA 111:3044–3049

Blount ZD, Borland CZ, Lenski RE (2008) Historical contingency and the evolution of a key innovation in an experimental population of Escherichia coli. Proc Natl Acad Sci USA 105: 7899–7906

Bonekamp F, Clemmesen K, Karlstrom O et al (1984) Mechanism of UTP-modulated attenuation at the pyrE gene of Escherichia coli: an example of operon polarity control through the coupling of translation to transcription. EMBO J 3:2857–2861

Burgard AP, Pharkya P, Maranas CD (2003) Optknock: a bilevel programming framework for identifying gene knockout strategies for microbial strain optimization. Biotechnol Bioeng 84: 647–657

Carpenter AC, Feist AM, Harrison FSM et al (2023) Have you tried turning it off and on again? Oscillating selection to enhance fitness-landscape traversal in adaptive laboratory evolution experiments. Metab Eng Commun 17:e00227

Chen K, Gao Y, Mih N et al (2017) Thermosensitivity of growth is determined by chaperone-mediated proteome reallocation. Proc Natl Acad Sci USA 114:11548–11553

Cheng KK, Lee BS, Masuda T et al (2014) Global metabolic network reorganization by adaptive mutations allows fast growth of Escherichia coli on glycerol. Nat Commun 5:3233

Cho BK, Kim D, Knight EM et al (2014) Genome-scale reconstruction of the sigma factor network in Escherichia coli: topology and functional states. BMC Biol 12:4

Choe D, Lee JH, Yoo M et al (2019) Adaptive laboratory evolution of a genome-reduced Escherichia coli. Nat Commun 10:935

Choe D, Kim K, Kang M et al (2022) Synthetic 3'-UTR valves for optimal metabolic flux control in Escherichia coli. Nucleic Acids Res 50:4171–4186

Chou HH, Chiu HC, Delaney NF et al (2011) Diminishing returns epistasis among beneficial mutations decelerates adaptation. Science 332:1190–1192

Conrad TM, Joyce AR, Applebee MK et al (2009) Whole-genome resequencing of Escherichia coli K-12 MG1655 undergoing short-term laboratory evolution in lactate minimal media reveals flexible selection of adaptive mutations. Genome Biol 10:R118

Conrad TM, Frazier M, Joyce AR et al (2010) RNA polymerase mutants found through adaptive evolution reprogram Escherichia coli for optimal growth in minimal media. Proc Natl Acad Sci USA 107:20500–20505

Cooper TF (2007) Recombination speeds adaptation by reducing competition between beneficial mutations in populations of Escherichia coli. PLoS Biol 5:e225

Du C, Li Y, Xiang R et al (2022) Co-utilization of multiple lignocellulose-derived sugars and inhibitors by the robust mutant Kluyveromyces marxianus through adaptive laboratory evolution. Fuel 315:122816

Fait A, Seif Y, Mikkelsen K et al (2022) Adaptive laboratory evolution and independent component analysis disentangle complex vancomycin adaptation trajectories. Proc Natl Acad Sci USA 119: e2118262119

Fong SS, Marciniak JY, Palsson BO (2003) Description and interpretation of adaptive evolution of Escherichia coli K-12 MG1655 by using a genome-scale in silico metabolic model. J Bacteriol 185:6400–6408

Fong SS, Burgard AP, Herring CD et al (2005) In silico design and adaptive evolution of Escherichia coli for production of lactic acid. Biotechnol Bioeng 91:643–648

Freedberg WB, Kistler WS, Lin EC (1971) Lethal synthesis of methylglyoxal by Escherichia coli during unregulated glycerol metabolism. J Bacteriol 108:137–144

Friesen ML, Saxer G, Travisano M et al (2004) Experimental evidence for sympatric ecological diversification due to frequency-dependent competition in Escherichia coli. Evolution 58:245–260

Gibson TC, Scheppe ML, Cox EC (1970) Fitness of an Escherichia coli mutator gene. Science 169: 686–688

Giraud A, Matic I, Tenaillon O et al (2001a) Costs and benefits of high mutation rates: adaptive evolution of bacteria in the mouse gut. Science 291:2606–2608

Giraud A, Radman M, Matic I et al (2001b) The rise and fall of mutator bacteria. Curr Opin Microbiol 4:582–585

Gleizer S, Ben-Nissan R, Bar-On YM et al (2019) Conversion of Escherichia coli to generate all biomass carbon from CO(2). Cell 179(1255–1263):e1212

Good BH, Mcdonald MJ, Barrick JE et al (2017) The dynamics of molecular evolution over 60,000 generations. Nature 551:45–50

Gresham D, Dunham MJ (2014) The enduring utility of continuous culturing in experimental evolution. Genomics 104:399–405

Gurdo N, Volke DC, Nikel PI (2022) Merging automation and fundamental discovery into the design-build-test-learn cycle of nontraditional microbes. Trends Biotechnol 40:1148–1159

Harwani D, Begani J, Barupal S et al (2022) Adaptive laboratory evolution triggers pathogen-dependent broad-spectrum antimicrobial potency in Streptomyces. J Genet Eng Biotechnol 20:1

Herring CD, Raghunathan A, Honisch C et al (2006) Comparative genome sequencing of Escherichia coli allows observation of bacterial evolution on a laboratory timescale. Nat Genet 38:1406–1412

Hua Q, Joyce AR, Palsson BO et al (2007) Metabolic characterization of Escherichia coli strains adapted to growth on lactate. Appl Environ Microbiol 73:4639–4647

Hutchison CA 3rd, Chuang RY, Noskov VN et al (2016) Design and synthesis of a minimal bacterial genome. Science 351:aad6253

Ibarra RU, Edwards JS, Palsson BO (2002) Escherichia coli K-12 undergoes adaptive evolution to achieve in silico predicted optimal growth. Nature 420:186–189

Ingolia NT, Brar GA, Rouskin S et al (2012) The ribosome profiling strategy for monitoring translation in vivo by deep sequencing of ribosome-protected mRNA fragments. Nat Protoc 7:1534–1550

Jakociunas T, Pedersen LE, Lis AV et al (2018) CasPER, a method for directed evolution in genomic contexts using mutagenesis and CRISPR/Cas9. Metab Eng 48:288–296

Jasmin JN, Kassen R (2007) Evolution of a single niche specialist in variable environments. Proc Biol Sci 274:2761–2767

Jayakody LN, Chinmoy B, Turner TL (2022) Trends in valorization of highly-toxic lignocellulosic biomass derived-compounds via engineered microbes. Bioresour Technol 346:126614

Jee J, Rasouly A, Shamovsky I et al (2016) Rates and mechanisms of bacterial mutagenesis from maximum-depth sequencing. Nature 534:693–696

Jensen KF (1993) The Escherichia coli K-12 "wild types" W3110 and MG1655 have an rph frameshift mutation that leads to pyrimidine starvation due to low pyrE expression levels. J Bacteriol 175:3401–3407

Jeong Y, Kim JN, Kim MW et al (2016) The dynamic transcriptional and translational landscape of the model antibiotic producer Streptomyces coelicolor A3(2). Nat Commun 7:11605

Jin C, Hou W, Yao R et al (2019) Adaptive evolution of Gluconobacter oxydans accelerates the conversion rate of non-glucose sugars derived from lignocellulose biomass. Bioresour Technol 289:121623

Jin S, Kang S, Bae J et al (2022) Development of CO gas conversion system using high CO tolerance biocatalyst. Chem Engineer J 449:137678

Johnson DS, Mortazavi A, Myers RM et al (2007) Genome-wide mapping of in vivo protein-DNA interactions. Science 316:1497–1502

Johnston CW, Badran AH, Collins JJ (2020) Continuous bioactivity-dependent evolution of an antibiotic biosynthetic pathway. Nat Commun 11:4202

Jumper J, Evans R, Pritzel A et al (2021) Highly accurate protein structure prediction with AlphaFold. Nature 596:583–589

Kamionka M (2011) Engineering of therapeutic proteins production in Escherichia coli. Curr Pharm Biotechnol 12:268–274

Kang M, Kim K, Choe D et al (2019) Inactivation of a mismatch-repair system diversifies genotypic landscape of Escherichia coli during adaptive laboratory evolution. Front Microbiol 10:1845

Kang S, Song Y, Jin S et al (2020) Adaptive laboratory evolution of Eubacterium limosum ATCC 8486 on carbon monoxide. Front Microbiol 11:402

Kassen R (2002) The experimental evolution of specialists, generalists, and the maintenance of diversity. J Evol Biol 15:173–190

Kavvas ES, Long CP, Sastry A et al (2022) Experimental evolution reveals unifying systems-level adaptations but diversity in driving genotypes. mSystems 7:e0016522

Kim D, Seo SW, Gao Y et al (2018) Systems assessment of transcriptional regulation on central carbon metabolism by Cra and CRP. Nucleic Acids Res 46:2901–2917

Kim K, Hou CY, Choe D et al (2022) Adaptive laboratory evolution of Escherichia coli W enhances gamma-aminobutyric acid production using glycerol as the carbon source. Metab Eng 69:59–72

Kim K, Kang M, Cho SH et al (2023) Minireview: engineering evolution to reconfigure phenotypic traits in microbes for biotechnological applications. Comput Struct Biotechnol J 21:563–573

Knoppel A, Knopp M, Albrecht LM et al (2018) Genetic adaptation to growth under laboratory conditions in Escherichia coli and Salmonella enterica. Front Microbiol 9:756

Kryazhimskiy S, Rice DP, Desai MM (2012) Population subdivision and adaptation in asexual populations of Saccharomyces cerevisiae. Evolution 66:1931–1941

Kvitek DJ, Sherlock G (2011) Reciprocal sign epistasis between frequently experimentally evolved adaptive mutations causes a rugged fitness landscape. PLoS Genet 7:e1002056

Lacroix RA, Sandberg TE, O'brien EJ et al (2015) Use of adaptive laboratory evolution to discover key mutations enabling rapid growth of Escherichia coli K-12 MG1655 on glucose minimal medium. Appl Environ Microbiol 81:17–30

Le Vo TD, Ko JS, Park SJ et al (2013) Efficient gamma-aminobutyric acid bioconversion by employing synthetic complex between glutamate decarboxylase and glutamate/GABA antiporter in engineered Escherichia coli. J Ind Microbiol Biotechnol 40:927–933

Leclerc JE, Li B, Payne WL et al (1996) High mutation frequencies among Escherichia coli and Salmonella pathogens. Science 274:1208–1211

Lee JK, Kim S, Kim W et al (2019a) Efficient production of d-lactate from methane in a lactate-tolerant strain of Methylomonas sp. DH-1 generated by adaptive laboratory evolution. Biotechnol Biofuels 12:234

Lee SY, Kim HU, Chae TU et al (2019b) A comprehensive metabolic map for production of bio-based chemicals. Nat Catalysis 2:18–33

Leiby N, Marx CJ (2014) Metabolic erosion primarily through mutation accumulation, and not tradeoffs, drives limited evolution of substrate specificity in Escherichia coli. PLoS Biol 12: e1001789

Lenski RE, Rose MR, Simpson SC et al (1991) Long-term experimental evolution in Escherichia coli. I. Adaptation and divergence during 2,000 generations. Am Nat 138:1315–1341

Liang S, Jiang W, Song Y et al (2020) Improvement and metabolomics-based analysis of d-lactic acid production from agro-industrial wastes by Lactobacillus delbrueckii submitted to adaptive laboratory evolution. J Agric Food Chem 68:7660–7669

Lloyd CJ, Ebrahim A, Yang L et al (2018) COBRAme: a computational framework for genome-scale models of metabolism and gene expression. PLoS Comput Biol 14:e1006302

Loh E, Salk JJ, Loeb LA (2010) Optimization of DNA polymerase mutation rates during bacterial evolution. Proc Natl Acad Sci USA 107:1154–1159

Luo H, Hansen ASL, Yang L, a. (2019) Coupling S-adenosylmethionine-dependent methylation to growth: design and uses. PLoS Biol 17:e2007050

Maeda T, Iwasawa J, Kotani H et al (2020) High-throughput laboratory evolution reveals evolutionary constraints in Escherichia coli. Nat Commun 11:5970

Mcmanus CJ, May GE, Spealman P et al (2014) Ribosome profiling reveals post-transcriptional buffering of divergent gene expression in yeast. Genome Res 24:422–430

Mitchell A, Romano GH, Groisman B et al (2009) Adaptive prediction of environmental changes by microorganisms. Nature 460:220–224

Moger-Reischer RZ, Glass JI, Wise KS et al (2023) Evolution of a minimal cell. Nature 620:122–127

Morgenthaler AB, Kinney WR, Ebmeie CC et al (2019) Mutations that improve efficiency of a weak-link enzyme are rare compared to adaptive mutations elsewhere in the genome. elife 8

Mundhada H, Seoane JM, Schneider K et al (2017) Increased production of L-serine in Escherichia coli through adaptive laboratory evolution. Metab Eng 39:141–150

Nahum JR, Godfrey-Smith P, Harding BN et al (2015) A tortoise–hare pattern seen in adapting structured and unstructured populations suggests a rugged fitness landscape in bacteria. Proc Natl Acad Sci USA 112:7530–7535

Novotny MJ, Frederickson WL, Waygood EB et al (1985) Allosteric regulation of glycerol kinase by enzyme IIIglc of the phosphotransferase system in Escherichia coli and Salmonella typhimurium. J Bacteriol 162:810–816

Nyerges A, Vinke S, Flynn R et al (2023) A swapped genetic code prevents viral infections and gene transfer. Nature 615:720–727

Orth JD, Thiele I, Palsson BO (2010) What is flux balance analysis? Nat Biotechnol 28:245–248

Overbeck TJ, Welker DL, Hughes JE et al (2017) Transient MutS-based hypermutation system for adaptive evolution of Lactobacillus casei to low pH. Appl Environ Microbiol 83

Park MK, Lee SH, Yang KS et al (2014) Enhancing recombinant protein production with an Escherichia coli host strain lacking insertion sequences. Appl Microbiol Biotechnol 98:6701–6713

Peabody VG, Li H, Kao KC (2017) Sexual recombination and increased mutation rate expedite evolution of Escherichia coli in varied fitness landscapes. Nat Commun 8:2112

Pham VD, Lee SH, Park SJ et al (2015) Production of gamma-aminobutyric acid from glucose by introduction of synthetic scaffolds between isocitrate dehydrogenase, glutamate synthase and glutamate decarboxylase in recombinant Escherichia coli. J Biotechnol 207:52–57

Phaneuf PV, Gosting D, Palsson BO et al (2019) ALEdb 1.0: a database of mutations from adaptive laboratory evolution experimentation. Nucleic Acids Res 47:D1164–D1171

Pontrelli S, Fricke RCB, Sakurai SSM et al (2018) Directed strain evolution restructures metabolism for 1-butanol production in minimal media. Metab Eng 49:153–163

Quandt EM, Deatherage DE, Ellington AD et al (2014) Recursive genomewide recombination and sequencing reveals a key refinement step in the evolution of a metabolic innovation in Escherichia coli. Proc Natl Acad Sci USA 111:2217–2222

Radek A, Tenhaef N, Muller MF et al (2017) Miniaturized and automated adaptive laboratory evolution: evolving Corynebacterium glutamicum towards an improved d-xylose utilization. Bioresour Technol 245:1377–1385

Raman S, Rogers JK, Taylor ND et al (2014) Evolution-guided optimization of biosynthetic pathways. Proc Natl Acad Sci USA 111:17803–17808

Robertson WE, Funke LFH, De La Torre D et al (2021) Sense codon reassignment enables viral resistance and encoded polymer synthesis. Science 372:1057–1062

Rugbjerg P, Myling-Petersen N, Porse A et al (2018a) Diverse genetic error modes constrain large-scale bio-based production. Nat Commun 9:787

Rugbjerg P, Sarup-Lytzen K, Nagy M et al (2018b) Synthetic addiction extends the productive life time of engineered Escherichia coli populations. Proc Natl Acad Sci USA 115:2347–2352

Sandberg TE, Long CP, Gonzalez JE et al (2016) Evolution of E. coli on [U-13C]glucose reveals a negligible isotopic influence on metabolism and physiology. PLoS One 11:e0151130

Sandberg TE, Lloyd CJ, Palsson BO et al (2017) Laboratory evolution to alternating substrate environments yields distinct phenotypic and genetic adaptive strategies. Appl Environ Microbiol 83

Sandberg TE, Salazar MJ, Weng LL et al (2019) The emergence of adaptive laboratory evolution as an efficient tool for biological discovery and industrial biotechnology. Metab Eng 56:1–16

Sastry AV, Gao Y, Szubin R et al (2019) The Escherichia coli transcriptome mostly consists of independently regulated modules. Nat Commun 10:5536

Saxer G, Doebeli M, Travisano M (2010) The repeatability of adaptive radiation during long-term experimental evolution of Escherichia coli in a multiple nutrient environment. PLoS One 5: e14184

Schwentne A, Feith A, Munch E et al (2018) Metabolic engineering to guide evolution - creating a novel mode for L-valine production with Corynebacterium glutamicum. Metab Eng 47:31–41

Seok JY, Han YH, Yang JS et al (2021) Synthetic biosensor accelerates evolution by rewiring carbon metabolism toward a specific metabolite. Cell Rep 36:109589

Seong W, Han GH, Lim HS et al (2020) Adaptive laboratory evolution of Escherichia coli lacking cellular byproduct formation for enhanced acetate utilization through compensatory ATP consumption. Metab Eng 62:249–259

Shendure J, Balasubramanian S, Church GM et al (2017) DNA sequencing at 40: past, present and future. Nature 550:345–353

Solopova A, Van Gestel J, Weissing FJ et al (2014) Bet-hedging during bacterial diauxic shift. Proc Natl Acad Sci USA 111:7427–7432

Spinck M, Piedrafita C, Robertson WE et al (2023) Genetically programmed cell-based synthesis of non-natural peptide and depsipeptide macrocycles. Nat Chem 15:61–69

Sprouffske K, Aguilar-Rodriguez J, Sniegowski P et al (2018) High mutation rates limit evolutionary adaptation in Escherichia coli. PLoS Genet 14:e1007324

Szappanos B, Fritzemeier J, Csorgo B et al (2016) Adaptive evolution of complex innovations through stepwise metabolic niche expansion. Nat Commun 7:11607

Thornburg ZR, Bianchi DM, Brier TA et al (2022) Fundamental behaviors emerge from simulations of a living minimal cell. Cell 185:345–360 e328

Thyer R, Shroff R, Klein DR et al (2018) Custom selenoprotein production enabled by laboratory evolution of recoded bacterial strains. Nat Biotechnol 36:624–631

Tokuyama K, Toya Y, Horinouchi T et al (2018) Application of adaptive laboratory evolution to overcome a flux limitation in an Escherichia coli production strain. Biotechnol Bioeng 115: 1542–1551

Umeyama T, Okada S, Ito T (2013) Synthetic gene circuit-mediated monitoring of endogenous metabolites: identification of GAL11 as a novel multicopy enhancer of s-adenosylmethionine level in yeast. ACS Synth Biol 2:425–430

Van Cleve J, Weissman DB (2015) Measuring ruggedness in fitness landscapes. Proc Natl Acad Sci USA 112:7345–7346

Vaud S, Pearcy N, Hanzevacki M et al (2021) Engineering improved ethylene production: leveraging systems biology and adaptive laboratory evolution. Metab Eng 67:308–320

Wang Z, Gerstein M, Snyder M (2009) RNA-Seq: a revolutionary tool for transcriptomics. Nat Rev Genet 10:57–63

Wang G, Li Q, Zhang Z et al (2023) Recent progress in adaptive laboratory evolution of industrial microorganisms. J Ind Microbiol Biotechnol 50

Wannier TM, Kunjapur AM, Rice DP et al (2018) Adaptive evolution of genomically recoded Escherichia coli. Proc Natl Acad Sci USA 115:3090–3095

Wong BG, Mancuso CP, Kiriakov S et al (2018) Precise, automated control of conditions for high-throughput growth of yeast and bacteria with eVOLVER. Nat Biotechnol 36:614–623

Woods RJ, Barrick JE, Cooper TF et al (2011) Second-order selection for evolvability in a large Escherichia coli population. Science 331:1433–1436

Wright S (1932) The roles of mutation, inbreeding, crossbreeding and selection in evolution, Proceedings of the sixth international congress of genetics. Brooklyn Botanic Garden, Brooklyn, NY

Wu Y, Jameel A, Xing XH et al (2022) Advanced strategies and tools to facilitate and streamline microbial adaptive laboratory evolution. Trends Biotechnol 40:38–59

Yang J, Seo SW, Jang S et al (2013) Synthetic RNA devices to expedite the evolution of metabolite-producing microbes. Nat Commun 4:1413

You L, Arnold FH (1996) Directed evolution of subtilisin E in Bacillus subtilis to enhance total activity in aqueous dimethylformamide. Protein Eng 9:77–83

Yu S, Zha Q, Miao X et al (2013) Enhancement of lipid production in low-starch mutants Chlamydomonas reinhardtii by adaptive laboratory evolution. Bioresour Technol 147:499–507

Yu T, Zhou YJ, Huang M et al (2018) Reprogramming yeast metabolism from alcoholic fermentation to lipogenesis. Cell 174(1549–1558):e1514

Zheng B, Ma X, Wang N et al (2018) Utilization of rare codon-rich markers for screening amino acid overproducers. Nat Commun 9:3616

Zhu L, Cai Z, Zhang Y et al (2014) Engineering stress tolerance of Escherichia coli by stress-induced mutagenesis (SIM)-based adaptive evolution. Biotechnol J 9:120–127

Zhu C, You X, Wu T et al (2022) Efficient utilization of carbon to produce aromatic valencene in Saccharomyces cerevisiae using mannitol as the substrate. Green Chem 24:4614–4627

Zwaig N, Lin EC (1966) Feedback inhibition of glycerol kinase, a catabolic enzyme in Escherichia coli. Science 153:755–757

The Non-coding Genome and Network Biology

Ryan Christopher Kelly, Rebecca Anne Morgan, Madison Brown, Ian Overton, and Gary Hardiman

Contents

Abstract A non-coding RNA (ncRNA) is a class of RNA that does not undergo translation. The DNA sequence encoding a non-coding RNA referred to as an RNA gene. According to the RNA world hypothesis, early life relied fundamentally on RNA, with subsequent evolution transferring information storage to the DNA a more robust molecule and catalytic functions to proteins. Initially, RNA was considered to primarily serve as an intermediary between genes and proteins despite its critical roles in primeval processes including RNA splicing and translation. However, recent discoveries challenge this assumption, revealing that a significant portion of the

R. C. Kelly
Faculty of Medicine, Health and Life Sciences, School of Biological Sciences, and Institute for Global Food Security (IGFS), Queen's University Belfast, Belfast, UK

National Cancer Institute, Frederick, MD, USA

R. A. Morgan · M. Brown · G. Hardiman (✉)
Faculty of Medicine, Health and Life Sciences, School of Biological Sciences, and Institute for Global Food Security (IGFS), Queen's University Belfast, Belfast, UK
e-mail: g.hardiman@qub.ac.uk

I. Overton
Faculty of Medicine, Health and Life Sciences, Patrick G. Johnston Centre for Cancer Research, Queen's University Belfast, Belfast, UK

J. Barciszewski (ed.), *Systems Biology II*, RNA Technologies 15,
https://doi.org/10.1007/978-3-031-62178-9_7

genome in higher organisms is transcribed across different developmental stages. Additionally, the identification of novel non-coding RNAs suggests that RNA evolved in parallel with DNA and protein. The Encyclopedia of DNA Elements (ENCODE) consortium has reported that 67% of genes are non-coding, while 33% encode proteins. Among all genes, 27% are comprised of long non-coding RNAs (lncRNAs), 16% consist of small non-coding RNAs and 24% are pseudogenes, which are transcribed but do not produce proteins. Notable types of non-coding RNAs include ribosomal RNAs (rRNAs), transfer RNAs (tRNAs), microRNAs, small nucleolar RNAs (snoRNAs), small nuclear RNAs (snRNAs), small interfering RNAs (siRNAs), PIWI-interacting RNAs (piRNAs), extracellular RNAs (exRNAs), small Cajal body-specific RNAs (scaRNAs) and long ncRNAs (lncRNAs). Typically, ncRNAs longer than 200 nucleotides are categorised as lncRNAs, present at levels much lower than total mRNA. Previously considered "junk RNA", lncRNAs have gained recognition as RNA molecules possessing functional attributes through initiatives like Functional Annotation of the Mammalian Genome (FANTOM). Like mRNAs, lncRNAs undergo transcription by the RNA polymerase II enzyme, followed by polyadenylation, capping and splicing. However, lncRNAs usually possess intron/exon structures; are expressed at lower levels, often in a tissue-specific manner; and exhibit poor evolutionary conservation compared to mRNAs. The most extensively studied category of ncRNAs is microRNAs (miRNAs), small ncRNAs around 22 nucleotides in length. In animals, miRNAs regulate post-transcriptional gene expression by controlling mRNA translation into proteins. This chapter discusses software toolkits developed to elucidate the regulatory roles of ncRNAs.

Keywords lncRNAs · miRNAs · Prostate cancer · Networks · ENCODE

1 Introduction

The completion of the Human Genome Project (HGP) in 2001 remains one of the most important milestones in the field of biology, the findings of which provided invaluable insight into the complexity of the human genome and, with it, a wealth of knowledge which remains useful today (Lander et al. 2001). Despite its scale of over three billion bases, a major finding was that only 1% of the DNA sequence of the human genome encodes protein, with the remaining 99% aptly named "non-coding" or "junk" DNA. Given its lack of known functionality, this junk DNA was believed to be the molecular scar of evolution; sequence which has progressively lost function over the years and now exists with no biological relevance (McLysaght and Guerzoni 2015). The observation of non-coding DNA had been reported as early as the 1970s, with studies providing descriptions of the human genome being pervasively transcribed into non-coding transcripts, but little was done to understand non-coding portions of the genome (Comings 1972). Many hypothesised the functional significance of non-coding DNA, but little progress was achieved in substantiating these claims (Orgel and Crick 1980).

More recently, the availability of next-generation sequencing (NGS) technologies, particularly RNA sequencing (RNA-seq), has led to a significant increase in the number of publications aiming to better characterise the non-coding transcriptome of the human genome. The past two decades has seen a significant investment into understanding the roles of long non-coding RNAs (lncRNAs) which comprise a substantial proportion of the total non-coding mass (Hangauer et al. 2013). The efforts made to annotate and understand lncRNAs have proven fruitful with estimates of human lncRNAs ranging from a more conservative 10,000 to over 200,000 transcripts (Pinkney et al. 2020). These investigations have aided clarity regarding the mechanisms by which lncRNAs act and their functional relevance during development and disease (Jarroux et al. 2017).

With these investigations, it is now generally considered that lncRNAs possess important roles in the initiation and progression of human disease. Indeed, reports have provided evidence to support the roles of lncRNAs in driving cancer, Alzheimer's disease and diabetes, among others, with many computational algorithms having been developed to predict such associations. Despite progress, lncRNAs remain elusive as many lack known function, let alone disease associations. This contrast is particularly stark when one considers cancer, as only a fraction of the total set of "cancer genes" have known and well-defined functions with this proportion diminishing considerably when considering cancer-associated lncRNAs, albeit progression is being made (Elliott and Larsson 2021). With this, what has been shown is that lncRNAs possess a remarkable ability to influence disease across multiple levels of biological complexity through their influence over a diverse array of molecular components.

In light of the above, it is important to consider that components of any biological system have functional interdependence with one another which results in the formation of a highly complex interaction network, sometimes referred to as an "interactome" (Vidal et al. 2011). With this, the introduction of network theory to biology has provided a conceptual framework from which to describe, characterise, probe, compare, and understand these complex interaction networks. Furthermore, this framework is highly versatile, facilitating a multi-omics approach with the incorporation of prior knowledge further increasing the quality of networks, thereby improving the validity of any inference made based on the analysis of the network. The inclusion of prior knowledge in network biology is increasingly relevant due to its application to disease science (Menche et al. 2015).

It has been observed that networks within biological systems operate in a structured manner, guided by a fundamental set of organising principles (Barabási et al. 2011). Examining disease pathogenesis through the lens of network theory has aided our understanding of the cellular and molecular components that contribute towards disease progression. Much is known regarding disease genes and the methods by which they enhance pathogenesis in each context; yet only a fraction of genes in the human genome are known to contribute towards disease directly (Claussnitzer et al. 2020). As a result, the association between much of the genome—coding and non-coding—and pathology is yet to be uncovered; however, there is clear evidence to suggest that the application of network science is a suitable means by which to tackle this monumental issue as it is capable of not only identifying associations but prioritising network components that are most likely to be contributing towards an aberrant phenotype.

2 Annotation and Classification of lncRNAs

Genetic annotation of both coding and non-coding genes is a hierarchical process: these annotations include a locus on a particular chromosome, within which there potentially exist several overlapping transcripts with themselves containing exons (Uszczynska-Ratajczak et al. 2018). Given the high evolutionary plasticity of lncRNA sequences, their annotation does not rely on sequence conservation but on evidence of physical transcription. For this purpose, the rise in popularity of RNA-seq has aided in the discovery of putative lncRNAs mapping to unannotated regions of the genome (Guttman et al. 2009). Despite there being no standardised criteria for lncRNA annotation and/or identification, most works adhere to a similar definition in which lncRNAs are operationally defined as RNAs longer than 200 nucleotides in length and lack an open reading frame within their sequence (Kopp and Mendell 2018).

As modern understanding of lncRNAs is far from comprehensive, classifying such molecules is a monumental task made more difficult by the fact that little is known about the relationship between lncRNA sequence and function. Often, sequence-function relationships are ideal for basic classification and commonplace for protein-coding genes (PCGs), but as these are mostly uninformative for lncRNAs, they are instead classed based on their proximity to PCGs. Major consortia groups like GENCODE are utilising this proximity model as the basis for a nomenclature for lncRNAs and in doing so are redefining the classical perspective of transcriptional loci in mammalian systems (Mattick et al. 2023).

It is now appreciated that transcripts, both coding and non-coding, exist within complex loci that produce a plethora of differing isoforms (Gerstein et al. 2007; Zerbino et al. 2020). With this, the proximity of lncRNAs to PCGs suggests similarities in transcriptional machinery and regulation. With some exceptions, most eukaryotic lncRNAs are produced through the activity of RNA polymerase II in addition to being capped at their 5′ end and spliced (Bunch 2018). It is important to consider that while many annotated lncRNAs exhibit polyadenylation, this tendency is most likely due to the polyA isolation methods used in modern RNA-seq library preparation methods (Chao et al. 2019). Similarities extend to the epigenome where lncRNAs, like mRNAs present similar promoter and gene body methylation enrichment (Guttman et al. 2009). Yet, the resemblance only extends so far, and much of the underlying lncRNA biogenesis mechanisms remain to be elucidated.

While there is no empirical classification for lncRNAs, three major subdivisions are commonly used to classify lncRNA based on their orientations with respect to their corresponding PCG, but it is generally seen that this description is an oversimplification (Jarroux et al. 2017). Original classifications of lncRNAs only included intergenic transcripts, but now, further research has revealed that lncRNAs can comprise overlapping, intronic, and natural antisense transcripts, among others (Laurent et al. 2015). A vast number of putative lncRNAs have been identified through high-throughput methods, and thus there has been a commensurate increase in the number of publicly available lncRNA catalogues/databases which, for the most part, can be subdivided into two main types: automatic and manually curated (Pinkney et al. 2020).

Automatic databases such as the MiTranscriptome or FANTOM databases benefit from much greater speed and reduced curation costs, but the annotations retained within are of comparably lower quality to lncRNAs held in manually curated databases (Ramilowski et al. 2020). GENCODE, RefSeq, and RefSeq Functional Elements are examples of manually curated databases that store lncRNA annotations. These require greater time, effort and cost to maintain but possess much higher-quality lncRNA annotations (Farrell et al. 2022). Importantly, annotations retained within the GENCODE database have been integrated into Ensembl, thereby facilitating comparisons, and thus large consortia-driven projects have primarily relied on the GENCODE database.

The results from the Matched Annotation from the NCBI and EMBL-EBI (MANE) collaboration highlight how close we are to achieving one of the most important milestones of the HGP, a fully annotated human genome (Morales et al. 2022). As we continue to unravel the complex relationship between the coding and non-coding genomes, we build upon this established framework to elucidate the roles of genomic loci in the context of disease.

3 lncRNAs and Cancer

Over the past decade, the cast of molecular players involved in the onset and progression of disease has grown substantially in recent years. Notably, within cancer biology, this evolution has been shaped by the inclusion of non-coding RNAs and the evidence emphasising the crucial roles played by lncRNAs. In many cases, the expression of the lncRNA alone provides insight into its potential as a therapeutic biomarker. For example, a multiphase study from Guo and colleagues highlighted lncRNA-GC1 as an aid in early detection and disease trajectory for patients with gastric cancers (Guo et al. 2020). In more straightforward investigations, researchers have gathered expression data from disease and control samples. Using differential expression analysis alone, they are able to identify a large number of lncRNAs that have a potential as therapeutic or prognostic biomarkers (Iaccarino and Klapper 2021). However, simple computational definitions of significance are underwhelming and can lead to a large number of potential candidates, many of which lack clinical value; therefore, it is crucial to employ functional investigations to validate potential candidates prior to any form of clinical investigation.

Among the limited subset of lncRNAs with established functions, the majority have been experimentally validated within the context of cancer. A notable example is *Malat1*, initially identified as a prognostic marker for lung cancers due to its expression, and recent studies have outlined its broader tumour intrinsic and extrinsic functions (Ji et al. 2003). In models of bladder cancer, the dynamic expression patterns of *Malat1* were found to be important in determining its invasive capacity (Zhu et al. 2023), while in models of breast cancer, *Malat1* was found to suppress T-cell-mediated anti-tumour immunity (Kumar et al. 2024). Similarly, the cancer immunogenic lncRNA *LIMIT* was found by Li and colleagues to enhance anti-tumour immune programmes through a signalling axis leading to greater expression of MHC-I in tumour cells and antigen-presenting cells (Li et al. 2021).

The influence of lncRNAs is often indirect, and it is common to see their actions influencing cancer pathogenesis through a signalling axis or target gene/protein. For example, PSPC1 (paraspeckle component 1) has been reported as a master regulator of EMT-related transcription factors (Yeh et al. 2018), and its expression has been described as a negative prognostic marker for multiple different cancer types. By reducing expression and stability of PSPC1, Zhan and colleagues found that the lncRNA *LOC105369504* is capable of reducing colorectal cancer cell pathogenesis (Zhan et al. 2023). The influence of lncRNAs has been known to significantly shape the tumour-suppressive effects of the p53 signalling pathway (Zhang et al. 2023a). For example, *TLIR* was found to bind and supress p53-mediated apoptosis in lung cancer cell lines, and while this highlights *TLIR* as a negative regulator of p53, the question remains as to its importance in cancers where p53 has obtained an oncogenic mutation (Iwai et al. 2022). In their proposed model, Zhang and colleagues suggest that the lncRNA *FTO-IT1* is capable of promoting prostate cancer progression by reducing the stability of specific p53 target genes, thereby perturbing the p53 pathways tumour-suppressive influence (Zhang et al. 2023a).

Several lncRNAs have been linked to chemotherapeutic response in different cancers. LncRNA *ZFN674-AS1* was found to be elevated in neuroblastoma patients with limited response to chemotherapy (Zhao et al. 2024). In colorectal cancers, CACClnc has been shown to function as a predictive biomarker as its expression in exosmal plasma was found higher in patients less susceptible to common chemotherapeutics (Zhang et al. 2023b). By sponging ATG3 through *miR-622*, the lncRNA *SP100-AS1* was found by Zhou and colleagues to be upregulated in radioresistant patients with CRC (Zhou et al. 2023).

The lncRNA *HOTAIR* is often found dysregulated in cancer, and in breast cancers, Gupta et al. noted that by associating with the EZH2 component of polycomb repressive complex 2, *HOTAIR* alters chromatin structure through methylation resulting in an perturbed transcriptional profile which facilitates a more invasive phenotype (Zhang et al. 2014b; Gupta et al. 2010). More recent reports have noted heightened levels of *HOTAIR* enhance chemotherapeutic resistance, metastasis and poor prognosis in hormone receptor-positive breast cancers with many suggesting it has utility as a diagnostic or predictive biomarker for multiple cancers (Mozdarani et al. 2020).

While certain lncRNAs have proven consistently insightful, advances and progressive functional annotation of novel lncRNAs are continually reshaping our understanding of the intricate relationship between the non-coding genome and cancer biology. This ongoing progress not only expands our understanding of cancer but highlights the potential for the future translation of research towards the clinic.

4 lncRNAs and Databases

The growing interest in non-coding biology coupled with advances and improved accessibility to NGS technologies has expanded the catalogue of annotated non-coding RNAs and their roles in biological processes. Numerous non-coding RNA databases have emerged in recent years aiming to compile and curate the vast array of published information and to act as an important resource for the wider scientific community (Ma and Zhang 2023). The databases themselves are often developed by various labs and collaborative efforts, leading to distinct purposes and varying criteria for inclusion.

GENCODE, a common reference base for genes and transcripts, has manually curated information on around 19,000 human and 14,000 mouse lncRNAs (Frankish et al. 2023). Similarly broad but with greater flexibility, the FANTOM project has aimed to provide functional characterisation of the mammalian genome and has provided functional annotations to many of the lncRNAs catalogued by GENCODE (Ramilowski et al. 2020). The value in such databases is restricted by their limited accessibility to the average lab. As a result, numerous publications have consolidated and condensed information into digestible units which are more useful for the typical research group.

Through an exhaustive review of over 25,000 publications, Zhou et al. have recently released EVLncRNA version 3.0 (https://www.sdklab-biophysics-dzu.net/EVLncRNAs3/), a manually curated database comprising information on 6195 experimentally validated lncRNAs and their roles in over 1700 diseases across 162 species (Zhou et al. 2024). Databases like EVLncRNA maintain the comprehensive nature of GENCODE while providing additional information, such as disease associations and interacting partners, thereby making it a more versatile resource for the scientific community. Similarly on its third iteration, the LncRNADisease v3.0 database (http://www.rnanut.net/lncrnadisease) hosted by the Cui lab holds curated information on over 13,000 lncRNA-disease associations between 6066 lncRNAs and 484 diseases (Lin et al. 2024). By retaining information on the response to drug treatment and development of resistance, the ncRNADrug database acts as an ideal clinical resource (Cao et al. 2024). Containing information on 9195 non-coding RNAs, of which 4145 are lncRNAs, the ncRNADrug database is well suited to address questions surrounding the roles of lncRNAs following drug treatment.

In general, the databases noted above are broad and applicable to the entire fields, but there is an emerging trend of databases/resources that have been developed with a narrower focus. For example, the lncR2metasta database is more restricted in that it comprises manually curated, experimentally validated lncRNAs which have been retained on account of their association with a particular metastasis-associated event (Zhang et al. 2021). BCLncRDB (http://sls.uohyd.ac.in/new/bclncrdb_v) is a manually curated database of 696 unique lncRNAs associated with breast cancer (Kumar et al. 2023). These databases may have value to some groups; however, broader databases, such as Lnc2Cancer or the lncRNA Modulator Atlas in Pan-Cancer database which both comprise information on lncRNAs across many cancers, are likely to be more valuable (Li et al. 2018b; Gao et al. 2021).

This leads to a wider perspective on the increasing volume of published lncRNA databases. While it is crucial to recognise that while certain databases hold a value on account of their disease-, species- or treatment-specific information, they often contain a large amount of overlapping information. This raises a question of whether a concerted effort is needed to ensure the most efficient use of a lncRNA database as a scientific resource. Despite these concerns, the expanding complexity and landscape of these databases are testament to our ever-growing interest in lncRNAs and our goal of understanding their roles in disease.

5 Principles of Network Theory and Its Application to lncRNA Biology

Historically, reductionism has been the dominant approach for biological, biomedical and medical researchers and has provided invaluable clarity relating to the roles and functions of many cellular components (Kienle and Kiene 2012; Heng 2008). Such an approach is embodied by the decades-old central tenet of cancer research with the somatic mutation theory (SMT) which attributes cancer as a direct result of driver mutations causally implicated in tumorigenesis (Stratton et al. 2009; Montévil and Pocheville 2017). The consequences of this reductionist approach to cancer research are obvious when examining the therapeutic landscape, in which drugs targeting single molecular markers dominate treatment regimens. However ever-mounting reports of resistance, underwhelming approval rates following clinical evaluation and translational failures, lead many to argue that the reductionist approach is unsustainable.

There is clear evidence to suggest that the interaction network in cells, as it relates to lncRNAs, is extremely large and our current understanding of its role is limited by the methods by which we study and interpret it. Considering the importance of the lncRNA "interactome", network science provides a means from which to study and analyse such a complex system. For network biology, a standard nomenclature is adopted from mathematics, and these basic terms and concepts, as they related to lncRNA interaction networks, will be discussed below.

6 Network Nomenclature and Basic Concepts

In simple terms, a network is a group of points connected by lines. The technical terms for these components are "nodes" and "edges", respectively, where a node may represent anything from a gene to a webpage, and the edge connecting two nodes represents the relationship between them, be it genetic similarity or a hyperlink. As we currently exist in an explosive era of social media, a good case for explaining the use of networks, to simplify and visualise a given system, is a social network. Individuals serve as nodes, and the relationships between each individual, familial, and romantic friendship are represented by the edges. Similarly, using X as an example, one person's profile acts as a node, and each connection they make with other profiles, i.e. the "following" of another profile, are the edges. Clustering within these networks can reveal small communities, and the overlaps between them, as well as the concept of "centrality" where one profiles influence, i.e. the size of their following or celebrity, can be calculated using algorithms and statistics to assign a score of importance to each node within the network (Samanta et al. 2021). Centrality is a quantifiable measure of node and edge significance which allows for the identification of central nodes and therefore critical areas within a system. There are various different measures of centrality, but this specific type is known as "degree" centrality where the degree of a given profile is the number of edges, i.e. followers they have attached. A node with significantly higher degree than other nodes within a system is referred to as "hub" node (Samanta et al. 2021).

At its most basic, a network can be viewed as a catalogue of components consisting of *nodes* (or vertices) and their connections between one another, called *edges* (or *links*). In most biological networks, interactions or regulation are directional, and thus the corresponding network reflects this by providing directionality to its *edges*. A key property attributable to all nodes is *degree*, which is defined by the number of edges adjacent to a particular node. In directed networks, the *indegree* and *outdegree* are distinguished with the former referring to the number of edges, incident from a node, while the outdegree is the number of edges incident to the node. In a sense, *degree* is a descriptive characteristic in that the more connections an individual node possesses, the greater its influence, and therefore, it is important. More broadly, networks can be generally described through their *average degree* which provides a metric from which to judge the connectivity of a given system (Fig. 1).

Given the characteristics described above, networks have an architecture governed by a few simple principles. The work of Erdös and Rényi proposed a model for a random network which assumes a fixed number of nodes with pairwise connections occurring at a fixed probability (Erdös and Rényi 1960). Networks such as these were simple, displaying uniform characteristics with degree distributions following a Poisson model, i.e. exhibiting a positively skewed shape, as nodes with an overly high or low number of edges were exceedingly rare. Although elegant, the model lacks topological relevance to most real-world networks, and through modelling investigations, it was revealed that degree distribution of real-world networks

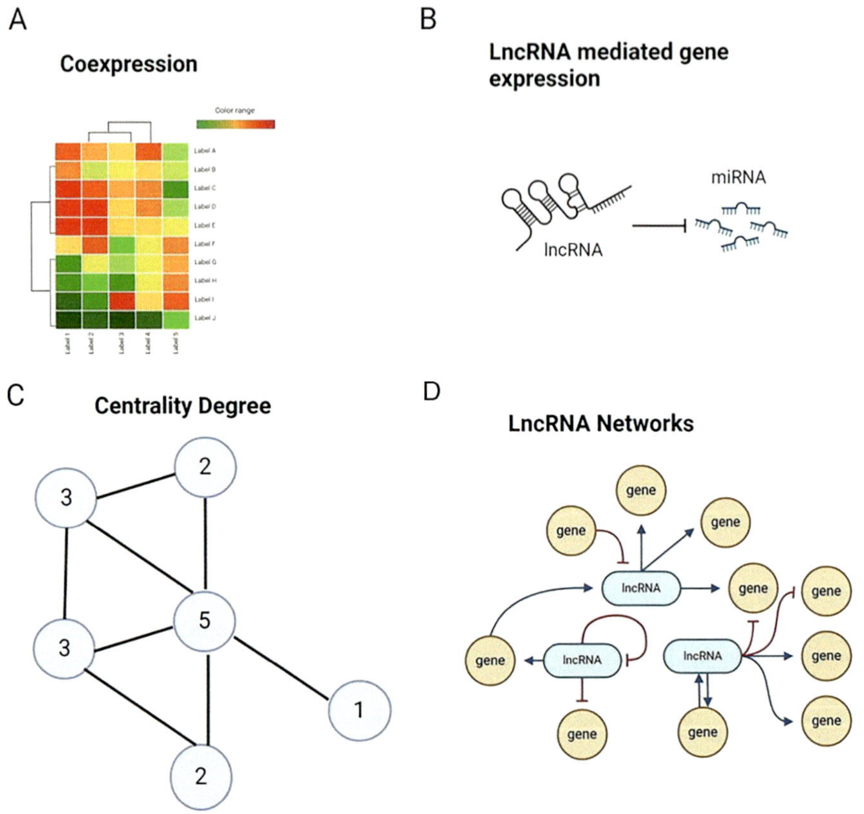

Fig. 1 How network science powers lncRNA research. (**a**) LncRNA and mRNA co-expression can be used for the extraction of important biology. (**b**) Long non-coding RNAs (lncRNAs) mediate gene expression through various mechanisms, despite not encoding proteins themselves. (**c**) At its simplest form, a network can be conceptualised as a collection of components, comprising nodes (or vertices), and the connections between them, referred to as edges (or vertices). In many biological networks, interactions or regulations are directional, and as a result, the network incorporates this by assigning directionality to its edges. The degree of centrality indicates the number of edges (represented by black lines) connected to a vertex (dots). The numerical value within each dot denotes the centrality degree of the corresponding vertex. (**d**) Schematic of potential interactions in a lncRNA-mRNA network (Created with BioRender.com)

follows a power-law distribution rather than a Poisson distribution. Unlike the regularity of the Poisson distribution, the updated scale-free models were much less uniform with most nodes possessing a small number of edges with only a select few possessing an exceeding high degree. These nodes with the highest degree, often defined as *hubs* or *authorities ceteris paribus*, have the greatest "importance" within their network.

The use of a network approach in identifying the underlying mechanisms of non-Mendelian diseases, where a mix of genetic mutations and environmental factors come into play, has shown a lot of potential thus far in that RNA sequencing techniques often produce data with low signal-to-noise ratios, and the use of networks seeks to correct this using "smoothing" techniques. Biological networks provide a means of reducing an otherwise extremely complex system into something simpler and more interpretable, with the added benefit of allowing for the visualisation of discernible patterns, where they exist. The reduction of these systems often means a lot of information is excluded, leaving only a basic connection between two entities to be interpreted, a feature that can be both a strength and a weakness of network approaches (Zhang et al. 2014a).

Advances in molecular profiling technologies have allowed for the curation of biological data from various levels with the drawback that differing profiling platforms result in heterogeneous datasets. The efficient combining of such datasets to catalogue different types of molecular interactions such as PPIs and biological pathways has presented a new computational challenge that may be solved using biological networks for integrating "omics" data with prior knowledge (Zhang et al. 2014a). However, lncRNA network analyses present several other limitations. Namely, lncRNA networks represent a snapshot in time of a biological process, an inaccurate or incomplete depiction of the overall picture. The expression of lncRNA genes is constantly up- and downregulated during disease progression, and therefore it is vital to consider temporal differences and disruptions in lncRNA-mRNA interactions when deciphering a given disease pathway.

A network algorithm is a computational method or procedure designed to analyse, manipulate or optimise networks, which are mathematical representations of interconnected objects or entities. These algorithms are used across various disciplines, including computer science, biology, sociology and engineering, to solve problems related to network structure, connectivity, dynamics and function. There are many types of network algorithms, each tailored to address specific tasks or challenges within the context of network analysis.

The PageRank algorithm is a well-established global standard for ranking nodes within a network based on their "popularity". Rank scores for each node are initially distributed equally, and then scores are altered as nodal interactions take place. "PageRank with Priors" describes the incorporation of "prior bias" into the algorithm to define the relative importance of each node. The "Hyperlink-Induced Topic Search (HITS) algorithm" also known as hubs and authorities works differently in determining importance, in that two different scores are assigned to each hub' also its authority score and its hub score. A node's authority is a calculation of how many other hubs are connected to it, whereas its hub score is indicative how many nodes it links to. HITS is commonly used for calculating the impact factor of scientific journals, where a journal with a high impact factor will have both high authority and hub value, i.e. the journal will have a high impact content (authority) and will be cited more frequently in other journals (hub) (Chen et al. 2007). HITS with Priors works similarly to PageRank with Priors in that there is the addition of prior knowledge or prior bias—towards nodes of known importance, before the algorithm is applied as well as the stochastic lengths of K. The PageRank and HITS algorithms without priors both have fixed lengths of K, where K is defined as a walk and a walk is defined as a path if no nodes are repeated in each sequence within a network. The

K-step Markov approach differs from both PageRank with Priors and HITS with Priors in that it calculates the relative probability that a system will spend time at any specific node given that it starts at a particular hub node and ends after K number of steps. Markov node clustering techniques can also be used to rank the importance of each node based on the sum of the node edge-weights within a cluster relative to the edge-weights of all nodes within a network. This method of clustering works by simulating stochastic flow within the network, and, in a biological network, it can identify clusters of functionally related genes. Another approach for determining node importance is the iterative ranking (IR) algorithm whereby each node is assigned an initial score which is then normalised according to the scores of surrounding nodes. These scores are updated after each sequential iteration. Naïve Bayesian protein networks operate by assigning each gene, protein or node with a likelihood score based on the probability that a given gene or protein is associated with a specific biological pathway. These scores are then combined to produce a weighted Bayesian estimate that the genes or proteins are linked to the same biological process (Lee et al. 2004, 2011).

7 Networks and Disease

Untangling the complex interplay between cellular networks and disease is essential in unravelling and exposing potential targets and biomarkers of clinical relevance (Menche et al. 2015). Mapping such disease-relevant interactions is conducted through network science, simplifying the complex disease interactome into tangible components such as a protein-protein interaction or gene regulatory network (Halu et al. 2019). The implication is that the interconnectedness of biological interaction networks extends the impact of a single mutation or abnormality beyond the activity of the affected component, potentially influencing all interaction partners. Importantly, such disease proteins or genes are not scattered throughout a network but cluster into subnetworks forming disease "modules". This concept extends as overlaps between disease modules suggest that perturbations related to disease A will share characteristics with disease B, resulting in similar clinicopathological characteristics, thereby exposing potential therapeutic targets. Molecular networks have a high degree of modularity, and functional similarities culminate in network modules containing nodes relating to the same biological function (Choobdar et al. 2019). Based on this concept, linkage, disease module and diffusion-based methods can be used to identify and predict candidate disease genes/proteins (Barabási et al. 2011). Although more modern disease-candidate prioritisation uses more complex methods, integrating multiple data sources including disease phenotype, known disease-associated lncRNA, and disease gene/protein data, as in the case of LncPriCNet, to better identify candidate disease nodes (Yao et al. 2017).

Differential gene expression data can be used to derive marker genes which indicate various human diseases as well as their up- or downregulation during disease progression or following treatment (Fig. 1). This data can also be aligned to highlight differences between wild-type and mutated genes, control and disease genes and differences between disease subtypes, commonly using a transcriptomic approach which allows thousands of genes to be analysed simultaneously. A few issues arise in the generation of such data due to its large volume, gene mapping, storage and retrieval, to name a few [35]. When it comes to DEGs, it is extremely important to bear in mind that small changes in expression can have significant implications on disease mechanisms, ignoring this fact can lead to largely incorrect interpretations of gene networks. It is also worthy of note that current network tools are usually only capable of constructing an interpretable visualisation up to hundreds or a couple of thousand nodes, a limitation that requires a significant amount of scaling down and filtering out of gene expression data by certain biological features, with the risk of losing important information. This converts the differential gene expression data into a more easily interpreted object while allowing for its visualisation using network software. Networks based on upregulated gene expression within disease samples can be compared to a network of downregulated genes which, in theory, could just as likely possess a significant biological impact such as regulation of apoptosis or proliferation in cancer, for example (Jean-Quartier et al. 2020).

Differential expression analysis (DEA) allows for the comparison of differences as well as similarities among datasets of selected disease types which may highlight unique characteristics of disease subtypes and novel molecular targets for treatment. Hub genes with a high degree are often helpful in evaluating key signalling points within a sample which can be of great value in understanding the molecular mechanisms of disease. However, well-studied genes and proteins are likely to have an increased number of interactions in comparison to uncharacterised proteins, and therefore a low degree is not always indicative of a low number of interactions; these outliers could still hold valuable and relevant biological functions within a disease pathway. Hub nodes within clusters in a network may also point to protein complexes since multiple proteins function within interacting aggregates. These interactions represent biological functions and can be indicative of signalling steps in disease progression (Jean-Quartier et al. 2020).

8 lncRNA-mRNA Differential Co-expression Network Analysis

For lncRNA, numerous computational methods have been developed to identify disease-associated lncRNAs. LncDisease developed by Cui and colleagues is a sequence-based tool which predicts lncRNA-disease associations based on lncRNA-miRNA interactions enriched by the use of miRNA-disease-associated datasets based on the TAM tool (Lu et al. 2010; Wang et al. 2016). However, a more popular approach is to use lncRNA-mRNA co-expression methods which have proven useful in studies of cancer (Xing et al. 2019). In co-expression networks, correlation coefficients (Pearson or Spearman) are obtained between pairs of differentially expressed lncRNAs and mRNAs from which correlation partners, above a defined threshold, are used to construct the network. These methods are effective but describe static environments; therefore, more modern investigations utilise differential interaction networks where the strongest interactions are those that are altered due to the biological condition or perturbation. Importantly, this allows for a direct investigation of the consequences of perturbations and thus is highly useful in the study of cancer (He et al. 2019).

For instance, Guo et al. generated a lncRNA-mRNA differential co-expression network analysis based on expression data from early- and late-stage ovarian cancer samples which produced a cancer lncRNA-mRNA network consisting of 295 lncRNAs and 2366 mRNAs in its major component (most connected cluster) (Guo et al. 2015). The authors further analysed this subcluster based on topological properties such as degree and centrality which revealed two major hub lncRNAs significantly associated with patient outcome. Functional analysis revealed both lncRNAs, *RP11-284N8.3.1* and *AC104699.1.1*, have protective characteristics and roles in regulating immune and cell cycle processes. This study, using a differential co-expression network, was able to identify specific, clinically relevant lncRNAs and define their function. With this, the potential for network science is apparent in cancer research, and thus there is a necessity to develop functional network analysis tools to accomplish the types of analysis above in a simple and intuitive manner (Zhang et al. 2014a).

9 LncRNA-Centric Network Analysis Tools

The inference of lncRNA-mRNA interactions assumed in the construction of co-expression networks is important in developing an accurate analysis tool. Simple network construction is capable of using sequence-based complementary base pairing and is used by many of the prediction algorithms which aim to identify lncRNA targets; however, lncRNAs are long and known to possess complex tertiary structures as seen with *MALAT1*, and thus these methods may not be comprehensive (Zhang et al. 2019). Alternatively, networks can be constructed based on expression data which, as mentioned before, assume interaction based on lncRNA-mRNA expression correlation as determined by a statistical equation (Zhang et al. 2019). But as interactions are determined by statistics alone, they are likely to contain numerous false positives. However, inferred interactions from both expression and prediction algorithms can be confirmed using experimentally validated lncRNA databases such as LIVE or LongHorn (An et al. 2019; Chiu et al. 2018). Alternatively, networks can be constructed solely from experimentally validated interaction data, thereby allowing for the creation of highly reliable networks leading to more accurate inferences.

As analyses can involve multiple networks, such as the case during a time-course analysis, methods are required which compare created networks using appropriate statistical tests. Several methods have been developed to determine where correlation networks are statistically different across conditions. The CoDiNA and DiffCorr packages evaluate node-pair correlations to determine if they differ between network conditions, whereas the DINGO and CoGA methods conduct a comparison based on a set of predetermined attributes. More recently BioNetStat was developed to conduct comparative analysis of two or more differential co-expression networks and is based on the CoGA method (Jardim et al. 2019).

The weighted gene co-expression network analysis (WGCNA) provides insights into lncRNA-mRNA networks. This method is limited as despite its popularity, sample size and data input requirements often present severe limitations (Langfelder and Horvath 2008). Although alternative methods are capable of conducting similar analyses to WGCNA, they are often substantially more technical and less popular meaning there is often less support for their use (Petereit et al. 2016). Unlike many available tools, the WGCNA package is provided with an extremely in-depth tutorial, but it could still be considered computationally intensive, time-consuming and not readily accessible to all end users. However, alternative web-based methods such as webCEMiTool implement WGCNA, thereby overcoming potential user technical limitations (Cardozo et al. 2019).

In stark contrast to the network-based methods for miRNA-centric analysis, there is a paucity of tools available for the analysis and interpretation of lncRNA-mRNA networks and perhaps due to the complex regulatory action of lncRNAs. The canonical function of miRNAs—their upregulation leads to a downregulation of their interaction partners—lends itself to the creation of simple tools such as miRNet or miRmapper which are capable of exploiting this function for analysis (Chang et al. 2020; Da Silveira et al. 2018).

Some lncRNA-centric network analysis methods do exist. For example, the CLING method (http://bio-bigdata.hrbmu.edu.cn/cling/), an integrative framework for the extraction of candidate lncRNAs through the analysis of multiple networks developed by Zhang and colleagues, is one such method applicable in this context (Zhang et al. 2020). However, the tool is utilised pre-created networks from which a user-input list of lncRNAs is compared against and prioritised. This limits the flexibility of the tool and could severely restrict results as CLING, at present, only makes use of nine different reference networks for prioritisation.

The NetConfer tool similarly works through a web-server (https://web.rniapps.net/netconfer) and facilitates the analysis of multiple user-provided networks (Nagpal et al. 2020). Although not lncRNA-specific, the tool takes simple network inputs—a two-column data frame comprising nodes and their targets—and uses this information to create networks which can then be analysed. This tool is highly useful but limited in one major aspect, i.e. it requires more than one network and thus is not suitable in studies where single networks require analysis.

The GDCRNATools R Bioconductor package provides access to the RNA-seq data retained within the Genomic Data Commons (GDC) portal and places an emphasis on the analysis of this data for the creation and interpretation of competing endogenous RNA networks (e.g. lncRNA-miRNA-mRNA networks). This method uses cancer-based RNA-seq data that is extremely valuable when investigating lncRNAs in the context of this disease. However the method provides no means as to integrate user-generated data, and it's extremely memory intensive and, in some respects, quite technical (Li et al. 2018a). Moreover, to actually conduct in-depth analysis of the networks generated by the package, the user is required to use the Cytoscape tool, which again adds to the technical limitations of the tool (Shannon et al. 2003).

10 Conclusions

Although initially considered as "junk" or "noise" in the genome, recent research has unveiled the diverse functions and importance of lncRNAs in regulating cellular processes and contributing to human health and disease. They are involved in various biological processes, including development, differentiation, immune response and cancer progression. Due to their regulatory roles, dysregulation of lncRNAs has been implicated in numerous diseases, making them potential targets for therapeutic intervention and diagnostic biomarkers. Prediction tools and network biology play a crucial role in understanding the complex interactions involving long non-coding RNAs (lncRNAs). While exploratory studies have hinted at the vast interactome associated with lncRNAs, functional validation lags, leaving many lncRNAs with ambiguous roles. Databases compile interaction data from various sources, offering insight into potential lncRNA interactions. Despite advances in these databases, the accuracy and completeness of interaction information remain challenging due to limited functional validation. Prediction tools serve as a valuable resource in the absence of experimental evidence, albeit with varying levels of accuracy. Despite the progress made in prediction methodologies, the complexity of the lncRNA interactome presents ongoing challenges in accurately predicting and understanding lncRNA functions and interactions, highlighting the need for further advancements in predictive algorithms and experimental validation techniques.

References

An G, Sun J, Ren C et al (2019) LIVE: a manually curated encyclopedia of experimentally validated interactions of lncRNAs. Database 2019

Barabási A-L, Gulbahce N, Loscalzo J (2011) Network medicine: a network-based approach to human disease. Nat Rev Genet 12:56–68

Bunch H (2018) Gene regulation of mammalian long non-coding RNA. Mol Genet Genomics 293: 1–15

Cao X, Zhou X, Hou F et al (2024) ncRNADrug: a database for validated and predicted ncRNAs associated with drug resistance and targeted by drugs. Nucleic Acids Res 52:D1393–D1399

Cardozo LE, Russo PS, Gomes-Correia B et al (2019) webCEMiTool: co-expression modular analysis made easy. Front Genet 10:146

Chang L, Zhou G, Soufan O et al (2020) miRNet 2.0: network-based visual analytics for miRNA functional analysis and systems biology. Nucleic Acids Res 48:W244–W251

Chao H-P, Chen Y, Takata Y et al (2019) Systematic evaluation of RNA-Seq preparation protocol performance. BMC Genomics 20:571

Chen P, Xie H, Maslov S et al (2007) Finding scientific gems with Google's PageRank algorithm. J Informetrics 1:8–15

Chiu H-S, Somvanshi S, Patel E et al (2018) Pan-cancer analysis of lncRNA regulation supports their targeting of cancer genes in each tumor context. Cell Rep 23:297–312. e212

Choobdar S, Ahsen ME, Crawford J et al (2019) Assessment of network module identification across complex diseases. Nat Methods 16:843–852

Claussnitzer M, Cho JH, Collins R et al (2020) A brief history of human disease genetics. Nature 577:179–189

Comings DE (1972) The structure and function of chromatin. In: Advances in human genetics. Springer, pp 237–431

Da Silveira WA, Renaud L, Simpson J et al (2018) miRmapper: a tool for interpretation of miRNA–mRNA interaction networks. Genes 9(9):458

Elliott K, Larsson E (2021) Non-coding driver mutations in human cancer. Nat Rev Cancer 21:500–509

Erdös P, Rényi A (1960) On the evolution of random graphs, vol 5. Publ Math Inst Hung Acad Science, pp 17–61

Farrell CM, Goldfarb T, Rangwala SH et al (2022) RefSeq functional elements as experimentally assayed nongenic reference standards and functional interactions in human and mouse. Genome Res 32:175–188

Frankish A, Carbonell-Sala S, Diekhans M et al (2023) GENCODE: reference annotation for the human and mouse genomes in 2023. Nucleic Acids Res 51:D942–D949

Gao Y, Shang S, Guo S et al (2021) Lnc2Cancer 3.0: an updated resource for experimentally supported lncRNA/circRNA cancer associations and web tools based on RNA-seq and scRNA-seq data. Nucleic Acids Res 49:D1251–D1258

Gerstein MB, Bruce C, Rozowsky JS et al (2007) What is a gene, post-ENCODE? History and updated definition. Genome Res 17:669–681

Guo Q, Cheng Y, Liang T et al (2015) Comprehensive analysis of lncRNA-mRNA co-expression patterns identifies immune-associated lncRNA biomarkers in ovarian cancer malignant progression. Sci Rep 5:17683

Guo X, Lv X, Ru Y et al (2020) Circulating exosomal gastric cancer–associated long noncoding RNA1 as a biomarker for early detection and monitoring progression of gastric cancer: a multiphase study. JAMA Surg 155:572–579

Gupta RA, Shah N, Wang KC et al (2010) Long non-coding RNA HOTAIR reprograms chromatin state to promote cancer metastasis. Nature 464:1071–1076

Guttman M, Amit I, Garber M et al (2009) Chromatin signature reveals over a thousand highly conserved large non-coding RNAs in mammals. Nature 458:223–227

Halu A, De Domenico M, Arenas A et al (2019) The multiplex network of human diseases. NPJ Syst Biol Appl 5:1–12

Hangauer MJ, Vaughn IW, McManus MT (2013) Pervasive transcription of the human genome produces thousands of previously unidentified long intergenic noncoding RNAs. PLoS Genet 9: e1003569

He R-Z, Luo D-X, Mo Y-Y (2019) Emerging roles of lncRNAs in the post-transcriptional regulation in cancer. Genes Dis 6:6–15

Heng HH (2008) The conflict between complex systems and reductionism. JAMA 300:1580–1581

Iaccarino I, Klapper W (2021) LncRNA as cancer biomarkers. Methods Mol Biol 2348:27–41

Jardim VC, Santos SS, Fujita A et al (2019) BioNetStat: a tool for biological networks differential analysis. Front Genet 10:594

Jarroux J, Morillon A, Pinskaya M (2017) History, discovery, and classification of lncRNAs. In: Long non coding RNA biology. Springer, pp 1–46

Jean-Quartier C, Jeanquartier F, Holzinger A (2020) Open data for differential network analysis in glioma. Int J Mol Sci 21:547

Ji P, Diederichs S, Wang W et al (2003) MALAT-1, a novel noncoding RNA, and thymosin β4 predict metastasis and survival in early-stage non-small cell lung cancer. Oncogene 22:8031–8041

Kienle G, Kiene H (2012) From reductionism to holism: systems-oriented approaches in cancer research. Glob Adv Health Med 1:68–77

Kopp F, Mendell JT (2018) Functional classification and experimental dissection of long noncoding RNAs. Cell 172:393–407

Kumar S, Agrawal A, Vindal V (2023) BCLncRDB: a comprehensive database of LncRNAs associated with breast cancer. Funct Integr Genomics 23:178

Kumar D, Gurrapu S, Wang Y et al (2024) LncRNA Malat1 suppresses pyroptosis and T cell-mediated killing of incipient metastatic cells. Nat Cancer 5:262. https://doi.org/10.1038/s43018-023-00695-9

Lander ES, Linton LM, Birren B et al (2001) Initial sequencing and analysis of the human genome

Langfelder P, Horvath S (2008) WGCNA: an R package for weighted correlation network analysis. BMC Bioinformatics 9:559

Laurent GS, Wahlestedt C, Kapranov P (2015) The landscape of long noncoding RNA classification. Trends Genet 31:239–251

Lee I, Date SV, Adai AT et al (2004) A probabilistic functional network of yeast genes. Science 306:1555–1558

Lee I, Blom UM, Wang PI et al (2011) Prioritizing candidate disease genes by network-based boosting of genome-wide association data. Genome Res 21:1109–1121

Li R, Qu H, Wang S et al (2018a) GDCRNATools: an R/Bioconductor package for integrative analysis of lncRNA, miRNA and mRNA data in GDC. Bioinformatics 34:2515–2517

Li Y, Li L, Wang Z et al (2018b) LncMAP: pan-cancer atlas of long noncoding RNA-mediated transcriptional network perturbations. Nucleic Acids Res 46:1113–1123

Li G, Kryczek I, Nam J et al (2021) LIMIT is an immunogenic lncRNA in cancer immunity and immunotherapy. Nat Cell Biol 23:526–537

Lin X, Lu Y, Zhang C et al (2024) LncRNADisease v3. 0: an updated database of long non-coding RNA-associated diseases. Nucleic Acids Res 52:D1365–D1369

Lu M, Shi B, Wang J et al (2010) TAM: a method for enrichment and depletion analysis of a microRNA category in a list of microRNAs. BMC Bioinformatics 11:419

Ma L, Zhang Z (2023) The contribution of databases towards understanding the universe of long non-coding RNAs. Nat Rev Mol Cell Biol 24:601–602

Mattick JS, Amaral PP, Carninci P et al (2023) Long non-coding RNAs: definitions, functions, challenges and recommendations. Nat Rev Mol Cell Biol 24:430–447

McLysaght A, Guerzoni D (2015) New genes from non-coding sequence: the role of de novo protein-coding genes in eukaryotic evolutionary innovation. Philos Trans R Soc Lond Ser B Biol Sci 370:20140332

Menche J, Sharma A, Kitsak M et al (2015) Disease networks. Uncovering disease-disease relationships through the incomplete interactome. Science 347:1257601

Montévil M, Pocheville A (2017) The Hitchhiker's guide to the cancer galaxy. How two critics missed their destination. Organisms J Biol Sci 1:37–48

Morales J, Pujar S, Loveland JE et al (2022) A joint NCBI and EMBL-EBI transcript set for clinical genomics and research. Nature 604:310–315

Mozdarani H, Ezzatizadeh V, Rahbar Parvaneh R (2020) The emerging role of the long non-coding RNA HOTAIR in breast cancer development and treatment. J Transl Med 18:152

Nagpal S, Baksi KD, Kuntal BK et al (2020) NetConfer: a web application for comparative analysis of multiple biological networks. BMC Biol 18:53

Orgel LE, Crick FH (1980) Selfish DNA: the ultimate parasite. Nature 284:604–607

Petereit J, Smith S, Harris FC et al (2016) Petal: co-expression network modelling in R. BMC Syst Biol 10:181–195

Pinkney HR, Wright BM, Diermeier SD (2020) The lncRNA toolkit: databases and in silico tools for lncRNA analysis. Non-coding RNA 6:49

Ramilowski JA, Yip CW, Agrawal S et al (2020) Functional annotation of human long noncoding RNAs via molecular phenotyping. Genome Res 30:1060–1072

Samanta S, Dubey VK, Sarkar B (2021) Measure of influences in social networks. Appl Soft Comput 99:106858

Shannon P, Markiel A, Ozier O et al (2003) Cytoscape: a software environment for integrated models of biomolecular interaction networks. Genome Res 13:2498–2504

Stratton MR, Campbell PJ, Futreal PA (2009) The cancer genome. Nature 458:719–724

Uszczynska-Ratajczak B, Lagarde J, Frankish A et al (2018) Towards a complete map of the human long non-coding RNA transcriptome. Nat Rev Genet 19:535–548

Vidal M, Cusick ME, Barabási A-L (2011) Interactome networks and human disease. Cell 144: 986–998

Wang J, Ma R, Ma W et al (2016) LncDisease: a sequence based bioinformatics tool for predicting lncRNA-disease associations. Nucleic Acids Res 44:e90–e90

Xing W, Qi Z, Huang C et al (2019) Genome-wide identification of lncRNAs and mRNAs differentially expressed in non-functioning pituitary adenoma and construction of an lncRNA-mRNA co-expression network. Biol Open 8:bio037127

Yao Q, Wu L, Li J et al (2017) Global prioritizing disease candidate lncRNAs via a multi-level composite network. Sci Rep 7:39516

Yeh H-W, Hsu E-C, Lee S-S et al (2018) PSPC1 mediates TGF-β1 autocrine signalling and Smad2/3 target switching to promote EMT, stemness and metastasis. Nat Cell Biol 20:479–491

Zerbino DR, Frankish A, Flicek P (2020) Progress, challenges, and surprises in annotating the human genome. Annu Rev Genomics Hum Genet 21:55–79

Zhan T, Cheng X, Zhu Q et al (2023) LncRNA LOC105369504 inhibits tumor proliferation and metastasis in colorectal cancer by regulating PSPC1. Cell Death Discov 9:89

Zhang B, Tian Y, Zhang Z (2014a) Network biology in medicine and beyond. Circ Cardiovasc Genet 7:536–547

Zhang J, Zhang P, Wang L et al (2014b) Long non-coding RNA HOTAIR in carcinogenesis and metastasis. Acta Biochim Biophys Sin 46:1–5

Zhang J, Le TD, Liu L et al (2019) Inferring and analyzing module-specific lncRNA–mRNA causal regulatory networks in human cancer. Brief Bioinform 20:1403–1419

Zhang J, Gao Y, Wang P et al (2020) CLING: candidate cancer-related lncRNA prioritization via integrating multiple biological networks. Front Bioeng Biotechnol 8:138

Zhang S, He X, Zhang R et al (2021) LncR2metasta: a manually curated database for experimentally supported lncRNAs during various cancer metastatic events. Brief Bioinform 22:bbaa178

Zhang J, Wei J, Sun R et al (2023a) A lncRNA from the FTO locus acts as a suppressor of the m6A writer complex and p53 tumor suppression signaling. Mol Cell 83:2692–2708.e2697

Zhang X, Ma D, Xuan B et al (2023b) LncRNA CACClnc promotes chemoresistance of colorectal cancer by modulating alternative splicing of RAD51. Oncogene 42:1374–1391

Zhao K, Wang X, Jin Y et al (2024) LncRNA ZNF674-AS1 drives cell growth and inhibits cisplatin-induced pyroptosis via up-regulating CA9 in neuroblastoma. Cell Death Dis 15:5

Zhou Y, Shao Y, Hu W et al (2023) A novel long noncoding RNA SP100-AS1 induces radioresistance of colorectal cancer via sponging miR-622 and stabilizing ATG3. Cell Death Differ 30:111–124

Zhou B, Ji B, Shen C et al (2024) EVLncRNAs 3.0: an updated comprehensive database for manually curated functional long non-coding RNAs validated by low-throughput experiments. Nucleic Acids Res 52:D98–D106

Zhu N, Ahmed M, Li Y et al (2023) Long noncoding RNA MALAT1 is dynamically regulated in leader cells during collective cancer invasion. Proc Natl Acad Sci USA 120:e2305410120

Mesoscopic Patterns as Interfaces for Complex Systems Analysis

Marco Ledda and Marta Bertolaso

Contents

Abstract Complex systems science approach is characterized by the consideration that the phenomena under study are strongly interconnected entities, composed of many levels and subjected to reciprocal influences. Given that this topological arrangement of physical entities incorporates biological entities, we argue that the investigation of a given object, which belongs to and is itself a network in a given layer, intrinsically includes a lower level and an upper level with which is intertwined. The choice of the proper level of investigation will constitute the *interface level* where information about upper and lower levels is comprised, thus considering it as a correlation level.

Keywords Interface · Biocomplexity · Systems biology · Computational biology · Complex networks · Agent-based models · Biodynamics

M. Ledda (✉)
Physics Department, Università di Catania, Catania, Italy
e-mail: marco.ledda@dfa.unict.it

M. Bertolaso
Research Unit of Philosophy of Science and Human Development, Università Campus Bio-Medico di Roma, Rome, Italy
e-mail: m.bertolaso@unicampus.it

J. Barciszewski (ed.), *Systems Biology II*, RNA Technologies 15,
https://doi.org/10.1007/978-3-031-62178-9_8

1 Introduction

While it is commonly believed that the behavior of an entire system can be derived from the local behavior of its individual components through some dynamical laws, complexity science reveals that this is not always the case. Systems consisting of a large, but not excessively large (Weaver 1948), number of strongly interacting elements exhibit unique *emergent properties* (Artime and De Domenico 2022). These properties are referred to as emergent since they arise from the collective behavior of individual entities at a lower level. Such a hierarchy can be structured by various factors, including ontological, causal, organizational, and regulatory elements across different scales. Biocomplexity research has demonstrated that simple local rules can give rise to complex phenomena at higher levels of organization. For instance, the movement of a single cell can be triggered or inhibited by on/off or gradient signals, gene transcription can be regulated by the presence or absence of a specific transcription factor, and simple foraging habits can lead to complex trophic networks in ecology. Living organisms can be viewed as dynamic collections of matter, energy, and information. Systems biology, which encompasses fields such as computational biology and bioinformatics, investigates how organisms organize matter and utilize energy through the interpretation of information. This involves breaking down these entities into individual components to study their specific features before reconstructing their relationships with other entities (Kitano 2002; Bruggeman and Westerhoff 2007; Bandesh et al. 2018). As such, we believe that the systems biology methodology is well-suited for developing a systemic/relational epistemology geared toward studying complex biological systems (Alberghina and Westerhoff 2007; Boogerd et al. 2007).

Even more interesting in this light are the recent scientific advancements that, for example, combine models of cell fates as attractors of gene regulatory networks and emergence of tissue architecture from cell-cell interaction with an integrated approach (Gigante et al. 2023; Smart and Zilman 2023). Through theoretically consistent quantitative model of the emergence of complex biological organization at different scales, these evidences show the importance of the mesoscopic approach and of the interface notion in order to account for the interplay between the cell microscale and the macroscopic tissue organization in biological systems. Typically, when studying a biological phenomenon, researchers analyze correlations between different systems using statistical methods. This involves selecting a set of variables from both systems and observing whether changes in one set correlate with changes in the other. While this approach is effective for relatively simple systems with clear causal pathways, it lacks predictive efficiency and fails to adequately describe emergent properties in more complex systems that operate on different ontological levels. A new approach (Arora et al. 2020) is to not only consider the two systems but also the level at which different inputs from each system can be detected. This level is referred to as the *interface level* which emerges wherever measuring the state of one system intrinsically includes inputs from another system. This approach involves considering a more sophisticated causal and regulatory structure in which the interface level receives inputs from lower levels (efficient/mechanical causes) and higher levels (formal/information constraints) that allow for a range of possible interface states. This enables researchers to observe how the dynamics of the mesoscale interface level changes in response to complex upward and downward inputs, even in cases where information about lower-higher level associations is lacking (i.e., no correlations). Using tools from systems biology, such as networks, cellular automata, and agent-based computational models, researchers can observe how the microscale initial disposition of agents in space, which behave according to certain rules and the macroscale patterns representing the final disposition of agents in space, cannot fully inform us about mesoscale patterns (interface levels), which are dynamic aggregations of local rules and constraints (Takeuchi and Hogeweg 2012; An et al. 2009).

In this study, we delve into the implicit interplay of relationships among different ontological levels when investigating a specific aspect of reality. Section 2 introduces the notion of *mesoscale pattern* particularly in the context of networks and agent-based models. In Sect. 3, we put forth a novel epistemological perspective which posits, even in traditional scientific inquiry, that the subject of study is not an idealized entity but rather a complex system of interconnected relations. Additionally, we provide a preliminary outline of a formalization of this concept. Section 4 highlights our conclusions and sets the stage for future research directions.

2 Mesoscale Patterns

2.1 Networks and Cellular Automata (CA)

A complex network is a collection of interconnected nodes that follow a set of rules. These mathematical structures exhibit characteristics that correspond remarkably well to many real-world phenomena, such as social networks, ecological webs, virus spreading, and brain networks. Consequently, they have gained great relevance in biological studies. These networks provide a formal representation of biological relationships, such as metabolic pathways and cell-cell signaling, which can be analyzed using classical systems of ordinary or partial differential equations (ODEs/PDEs) to extract dynamics or through statistical invariants. Statistical invariants include quantities like node degree, degree distribution, short and mean path, power law or normal distribution of edges, and robustness (Barabási 2013; Boccaletti et al. 2006; Giuliani et al. 2014; Albert and Barabási 2002).

One of the earliest attempts to formalize a biological system using a network approach in the field of computational biology was the seminal work by Stuart Kauffman on random Boolean networks (RBN) (Kauffman 1969; Kauffman et al. 1993). In this work, the network was conceptualized as genes (nodes) connected through activation-suppression relations or regulatory interactions (Fig. 1a). A significant finding of Kauffman's study was that elements with only two possible states (activated or deactivated) based on randomly assigned interactions with other

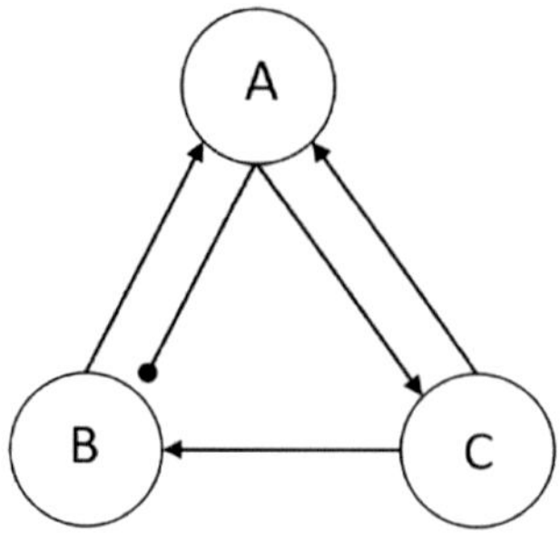

(a)

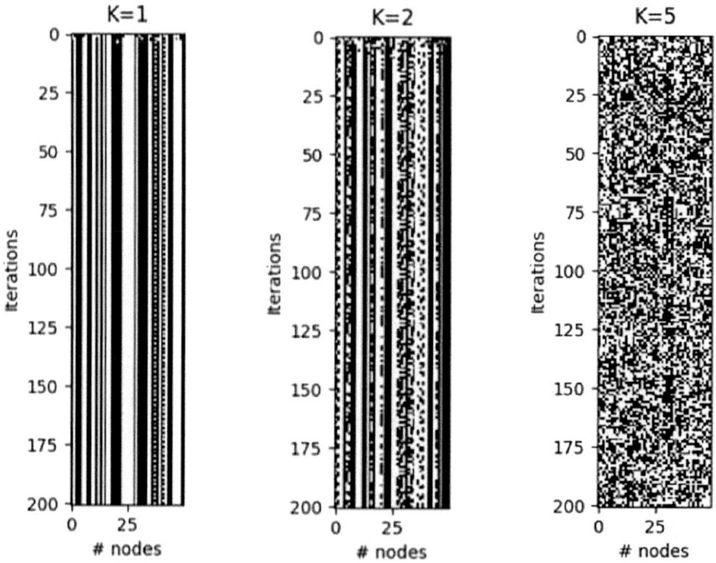

(b)

Fig. 1 (**a**) RBN with three nodes (A, B, C), whose interactions follow the Boolean rules A = (B ∧ C), B = (C ∧ ¬ A), C = (A). (**b**) Iterations run from the top to the bottom with each line reporting the state of every node in the network (black 1, white 0)

elements exhibited highly ordered behaviors resembling the structure of living organisms. For example, the number of attractors in the phase space was close to the number of different cell types in organisms. Another important finding was the existence of three dynamical regimes: *ordered*, *chaotic*, and *critical*, corresponding to different values of the number of connections (K) per node: $K < 2$, $K = 5$, and $K = 2$, respectively. These behaviors became evident when plotting the active state of each node over time, resulting in distinctive spatial patterns (Fig. 1b) (López-Díaz et al. 2023).

In recent years, the Boolean network (BN) model has experienced rapid growth and flourished in various forms, thanks in part to its relatively simple nature, making it suitable for modeling and implementing in computer simulations. It has found applications in many biological phenomena (Schwab et al. 2020) and has been used to tackle computational problems using bio-inspired techniques (Pontes-Filho et al. 2020).

Several interesting attempts have been made to describe biological systems using BNs. For instance, studies on cell fate, including growth, differentiation, and death, have shown that even with a small number of genes, cell regulatory networks self-organize into multiple cellular states. From a dynamical systems perspective, different connections within the network lead to different attractors (Huang and Ingber 2000; Huang et al. 2005). Modeling a network of 2773 proteins during neutrophil differentiation revealed that two different pathways eventually reach the same attractor, but at different times. Furthermore, modeling the yeast transcriptional network to identify interaction rules between promoter regions and genes demonstrated the robustness and redundancy of the network, as exemplified by "forcing structures" which are types of Boolean functions insensitive to multiple stimuli (Kauffman et al. 2003).

So far, we have examined networks structured in a single layer, representing different aspects of the same phenomenon. However, as depicted in Fig. 1, we can observe two levels: the microscopic level comprising nodes and their relations and the macroscopic level representing the dynamic behavior of the entire system over time. Even more relevant to the purpose of this work is the generalization of the traditional complex network approach, known as a multilayer network. If the notion of biological entities as nested networks holds true, then the multilayer approach provides the best framework for formalizing and representing these systems, which are inherently composed of interconnected subsystems. A multilayer network is defined as a pair $M = (G, C)$, where G is a family of graphs representing the layers of the multilayer network and C is the set of interconnections between the nodes across different layers (Fig. 2) (Boccaletti et al. 2014).

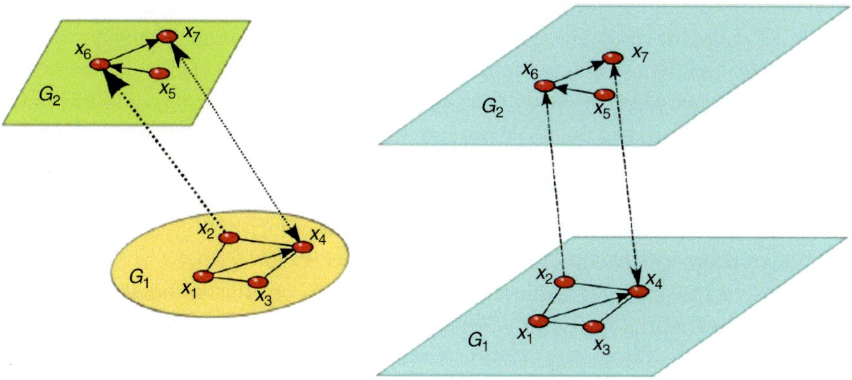

Fig. 2 Two representations of interconnected levels in a multilayer network

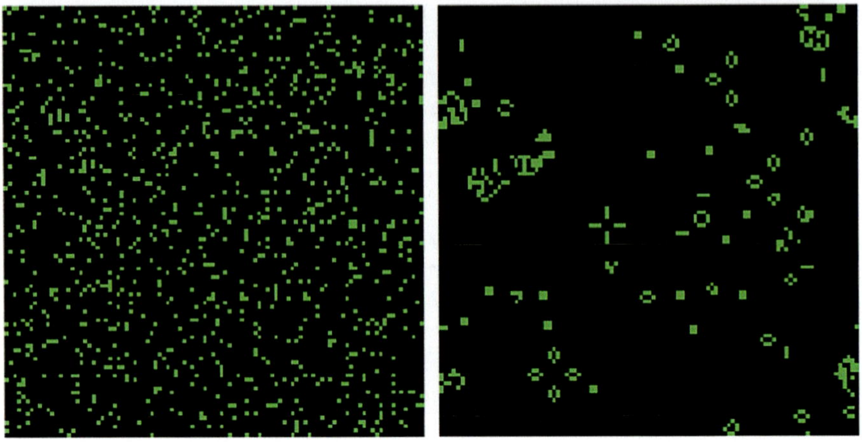

(a) (b)

Fig. 3 The world of Game of Life at two time steps: (**a**) initial conditions, (**b**) successive time steps. It is easily observable that, from a random initial disposition in space, clusters of stationary or moving cells emerge

The significance of this type of representation for biological systems lies in its ability to model not only the hierarchical subdivision of reality into ontological units but also different functions and behaviors. Numerous studies have demonstrated the relevance of this approach in various domains, including modeling human physiology and diseases, brain structure and functions, spreading phenomena, and ecological interactions (Chaudhuri and Srivastava 2022).

Another valuable modeling paradigm is cellular automata (CA), which can be seen as a subset of Boolean networks (BNs) with a distinctive feature of local interactions. CA are finite-state machines, where the elementary unit is a cell that follows certain rules based on its neighbors. In one-dimensional CA, the cell interacts with its right and left neighbors, while in two-dimensional CA, it interacts with its surrounding neighbors. One of the most well-known examples of CA is the Game of Life, devised by the brilliant mathematician John Conway. In this two-dimensional world divided into cells or patches (Fig. 3), each patch can be in one of two states: alive or dead. The Game of Life operates based on three simple rules: birth, survival, and death. According to these rules, an alive patch with two or three alive neighboring patches survives into the next generation. An alive patch with four or more alive neighbors dies from overpopulation, and a patch with one or no alive neighbors dies from isolation. Finally, an empty cell adjacent to exactly three alive neighbors becomes a birth cell (Gardner 1970).

Extensive research has explored the potential of cellular automata (CA) as a modeling framework, spanning various fields from molecular biology (Hogeweg 1989) to ecology (Hogeweg 1988). These studies have demonstrated that the fundamental concept of locality of interactions in CA is highly insightful for modeling biological systems. Locality, along with asynchronous updating of the system state, provides a more realistic representation of how a real system operates when performing environmental computations.

In synchronous updating, it is assumed that there is a universal timing, and each unit of the grid is updated simultaneously. This approach mixes all the outputs and makes it difficult to determine which unit performs a specific task. On the other hand, asynchronous updating introduces a more realistic procedure. A random unit is chosen to act first, followed by its neighbors and the neighbors of those neighbors, until every unit in the grid has performed the next state function. This approach allows for the observation of interesting spatial patterns that can spread throughout the model world. However, it should be noted that the biological significance of these patterns depends on the assumptions made in the model (Fig. 4).

2.2 Networks and CA Merged Together

So far, we have discussed complex networks and cellular automata (CA) as related modeling frameworks. In this section, we will demonstrate how they can work together.

The locality constraints inherent in CA behaviors give rise to a unique phenomenon known as the formation of mesoscale patterns in space. While information in classical networks can propagate throughout the entire network topology (i.e., long-range relations), in CA, we observe that the simple deterministic rules at the local level, specifically within the neighborhood, generate effects that are confined to restricted areas where the agents interact.

Adopting a mesoscale approach means that we cannot rely solely on a purely bottom-up perspective, adhering to the reductionist dream, nor can we rely solely on a top-down perspective, which is characteristic of the holistic paradigm. Therefore, the most suitable choice is the middle-out approach, which emphasizes the importance of a middle-relational level between the macro and the micro. At this mesoscale level, we observe feedback from the mesoscale to the microscale behavior and reciprocal emergent behavior following these interactions (Hogeweg 2010; Bertolaso et al. 2014).

In this section, we present cases where the explicit integration of CA and networks has led to a hierarchical understanding of the system under study.

One significant phenomenon studied in biophysics and bioinformatics is the simulation of cell differentiation. Here, we focus on an in-depth research study investigating the emergence of self-coordinating patterns in cellular populations during development (Smart and Zilman 2023; Gigante et al. 2023). In this well-designed model, the authors utilize different techniques, including the power of statistical mechanics, to establish connections between three key aspects of cell development: gene regulatory networks, single-cell fate, and cell-cell interactions. Regulatory gene networks are mapped onto a Hopfield network, where each gene is represented by a binary-state vector $s = 1, -1$, following the traditional formulation of a Boolean network (BN). The network attractors represent stable phenotypes of

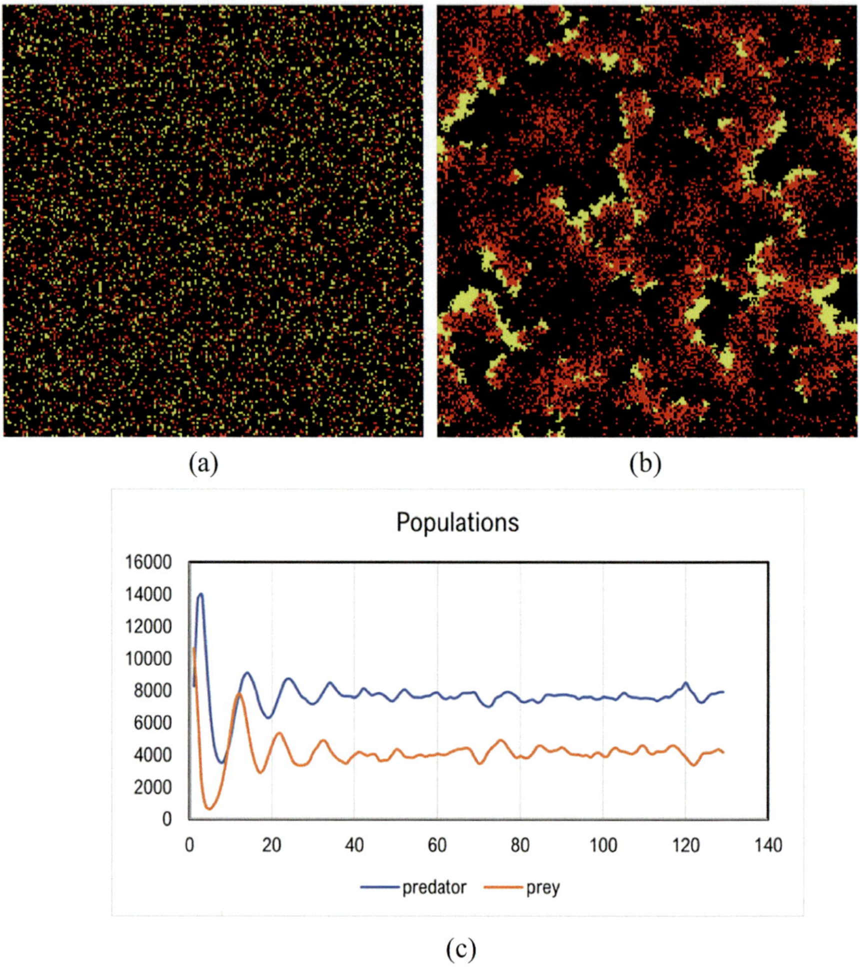

Fig. 4 A predator-prey model simulated with a 2D CA, with two panels representing (**a**) the initial disposition of predators (red patches) and preys (yellow patches) and (**b**) the long-run dynamic characterized by typical mesoscale patterns which are spirals. In (**c**) we can observe the population trend of both units

single cells. The model is then extended to incorporate multicell interactions using a commonly employed feature in the simulation of complex systems: a control parameter γ. This control parameter, γ, represents the strength of cell-cell interactions. When $\gamma \longrightarrow 0$, there is no relatedness between cells, and they behave as individual entities. However, when $\gamma > 0$ and surpasses a certain threshold, the cells exhibit collective behavior with distinct spatial patterns and signal-dependent interactions. The effects of various genes within the internal single-cell network are coupled with target genes in neighboring cells, mimicking a ligand-receptor relationship. The simulation results demonstrate that complexity emerges from the interaction between intracellular regulatory network behavior and extracellular signal interactions. The cells exhibit suboptimal behavior, characterized by high energy and being far from equilibrium, as they organize themselves into multicellular tissues.

Aside from the remarkable findings presented in this insightful study, we want to highlight the methodology employed, which involves multiple levels of simulation, description, and explanation of the system. When examining the figure (Fig. 5), we can discern three distinct levels: the microscopic level, where the authors program the individual cell behavior using the Hopfield network and establish interactions with neighboring cells; the mesoscopic level, where the lattice formed by the cells adopts various patterns based on the γ parameter; and the macroscopic level, where the fate of the entire system is represented by a fitness landscape exhibiting increasingly rugged morphology as γ increases.

Once a modeling framework begins to yield results, it becomes legitimate to question whether there is a privileged level at which to investigate the behavior of the system. In our opinion, this privileged level is the mesoscopic level. At this level, cellular automata (CA) exhibit their remarkable potential to demonstrate information flow, revealing the spatial arrangement of entities that are themselves networks of entities. These entities organize into higher-level structures, such as tissues, exhibiting behaviors distinct from those at lower levels.

The second case we would like to present concerns prebiotic evolution and the origin of life, as envisioned by the RNA world hypothesis (Takeuchi and Hogeweg 2009). According to the authors, multilevel selection is a crucial concept in the evolution of biological systems, and it becomes even more important in the context of the RNA world hypothesis. It has been shown that natural selection, acting solely on individual RNA replicators, is insufficient to explain the complexity of structures and behaviors observed in other RNA templates, such as parasite replicators.

In the model presented by the authors, multilevel selection can be either implicit or explicit. Implicit multilevel selection occurs through the spatial self-organization of replicator populations, leading to higher-level selection among emergent mesoscopic spatial patterns. Explicit multilevel selection is achieved through the compartmentalization of replicators by vesicles, which results in higher-level evolutionary dynamics among explicitly imposed mesoscopic entities.

The model is based on a stochastic cellular automaton (CA) implemented on a two-dimensional grid. Each square of the grid can be occupied by either R molecules (replicase molecules) or L molecules (parasite molecules). Empty squares represent space with resources for molecule replication, denoted as θ. The more complex molecules are labeled as C_R and C_L. The dynamics of the grid are governed by a reaction diffusion algorithm. At each time step, a random square is selected, followed by the selection of a neighboring square. Then, one of the possible reactions is executed with a certain probability, which is determined by the reaction rate multiplied by a constant α. The key feature of the reaction norms is that R molecules can replicate themselves, while L molecules can only be replicated by other L molecules. Furthermore, the parasite molecules have two states: folded, where catalysis is not allowed, and unfolded, where replication can occur.

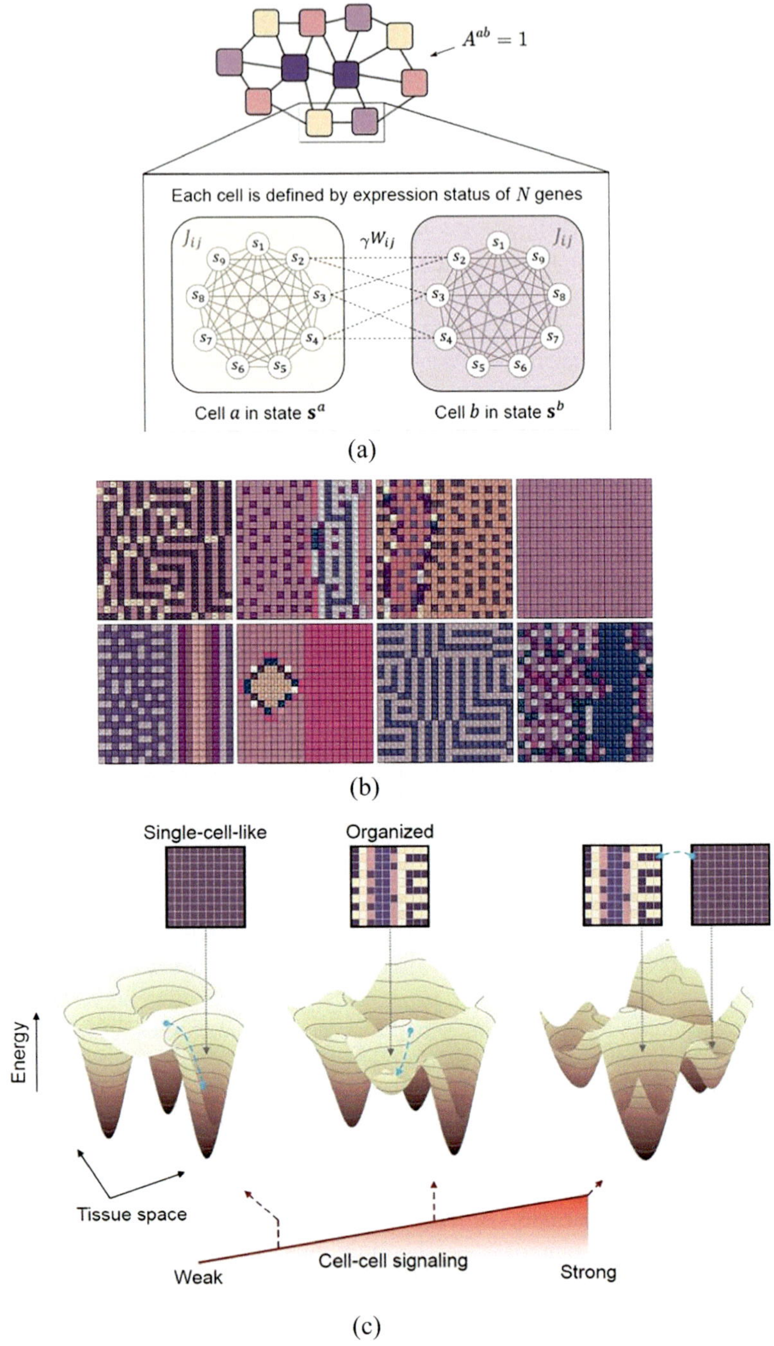

Fig. 5 The multicellular system development is characterized by three levels of description. (**a**) illustrates the gene network within each cell. (**b**) presents the mesoscale aggregation outcomes of different cell types. (**c**) depicts the fitness landscape based on cellular signal strength

The model focuses on two parameters: l and k_L. The former represents the portion of time during which a parasite is in the folded state and unable to replicate, while the latter represents the strength of the bond between replicators and parasites. These parameters serve as evolutionary traits in the model. The authors simulated three different scenarios: one with no spatial population structure (i.e., no local constraints), one with spatially imposed mesoscale entities (vesicles), and one with a spatial population structure free to organize itself in space. One important result obtained from the simulations is that the space-free model with ordinary differential equations (ODEs) is evolutionarily unstable. This is due to the eventual extinction of the system caused by the invasiveness of evolved parasites. This result is illustrated in the third panel of Fig. 6, which shows how all molecule phenotypes trace toward equilibrium as a function of l. With the addition of space in the self-organized model, the authors observed that parasites and replicators arrange themselves in growing traveling waves with a starting point and subsequent dynamics. In this scenario, parasites evolve to spend more time in the folded state, allowing replicators to maintain a population level that ensures the survival of the system. This can be seen in the second panel of Fig. 6. The compartment scenario is characterized by the imposition of mesoscale patterns in the form of vesicles. These vesicles constrain replicators and parasites within a bounded portion of space, with the volume of the compartment increasing or decreasing depending on the replication of internal molecules. Some molecules become extinct while others survive, demonstrating that even with superimposed mesoscale entities, the system reaches an evolutionary equilibrium. Similar to the model of multicellularity development, the RNA evolutionary space demonstrates how integrating the microscopic approach (encoding the dynamic rules of individual entities) with local constraints (reflecting multilevel behavior) leads to a hierarchy of levels. In this hierarchy, the privileged middle level allows for tracking the local disposition of individual entities, extrapolating attractors, and understanding general dynamic features of the entire system at the macroscale.

3 A Relational Onto-epistemology

The systems within the biosphere, as well as the biosphere itself, can be understood as hierarchical structures composed of entities and relations that span different ontological levels (Ellis 2012; List 2019; Tëmkin and Eldredge 2015). By *ontological* level, we refer to a distinct slice of the world where each element within that slice shares a common essential predicate. For example, a cell exists at the same ontological level as other individual cells, but it exists at a different level compared to the entire organism, and, similarly, a single nucleotide exists at a different level compared to a cell.

It is important to note another perspective from the philosophy of biology, which suggests that living systems are processes that constantly flow through time. According to this view, these processes are treated as temporarily clustered objects,

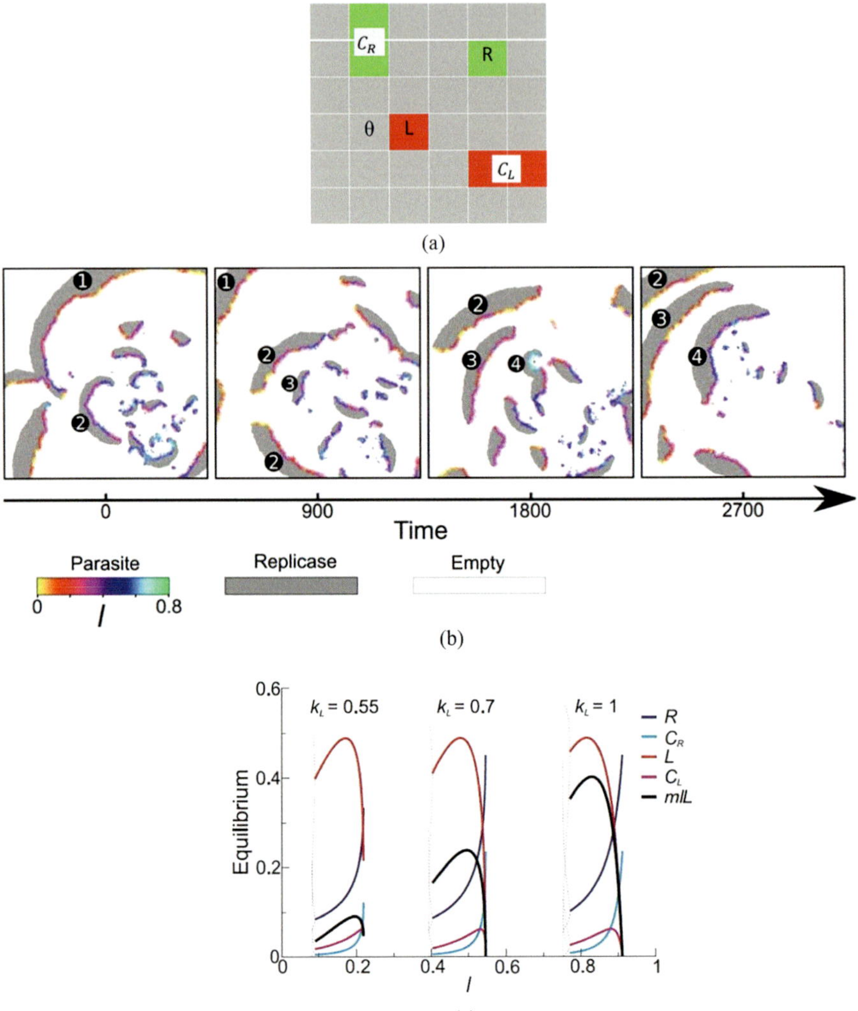

Fig. 6 The scenario of RNA replicators is delineated through distinct layers. (**a**) exhibits the two-dimensional grid of the cellular automaton (CA). (**b**) highlights the emergent waves in the scenario without pre-imposed mesoscale patterns. (**c**) displays the outcomes obtained through ordinary differential equation (ODE) analysis

such as species (Nicholson and Dupre 2018; Dupre 2020). It is not clear whether this metaphysical perspective is in conflict with a network approach to biological systems, as complex networks can also be seen as evolving relationships between nodes over time. Furthermore, each node can be considered as a complex network itself until a bottom level of reduction is reached, forming a continuous process of interacting nodes and rewiring networks.

Within such a hierarchical framework, systems can develop in two directions: within the same level and between levels. Using the example of cells, the mitosis of a cell population within a tissue represents a progression within the same level. In this case, cells perform individual tasks while interacting with neighboring cells (i.e., relations within the same level) without creating something fundamentally different from an individual cell. However, if we consider a functional task specific to the tissue, the organized behavior of individual cells leads to an outcome that is no longer an individual accomplishment. A tissue is not simply a collection of cells; it has its own distinct functions.

Considering that reality can be divided into nested layers with complex networks within them, we can also assert, following the perspective of nearly-decomposable systems (Simon 1996; Bechtel and Richardson 2010; Bertolaso 2013), that these networks, which structure a complex system, are decomposable until they reach a fundamental level that cannot be further divided. The differences between these levels can be attributed to the behavior of the agents within the network, which in turn relates to the variation of relations over time. In other words, the ontological characteristic that distinguishes the various levels is the behavioral dynamics, which are identical among agents at the same level but differ across multiple levels.

The molecular biology paradigm suggests a decomposable view, where entities are analyzed as isolated objects, followed by the recomposition of these entities as the sum of their individual parts. However, we argue that systems biology inherently implies a *relational ontology*, where the relations come before the entities themselves and entities are properly defined based on their relations (Bertolaso and Ratti 2018). In this view, the decomposition of hierarchies is not sufficient for a complete understanding of the relationships between levels. Once the description within the same level is reached, what is missed are the properties that emerge from interactions with other levels of description. An additive analysis of all levels considered in sequence will inevitably overlook the inter-level influences.

This leads to the problem of the *biological orchestra*—how can a single number or a snapshot of complex temporally arranged phenomena capture all the information needed for the description, prediction, or evolution of the system? This problem highlights the challenges of comprehensively capturing the intricacies and dynamics of biological systems and calls for a more nuanced approach that considers the interplay of multiple levels and their relationships. The issue is faced in a pioneering work on environmental health (Arora and Curtin 2021), where interdependence of relations and complexity, as a feature that must be extended to the individual, are the pillars for a new representation of the interplay between human and environment. The authors specify that their approach is in direct contrast with omics science, the most relevant methodological paradigm of modern biology, for several reasons. One reason is that they intend to focus on processes rather than measuring many things, and focusing on processes means to concentrate on trajectories of things over the time and space. Another crucial point is the interest in what Weaver called *organized complexity*, rather than *disorganized*, implying that complex systems are not independent to one another, instead, they operate together via a *biodynamic interface*, which is a correlation space where the variables of the interacting complex

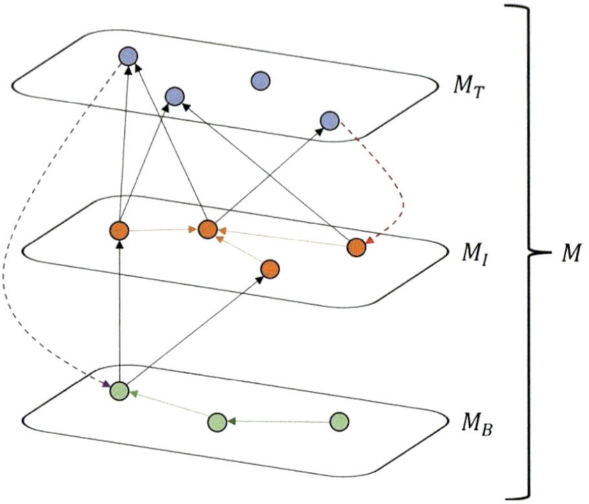

Fig. 7 Interface network with three layers divided by some predicate with connections in the same level and among different levels

systems covariate together. An important addition to that characteristic is that this interface emerges as a distinct level among the systems while remaining operationally independent. This means that the interface level itself is the appropriate focus of inquiry, as it "uses" or processes information from all systems involved.

3.1 Formalization

Given all the results so far exposed, we can try to sketch some formal definition of what an interface level would be. Let $\boldsymbol{M}$ be a *relational entity* with a number of layer >2. Let $\boldsymbol{M_T}$ be the higher level, $\boldsymbol{M_B}$ the lower level, and $\boldsymbol{M_I}$ the middle *interface* level (Fig. 7). If we divide $\boldsymbol{M}$ by causal predicates, the bottom-up relations can be represented by black lines, which signify efficient causation. On the other hand, the red and violet dotted lines represent top-down influences or formal constraints. When $\boldsymbol{M}$ is divided by an ontological predicate, the relations can be understood as inclusive or nested relations, which also encompass historical features. In this case, there is a time arrow from the bottom to the top. From an epistemological perspective, the top level is where we can find the first principles for deriving explanations. The bottom level involves instantiating the variables of the system with constants of nature. The middle level is where the physical constants and the laws of nature are processed, resulting in a dynamic level where variables can take on all the possible values allowed by the principles at the top level and provided by physical reality.

We can define an *interface level* if the following conditions are met:

1. The level being studied inherently includes a lower level and an upper level. This means that the entities and relations at the interface level encompass elements from both the lower and upper levels.

2. The dynamics and behavior of the entities at the interface level, as well as the spatial arrangement of the agents, cannot be predicted solely based on the rules and interactions of the lower and upper levels. There is emergent behavior that arises specifically at the interface level.
3. After a certain period of time in the dynamic of the network, the higher level exerts causal influences on the interface level, triggering a feedback loop from the lower-middle levels. This feedback loop establishes a bidirectional interaction between the interface level and the levels above and below it. Furthermore, another important feature of such a causal dynamic between levels is that relations are mostly asymmetric; specifically, there is an asymmetrical, but mutual, correspondence of causal relations between levels (Bertolaso 2016).
4. The variables in both the lower and upper levels are correlated through patterns that can be detected and studied at the interface level. These patterns reflect the relationships and interactions between entities from the lower and upper levels and provide insight into the dynamics of the system as a whole.

4 Conclusion

The objective of this work is to reconcile the reductionist perspective of biological entities with a holistic and systemic approach. The combination of complex network analysis and agent-based models offers a promising framework for addressing questions ranging from the origins of life to the development of multicellularity. The concept of the *relational entity*, introduced here, seeks to reinterpret classical biological entities such as genes, cells, organisms, populations, species, ecosystems, and the biosphere in a purely systemic manner. From a bottom-up standpoint, the biosphere represents the highest level of analysis as a complete system, with decreasing levels of reductionism as we delve deeper into the hierarchy of biological entities. For example, species can be regarded as systems, but with a greater inclination toward bottom-up analysis compared to the entire biosphere, implying a lower degree of complexity. This hierarchical approach can be extended to levels where fundamental sciences provide functional and structural explanations for biological phenomena. From a top-down perspective, the biosphere and other levels with organizational and relational properties hold the utmost significance in the study of biological systems. They provide crucial insights into macroscale effects and account for a certain level of complexity, but without an extension of complexity to the agents of lower levels. A more effective approach to studying interactions within complex biological systems should involve integrating the knowledge derived from a reductionist approach into a regulatory framework of the systemic approach. An interesting example of such approach is the study by Palazzi et al. (2019), which highlights the close relationship between constraints and functions, as respectively represented by nestedness at the macrolevel and modularity at the mesoscopic level.

Future research endeavors should focus on establishing a rigorous logical formalism for the *relational entity* or, alternatively, develop a mathematical framework that mirrors the characteristics of a multilayer network. This approach aims to ensure precision in the assumptions and deductions made regarding the structure and dynamics of the relational entity itself.

References

Alberghina L, Westerhoff HV (2007) Systems biology: definitions and perspectives, vol 61. Springer Science & Business Media
Albert R, Barabási AL (2002) Statistical mechanics of complex networks. Rev Mod Phys 74:47
An G, Mi Q, Dutta-Moscato J et al (2009) Agent-based models in translational systems biology. WIREs Syst Biol Med 1:159–171
Arora M, Curtin P (2021) Environmental biodynamics: a new science of how the environment interacts with human health. OUP
Arora M, Giuliani A, Curtin P (2020) Biodynamic interfaces are essential for human–environment interactions. BioEssays 42:2000017
Artime O, De Domenico M (2022) From the origin of life to pandemics: emergent phenomena in complex systems. Phil Trans R Soc A 380:20200410
Bandesh K, Dhar PK, Bharadwaj D (2018) A systems perspective of complex diseases: from reductionism to integration. Syst Biol 17–36
Barabási AL (2013) Network science. Phil Trans R Soc A 371:20120375
Bechtel W, Richardson RC (2010) Discovering complexity: decomposition and localization as strategies in scientific research. MIT press
Bertolaso M (2013) How science works. Choosing levels of explanation in biological sciences. Preface by Sandra D Mitchell Aracne Editrice
Bertolaso M (2016) Philosophy of cancer. Springer
Bertolaso M, Giuliani A, Filippi S et al (2014) The mesoscopic level and its epistemological relevance in systems biology. In: Recent advances in systems biology research. Nova Science Publishers, Inc, pp 19–36
Bertolaso M, Ratti E (2018) Conceptual challenges in the theoretical foundations of systems biology. Syst Biol:1–13
Boccaletti S, Bianconi G, Criado R et al (2014) The structure and dynamics of Multilayer networks. Phys Rep 544:1–122
Boccaletti S, Latora V, Moreno Y et al (2006) Complex networks: structure and dynamics. Phys Rep 424:175–308
Boogerd F, Bruggeman FJ, Hofmeyr JHS et al (2007) Systems biology: philosophical foundations. Elsevier
Bruggeman FJ, Westerhoff HV (2007) The nature of systems biology. Trends Microbiol 15:45–50
Chaudhuri S, Srivastava A (2022) Network approach to understand biological systems: from single to multilayer networks. J Biosci 47:55
Dupre J (2020) Life as process, vol 57. Russian Academy of Sciences, p 96
Ellis GF (2012) Top-down causation and emergence: some comments on mechanisms. Interface Focus 2:126–140
Gardner M (1970) The fantastic combinations of John Conway's new solitaire Game'Life. Sci Am 223:20–123
Gigante G, Giuliani A, Mattia M (2023) A novel network approach to multiscale biological regulation. Cell Syst 14:177–179
Giuliani A, Filippi S, Bertolaso M (2014) Why network approach can promote a new way of thinking in biology. Front Genet 5:83

Hogeweg P (1988) Cellular automata as a paradigm for ecological modeling. Appl Math Comput 27:81–100
Hogeweg P (1989) Local TT cell and TB cell interactions: a cellular automaton approach. Immunol Lett 22:113–122
Hogeweg P (2010) Multilevel cellular automata as a tool for studying bioinformatic processes. Simulating complex systems by cellular automata. Springer, pp 19–28
Huang S, Eichler G, Bar-Yam Y et al (2005) Cell fates as high-dimensional attractor states of a complex gene regulatory network. Phys Rev Lett 94:128701
Huang S, Ingber DE (2000) Shape-dependent control of cell growth, differentiation, and apoptosis: switching between attractors in cell regulatory networks. Exp Cell Res 261:91–103
Kauffman SA (1969) Metabolic stability and epigenesis in randomly constructed genetic nets. J Theor Biol 22:437–467
Kauffman S, Peterson C, Samuelsson B et al (2003) Random Boolean network models and the yeast transcriptional network. Proc Natl Acad Sci USA 100:14796–14799
Kauffman SA et al (1993) The origins of order: self-organization and selection in evolution. OUP, USA
Kitano H (2002) Computational systems biology. Nature 420:206–210
List C (2019) Levels: descriptive, explanatory, and ontological. Nouˆs 53:852–883
López-Dı́az AJ, Sánchez-Puig F, Gershenson C (2023) Temporal, structural, and functional heterogeneities extend criticality and antifragility in random Boolean networks. Entropy 25:254
Nicholson DJ, Dupre J (2018) Everything flows: towards a processual philosophy of biology. OUP
Palazzi MJ, Borge-Holthoefer J, Tessone CJ et al (2019) Macro-and mesoscale pattern interdependencies in complex networks. J R Soc Interface 16:20190553
Pontes-Filho S, Lind P, Yazidi A et al (2020) A neuro-inspired general framework for the evolution of stochastic dynamical systems: cellular automata, random Boolean networks and echo state networks towards criticality. Cogn Neurodyn 14:657–674
Schwab JD, Kühlwein SD, Ikonomi N et al (2020) Concepts in Boolean network modeling: what do they all mean? Comput Struct Biotechnol J 18:571–582
Simon HA (1996) The architecture of complexity. In: Managing in the modular age: architectures, networks, and organizations, pp 15–38
Smart M, Zilman A (2023) Emergent properties of collective gene-expression patterns in multicellular systems. Cell Rep Phys Sci 101247
Takeuchi N, Hogeweg P (2009) Multilevel selection in models of prebiotic evolution II: a direct comparison of compartmentalization and spatial self-organization. PLoS Comput Biol 5: e1000542
Takeuchi N, Hogeweg P (2012) Evolutionary dynamics of RNA-like replicator systems: a bioinformatic approach to the origin of life. Phys Life Rev 9:219–263
Tëmkin I, Eldredge N (2015) Networks and hierarchies: approaching complexity in evolutionary theory. In: Macroevolution: explanation, interpretation and evidence, pp 183–226
Weaver W (1948) Science and complexity. Am Sci 36:536–544

Recent Developments in Transcriptomic Technologies: Applications to Immunological Systems and Diseases

Bailey Andrew, Michal Zulcinski, Amber Emmett, and David R. Westhead

Contents

Abstract The immune system is immensely complex. A diverse repertoire of immune cells transits through blood to infiltrate tissues, and immune responses are driven through complex intercellular interactions. Therefore, the study of immunological diseases demands methods capable of disentangling the inherent complexity of the immune system. This chapter explores recent developments in RNA sequencing technologies, including single-cell RNA-seq (scRNA-seq), multi-omics integration, and spatial transcriptomics. Each of these technologies is regarded from the

B. Andrew
Faculty of Engineering and Physical Sciences, University of Leeds, Leeds, UK

M. Zulcinski
Faculty of Medicine and Health, University of Leeds, Leeds, UK

A. Emmett · D. R. Westhead (✉)
Faculty of Biological Sciences, University of Leeds, Leeds, UK
e-mail: D.R.Westhead@leeds.ac.uk

J. Barciszewski (ed.), *Systems Biology II*, RNA Technologies 15,
https://doi.org/10.1007/978-3-031-62178-9_9

perspective of how it has accelerated our understanding of immunological systems and diseases, with their advantages and limitations highlighted.

Keywords Transcriptome analysis · Single-cell RNA sequencing · Spatial transcriptomics · Multi-omics · Immunology

Abbreviations

ASTAR-seq	Assay for single-cell transcriptome and accessibility regions sequencing
BCR-seq	B cell receptor sequencing
CAR T cells	Chimeric antigen receptor T cells
CEL-seq	Cell expression by linear amplification and sequencing
CITE-seq	Cellular indexing of transcriptomes and epitopes sequencing
COTECH	Combined assay of transcriptome and enriched chromatin binding
CRS	Cytokine release syndrome
DLBCL	Diffuse large B cell lymphoma
DR-seq	Single-cell DNA-mRNA sequencing
DSP	Digital spatial profiling
Drop-seq	Droplet sequencing
ECCITE-seq	Expanded CRISPR-compatible cellular indexing of transcriptomes and epitopes by sequencing
FACS	Fluorescence-activated cell sorting
FFPE	Formalin-fixed paraffin-embedded
G&T-seq	Single-cell genome and transcriptome sequencing
ICANS	Immune effector cell-associated neurotoxicity syndrome
ISH	In situ hybridization
ISS	In situ sequencing
IVT	In vitro transcription
LCM	Laser capture microdissection
MACS	Magnetic-activated cell sorting
MATQ-seq	Multiple annealing and dC-tailing-based quantitative sequencing
PBMCs	Peripheral blood mononuclear cells
PCA	Principal component analysis
PCR	Polymerase chain reaction
PLAYR	Proximity ligation assay for RNA
REAP-seq	RNA expression and protein sequencing
ROI	Region of interest
SCAN-seq2	Single-cell amplification and sequencing
SMART-seq	Switching mechanism at the end of the 5′-end of the RNA transcript sequencing
SMRT	Single molecule real-time sequencing

SNARE-seq	Single nucleus chromatin accessibility and mRNA expression sequencing
SPLiT-seq	Split pool ligation-based transcriptome sequencing
TCR-seq	T cell receptor sequencing
TME	Tumor microenvironment
UMAP	Uniform Manifold Approximation and Projection
UMI	Unique molecular identifier
VASA-seq	Vast transcriptome analysis of single cells by dA-tailing sequencing
WM	Waldenström macroglobulinemia
cDNA	Complementary deoxyribonucleic acid
scChIP-seq	Single-cell chromatin immunoprecipitation sequencing
scM&T-seq	Single-cell methylome and transcriptome sequencing
scRNA-seq	Single-cell ribonucleic acid sequencing
scTRIO-seq	Single-cell triple omics sequencing
sciCAR-seq	Single-cell combinatorial indexing chromatin accessibility and mRNA sequencing
sciRNA-seq	Single-cell combinatorial indexing RNA sequencing
scifiRNA-seq	Single-cell combinatorial fluidic indexing
snRNA-seq	Single nucleus ribonucleic acid sequencing
tSNE	T-distributed stochastic neighbor embedding

1 Introduction

Single-cell RNA sequencing (scRNA-seq) has come a long way in the 15 years since its first use. Numerous advancements, such as the inclusion of unique molecular indexing, droplet-based microfluidics, and combinatorial indexing have all allowed an increase in throughput from the single digits to hundreds of thousands of cells while reducing biases. New technologies such as long-read sequencing have complemented the deficiencies of second-generation sequencing, enabling high-throughput detection of mutations. The inclusion of other "omics" has allowed the identification of cell subtypes at a higher resolution than possible with just scRNA-seq, and the inclusion of a spatial dimension has allowed these developments to be put into physical context.

The foundations for spatial omics were laid by the work of Doyle et al. in the 1990s, who aimed to spatially map biological function data onto images of tissue structure of embryos, to create multidimensional models of tissue morphology. A system named "spatial genomics" composed of morphological imaging, tissue rasterization, RNA amplification, cDNA microarray analysis, and mapping of spatial data was subsequently submitted as patent application in 2000 and eventually patented in 2009. This work was expanded by a technique for spatial visualization and quantitative analysis of the expression of transcripts in tissue sections, named "spatial transcriptomics" (Ståhl et al. 2016).

These advances have enabled new insights into disease pathogenesis, the tumor microenvironment, cell-type development, and therapy toxicities. They have done this at increasingly higher throughput, increasingly better sensitivity, and increasingly lower cost. With many new sequencing protocols being published every year, this progress shows no signs of slowing down.

This chapter gives an overview of the current state of single-cell RNA sequencing, including its synthesis with other modalities and the inclusion of spatial information. These developments, challenges, and limitations are placed in the context of immunology whenever possible. Numerous examples of how scRNA-seq has been used in immunology are described, ranging from the study of cancers and their therapies to the development and diversification of immune cells.

2 Methodological Developments in Transcriptomics Technologies

In this section, an overview of the innovations that arose in the past decade is depicted, including reference to how they have affected the field of immunology. For a comprehensive summary of scRNA-seq protocols, the reader is referred to the recent review by Huang et al. (2023).

2.1 Single-Cell RNA-seq in the Wet Lab

The primary pipeline of scRNA-seq in the wet lab consists of the following steps: tissue dissociation, cell separation, cell lysis, transcript amplification, and sequencing. Collectively, all but the last step comprises "library preparation." Errors in cell dissociation can interfere with cell separation as well as bias the cell types and transcripts recovered at the end of the process; lysis errors can also lead to the latter. Cell separation errors can lead to doublets (when two or more cells are mistakenly identified as a single cell, Fig. 1). Amplification can lead to amplification bias, while sequencing can incorrectly combine reads leading to fake transcripts (Fig. 2).

In the first step to scRNA-seq workflow (tissue dissociation), the extracellular structure binding the cells together needs to be destroyed so that RNA can be extracted from each cell at an individual level. This is often done by dissolving the bindings via enzymes, although nonenzymatic methods exist as well. One important consideration of tissue dissociation procedures is that they need to be performed faster than the timescale over which gene expression operates; otherwise, the dissociation procedure will affect the final gene counts. Similar concerns exist for cell lysis, and the best method for both will depend on the tissue being considered. Dissociation is not an issue in studies of peripheral blood mononuclear cells but is essential when working with solid tissue samples such as from lymph nodes. The

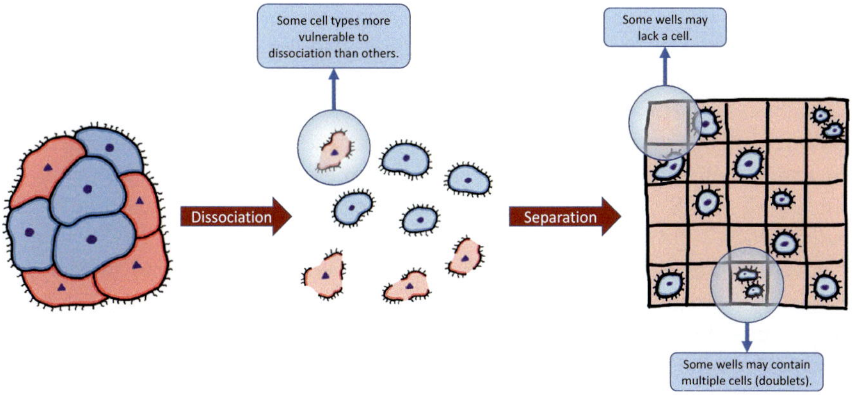

Fig. 1 The dissociation and separation stages of the scRNA-seq pipeline. A few common issues and biases are highlighted

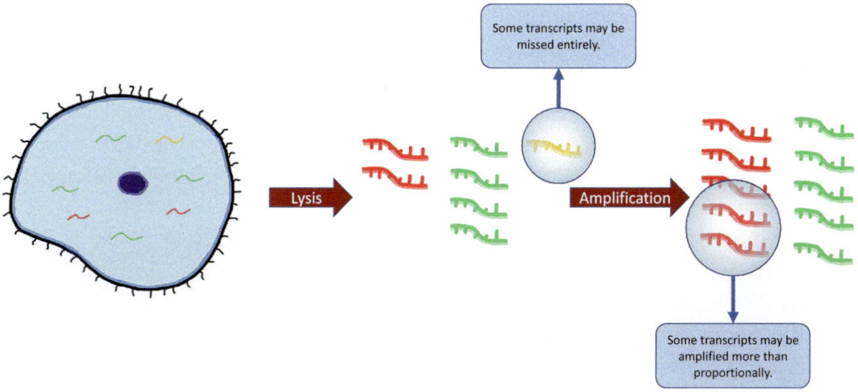

Fig. 2 The lysis and amplification stages of the scRNA-seq pipeline. A few common issues and biases are highlighted

method of dissociation has a disproportionate effect on the gene expression levels of immune cells (Denisenko et al. 2020).

One method to circumvent some of the difficulties of dissociation is single-nucleus RNA sequencing (snRNA-seq). It is easier to separate individual nuclei than it is to separate whole cells without damaging them, and they can be obtained from frozen samples (Habib et al. 2017). As a drawback, all cytoplasmic RNA is lost, and snRNA-seq appears more vulnerable to "zero inflation," the higher-than-expected number of undetected genes (Bakken et al. 2018). Furthermore, T, B, and NK cells tend to be underrepresented in snRNA-seq libraries (Denisenko et al. 2020).

Cell separation is a pivotal stage in scRNA-seq. The efficacy of separation techniques can be measured in terms of throughput (how quickly the cells can be sorted) and recovery/sensitivity (ratio of successfully separated cells to total cells in the sample). High-throughput methods are attractive due to their ability to handle large samples. The most common of these are fluorescence-activated cell sorting (FACS), magnetic-activated cell sorting (MACS), and microfluidics. Laser capture microdissection (LCM), a low-throughput method, preserves spatial information between cells. The dissociation stage can affect the suitability of these methods; the loss of cell surface protein expression due to enzymatic dissociation limits the efficacy of FACS (Hosic et al. 2016). These methods can be described as "plate-based."

Microfluidics separation methods can be generally divided into three types; droplet, pneumatic membrane valving, and hydrodynamic cell traps. Of these, droplet methods are the most suitable for high-throughput analysis (Gross et al. 2015). Over the past decade, droplet-based methods have become increasingly popular (Pan et al. 2022) and can be described as "ultrahigh throughput." Both Drop-Seq and 10x Genomics' Chromium platform, some of the most widely used platforms for scRNA-seq, use droplet-based methods. 10x Genomics in particular has had a great impact on immunology (Huang et al. 2023).

A major advantage of microfluidics over FACS is their low sample consumption; FACS, by contrast, requires at least 10,000 cells in the source sample and thus is not suitable for small-sample analysis. Furthermore, droplet-based methods allow for massively parallel cell processing, which is not traditionally possible. Due to the small size of each droplet, low-volume secretions of antibodies are able to reach detectable concentrations, allowing efficient sorting of cells of interest (Shembekar et al. 2016).

Amplification of transcripts is typically done by polymerase chain reaction (PCR), often after having been "reverse transcribed" into cDNA. Amplification is needed due to the low quantities of a given transcript in a single cell. Ideally, amplification will affect all transcripts equally. In such a case, measured transcript counts will be proportional to the true counts. When transcripts are disproportionately amplified, the final measurement is said to suffer from amplification bias.

One major downside of PCR is its exponential nature; as abundant transcripts are more likely to get amplified (thus further increasing their abundance), a feedback loop can be triggered. An alternative is in vitro transcription (IVT). However, IVT is slower than PCR and is thus less commonly used.

An important development in single-cell RNA sequencing is that of barcoding, first of the cells and then of the transcripts themselves ("unique molecular identifiers"—UMIs). The barcoding of individual cells allows more to be analyzed in parallel, increasing throughput. The recent development of combinatorial indexing (multiple barcodes per cell) in protocols such as the plate-based sciRNA-seq (Cao et al. 2017), the droplet-based scifiRNA-seq (Datlinger et al. 2021), and the (plateless, droplet-less) SPLiT-seq (Rosenberg et al. 2018) has led to the ability to profile hundreds of thousands of cells at the same time.

UMIs are used to nearly eliminate amplification bias, by allowing duplicate transcripts to be accurately mapped to each other. Due to read-length limitations of many sequencers, transcripts are typically fragmented post-amplification. As UMIs can only be added to the ends of a transcript, long transcripts will have many fragments without UMIs. This process makes it difficult to identify variations that occur in the middle of a transcript.

Despite the attractiveness of high-throughput library preparation methods, slower methods such as SMART-seq2 (Picelli et al. 2013), MATQ-seq (Sheng et al. 2017), and CEL-seq2 (Hashimshony et al. 2016) have higher sensitivity (Ding et al. 2020). The most sensitive low-throughput protocols can detect more than twice as many genes per cell as high-throughput protocols (Pan et al. 2022).

The final step of scRNA-seq in the wet lab is sequencing, which can be divided into three generations. The first generation was Sanger sequencing, but it was second-generation ("next-generation") sequencing platforms, such as those made by Illumina, Ion Torrent, and Pacific Biosciences, that enabled the development of scRNA-seq. Second-generation sequencing methods tend to have shorter but historically more accurate reads compared to the third generation (Hu et al. 2021) and are often built around the "sequencing by synthesis" methodology (Uhlen and Quake 2023).

Third-generation methods, such as those developed by Pacific Biosciences and Oxford Nanopore, tend to have higher read lengths but were less accurate than the generation before. Nanopore sequencing has achieved read lengths measured in the millions of base pairs, compared to second-generation sequencing whose read lengths were measured in the hundreds of base pairs. Third-generation methods also tend to be quicker than the second generation and do not necessarily need to undergo amplification. As the generations have increased, so has throughput; throughput has increased faster than Moore's law (Kharchenko 2021).

2.2 *Single-Cell RNA-seq in the Dry Lab*

After sequencing, the computational workflow for the scRNA-seq data can be split into two phases: preprocessing and analysis. The preprocessing includes all stages of the workflow to do with read assembly, quality control, and the creation of count matrices (the format which most downstream analyses rely upon). The analysis stage is more open ended but frequently involves dimensionality reduction, manifold projection, differential expression analysis, pseudotime analysis, and other downstream analyses.

Preprocessing is enabled by tools such as Cell Ranger (10× Genomics, Pleasanton, CA, USA) and STAR (Dobin et al. 2013). There are substantial differences between basecalling and read assembly for second- and third-generation sequencing. Basecalling for second-generation sequencing is less complicated than third-generation methods (especially nanopore methods). Third-generation methods have error rates more than an order of magnitude larger than second-generation methods. In contrast, read assembly is significantly easier for third-generation methods, due to their long (often full) read lengths.

The major third-generation methods, Pacific Biosciences' single molecule real-time (SMRT) sequencing and Oxford Nanopore's nanopore sequencing, require different basecalling techniques from each other. Of the two, nanopore sequencing is more complex. This complexity is reflected in the number of tools available for basecalling; there are nearly ten times as many nanopore-specific basecalling tools available compared to SMRT (Amarasinghe et al. 2020).

Nanopore basecalling is performed by neural networks. It is important to use a basecaller trained on data similar to that which is being collected; the standard Oxford Nanopore basecaller may perform worse on plant RNA due to this concern (Schmidt et al. 2017). Error correction can be done via consensus (comparing overlapping long reads) or by hybrid approaches (mixing short- and long-read methods).

The most popular methods for scRNA-seq analysis are the R package Seurat and the Python package Scanpy. Both are in active development, with Seurat about to release its fifth version. The focus on multimodal integration in the development of Seurat and the MUON framework, a member of the Scanpy ecosystem, demonstrates the growing importance of multi-omics.

A common step in scRNA-seq analysis is the use of principal component analysis (PCA) for dimensionality reduction (typically on just the most highly variable genes). This is a practical necessity due to the large number of unique genes measured in a set of cells but is also theoretically justified; the number of degrees of freedom of a cell's regulatory system is expected to be much lower than the number of genes (Kharchenko 2021). Due to the high proportion of zero counts, more sophisticated count-based models, such as zero-inflated factor analysis (Pierson and Yau 2015), may be more sensible. However, in practice, many analyses use just PCA (often as a preprocessing step to a nonlinear method). Deep learning techniques have also been applied (Svensson et al. 2020). Alternatively, one can treat the zeros as missing values and impute them (Zhang and Zhang 2020).

Another frequently included step in the analysis pipeline is the use of nonlinear dimensionality reduction techniques such as tSNE and, more recently, UMAP. Many dimensionality reduction methods, including tSNE and UMAP, are based on the k-nearest-neighbors graph. Clustering on these graphs is common, with the Leiden algorithm (Traag et al. 2019) recently supplanting Louvain (Blondel et al. 2008) due to the latter's tendency to find disconnected communities in the graph.

The development of tools such as Monocle allows the estimation of cell development trajectories, called pseudotime analysis (Trapnell et al. 2014). Additionally, analysis of pre-mRNA can provide information about the change in cell state of each cell, also called RNA velocity (La Manno et al. 2018). Together, these techniques mitigate the snapshot nature of scRNA-seq.

The explosion in single-cell sequencing has enabled and aided the development of several online cell atlases and repositories, such as the Single Cell Expression Atlas and the Cancer Single-Cell Expression Map. Immunology-specific atlases, such as JingleBells and the Tumor Immune Single Cell Hub (TISCH), have also been created. A major benefit of such atlases is the ability to use it as a reference for cell annotation of new data. Several methods exist, using both deep learning and statistical modeling techniques. The popular analysis package Seurat comes packaged with its own method.

2.3 Multi-omics

Single-cell transcriptomics has yielded unprecedented insight into the gene expression profiles of complex biological systems. Since the introduction of scRNA-seq, numerous other "omics" techniques have been adapted to achieve single-cell resolution. The past decade has seen the development of "omics" techniques which measure properties of DNA in single cells. These include single-cell genomics, epigenomics and proteomics methods. Through integrating multiple "omics" datatypes, researchers can gain a holistic insight into the combined biomolecular properties of single cells. This section introduces methods to quantify the DNA content, epigenomic modifications, and chromatin conformation of single cells as well as multi-omics techniques which simultaneously capture two or more "omics" layers from the same cell. Recent developments of a few selected techniques are also summarized.

Han et al. (2014) were the first to simultaneously sample genomic and transcriptomic data from the same cells, at single-cell precision. Building up on their work, the next year G&T-seq (Macaulay et al. 2015) and DR-seq (Dey et al. 2015) were introduced. Compared to the unimodal methods described in Sect. 2.2, these methods have very low throughput; G&T-seq has a throughput measured in hundreds of cells. More recent methods, such as Target-seq, have achieved throughput in the thousands.

The integration of epigenetic and transcriptomic data can take many forms, depending on the type of epigenetic information one wishes to profile. The first paired transcriptomic-epigenetic technique, scM&T-seq (Angermueller et al. 2016), allows the investigation of DNA methylation. scTRIO-seq (Hou et al. 2016) even allows the simultaneous profiling of transcriptomics, methylation, and genomics. As with transcriptomics + genomics techniques, transcriptomics + methylation suffers from low throughput. Another avenue of epigenetic exploration is chromatin accessibility, which has been combined with transcriptomics in high-throughput (>10,000 cells) methods such as sciCAR-seq (Cao et al. 2018) and SNARE-seq (Chen et al. 2019) as well as low-throughput high-sensitivity methods like ASTAR-seq (Xing et al. 2020). Commercial methods, such as 10x Genomics' Chromium Multiome, are available. Several other types of epigenetic data can be paired with transcriptomics; scChIP-seq (Grosselin et al. 2019) and COTECH (Xiong et al. 2021) enable the study of chromatin immunoprecipitation and chromatin occupancy, respectively. Paired transcriptomic-epigenetic technologies have been used to show how vaccines remodel the epigenetic makeup of the innate immune system.

Transcriptomic approaches have also been combined with proteomics. The initial techniques, such as PLAYR (Frei et al. 2016) did not allow transcriptome-wide analysis but were followed up by development of methods such as CITE-seq (Stoeckius et al. 2017) and REAP-seq (Peterson et al. 2017) that allow transcriptome-wide sequencing. These techniques specifically investigate cell surface proteins and have high throughput; recent developments in combinatorial indexing have enabled ultrahigh-throughput multi-omics techniques; epigenetic data has also been incorporated. Other types of proteins, such as nuclear proteins, have been paired with transcriptomics. These technologies have been used to investigate resistance to immune checkpoint inhibitors in cancer (Frangieh et al. 2021).

Of particular interest to immunology are T cell receptor sequencing (TCR-seq) and B cell receptor sequencing (BCR-seq). Both TCRs and BCRs can be reconstructed from scRNA-seq data. TCR sequencing can be combined with scRNA-seq, and paired analysis is commercially available through 10x Genomics. Trimodal single-cell TCR + RNA + protein analysis is possible via protocols such as ECCITE-seq (Mimitou et al. 2019). Paired single-cell BCR-RNA technologies have also found widespread use.

2.4 Spatial Transcriptomics

Since the first introduction of spatial transcriptomics by Ståhl et al. (2016), the method was adapted by other researchers and further modified which led to the development of spatial transcriptomic profilers based on the technology that involves in situ capturing of molecules, followed by ex situ sequencing. Other approaches that also can be used to generate spatially resolved transcriptome data include microdissection, in situ hybridization (ISH), and in situ sequencing (ISS). All these approaches predate the work of Ståhl et al. (2016), although they are not "omics" techniques, because they cannot collectively capture the whole transcriptome repertoire, but are only limited to a small part of it. Lastly, several in silico methods have been developed in recent years that allow reconstruction of spatial data with or without a preexisting reference map. The scope of this chapter focuses on omics methods (systems for spatial transcriptomics that can target the whole transcriptome) and are primarily based on in situ capturing protocol principles, as first devised by Ståhl et al. (2016). Currently, the most used systems for spatial transcriptomics that are commercially available as integrated platforms are Visium by 10x Genomics (Ståhl et al. 2016) and GeoMx (Merritt et al. 2020) by NanoString. The latest platform for "spatial molecular imaging" called CosMx (He et al. 2021), introduced by NanoString, is based on in situ hybridization (ISH) technique but only allows to quantify and visualize of up to 1000 RNA and 64 validated protein analytes.

The workflow of the Visium technology is comprised of six main steps and can be used for either fresh frozen or FFPE tissue. The first step involves tissue sectioning and placing it on a slide that contains capture probes with oligo(dT) primers. The specimen is then subjected to histological staining and imaging (step 2) followed by fixing and permeabilization on the slide (step 3). The released RNA can bind to capture probes, and barcoded cDNA is synthesized with Cy3-labeled nucleotides to construct libraries for sequencing (step 4). The barcodes are positional to indicate the exact X and Y position on the arrayed slide, which can be then mapped to the location on the tissue sample. The barcoded libraries can be then sequenced using a short-read sequencer (step 5), followed by the final step of data visualization and analysis using the 10X software tools (Ståhl et al. 2016).

Alike Visium, the GeoMx digital spatial profiling (DSP) technology developed by NanoString also offers an integrated system composed of hardware and software, which is combined with an NGS sequencer or nCounter for spatial profiling. The processing workflow is composed of five steps that are repeated for each ROI (region of interest), which are selected on the scanned tissue image. Before the selection of ROIs, a sectioned tissue is mounted on a slide and undergoes an antigen retrieval process, followed by incubation with a cocktail of primary PC-oligo-labeled antibodies and hybridization of RNA probes, as well as fluorescent staining to allow identification of tissue features and markers of interest (step 1). The slide is then placed on the instrument and subjected to fluorescent imaging which creates digital images that are used to guide the selection of ROIs (step 2). In the next step, the selected ROI is exposed to UV light which allows the oligonucleotides to be released in a region-specific manner (step 3), which is followed by their collection via microcapillary aspiration (step 4) and dispensing into a well of microtiter plate (step 5). Once all ROIs are collected in individual wells on a plate, the plate is digitally counted using the NanoString nCounter system or quantified on a standard NGS sequencer. The obtained reads can be then mapped back to each ROI that the oligos originated from (Merritt et al. 2020).

In comparison, the two systems described above are both based on a similar principle of using probes with positional barcoding, although the techniques of how the capturing of the target molecules is achieved are different. In addition, Visium only allows to profile RNA, while GeoMx DSP offers whole transcriptome and several panels of proteins that can be targeted.

3 Current Challenges in Single-Cell Transcriptomics and Related Technologies

This section encompasses current challenges encountered in scRNA-seq and adjacent techniques, as well as the progress that has been made toward their solutions. This is done with reference to the effect these challenges have on immunology.

3.1 Limitations

Despite the recent advances, scRNA-seq still suffers from high cost and low sensitivity. There are also numerous biases arising from the sequencing protocol and various challenges inherent in RNA data, which are discussed in Sect. 3.2. These challenges are largely shared by multi-omics and spatial technologies as well, although both come with their own unique limitations as well—primarily the lack of tools to analyze data.

Due to the abundance of new sequencing technologies, it is helpful to divide them into three categories: plate-based, droplet-based, and third generation. Plate-based methods such as CEL-seq2 (Hashimshony et al. 2016) have higher sensitivity compared to droplet-based methods such as Drop-seq (Macosko et al. 2015). CEL-seq2 can detect four times as many genes and twice as many cells as Drop-seq. Likewise, third-generation methods, they tend to have lower sensitivity than droplet-based methods, although this discrepancy has been reduced over time; recent protocols such as VASA-seq (Salmen et al. 2022) and SCAN-seq2 (Liao et al. 2023) have shown comparable sensitivity to plate-based methods such as SMART-seq3 (Hagemann-Jensen et al. 2020).

While combinatorial indexing protocols such as sciRNA-seq (Cao et al. 2017) have allowed even higher throughput, their decreased sensitivity has made them unsuitable for analysis of PBMCs (Ding et al. 2020) and clinical data (Hughes et al. 2020). Increased sensitivity can be achieved by increasing sequencing depth (i.e., the average number of times a nucleotide is sequenced), but only to a certain point—sensitivity tends to plateau for all protocols after reaching a depth of one million (Ziegenhain et al. 2017).

For detecting isoforms, an even larger sequencing depth would be needed (Hedlund and Deng 2018); it is more expensive to sequence deeper and thus is not always an option. Most genes express only one isoform in any given cell, and furthermore this choice of isoform is informative for subtyping (Li et al. 2020). However, simulation studies have raised doubts on the accuracy of isoform detection in scRNA-seq analyses due to the high sparsity of the data.

As throughput increases, computational complexity becomes an increasing limitation. This is especially true when considering multi-omics, in which common steps in scRNA-seq analysis cannot be directly mapped onto multi-omics analyses due to either insufficient or inadequate tools (Ma et al. 2022). The development of multi-omics-specific tools such as the MUON framework in Python and the addition of multi-omics techniques to Seurat indicate that the community is making progress in addressing these needs.

In particular, visualization becomes more difficult in the multi-omics context. While dimensionality reduction is a well-studied (but still active) field in the case of unimodal data, the multiple modalities of multi-omics data have their own challenges. There is a need for tools that can efficiently create a reasonable low-dimensional representation not only of the whole dataset but also simultaneously of each modality independently. In the case of scRNA-seq + scATAC-seq, the nature of the data means that both modalities of the data capture the cell at different stages in its development (Ma et al. 2020). This could affect pseudotime analysis if not taken into account.

Spatial RNA-seq suffers from even higher cost and low sensitivity. Current platforms, such as 10x Genomics' Visium and NanoString's GeoMx, cannot adequately capture data at the level of single cell (Li and Wang 2021). Spatial data is currently only practically feasible to acquire for 2D slices, whereas real biological interactions happen in 3D (Bressan et al. 2023). There are also fewer options to include multiple “omics” in an experiment with spatial data, although this will likely change as spatial data becomes more commonly collected.

3.2 *Statistical Solutions*

One of the most common types of biases in scRNA-seq data are amplification biases, biases caused due to the PCR amplification step of library preparation. This includes GC-bias (biased amplification based on transcript contents) and gene length bias. They are largely absent in UMI-based protocols and third-generation sequencing, although UMIs of insufficient length can lead to undercounting bias (Saunders et al. 2020).

Whether or not “zero inflation,” also called “gene dropout”—the overabundance of zero counts for genes in a given cell—can be considered purely due to amplification bias or not is a matter of considerable controversy. Failure to account for zero inflation can bias correlation and differential expression analyses. Traditionally, this is accounted for by modeling the data as being drawn from a zero-inflated count distribution such as the zero-inflated negative binomial distribution. However, nonzero-inflated models have been shown to adequately capture the measured zeros, and use of zero-inflated models can result in false conclusions (Silverman et al. 2020).

There is a growing consensus, especially in the case of UMI protocols, that zero-inflated models are unnecessary and even harmful (Jiang et al. 2022a). Rather, the zeros may better be explained due to the sample cell-type heterogeneity (Kim et al. 2020). The success of UMIs at removing zero inflation can be counterintuitive; UMIs do not increase the number of reads but rather allow the mapping of reads from the same original molecule to each other. Due to this behavior, a zero count observed via non-UMI protocols would remain a zero count if UMIs had been included. This leads to the conclusion that zero inflation is not in fact the overabundance of zeros, but rather the over counting of nonzero that makes the zeros seem more dramatic (Cao et al. 2021).

The efficacy of UMIs at removing zero inflation also depends on the origin of the zero. As UMIs preserve zeros, they cannot remove technical zeros caused by other stages in library preparation. One solution would be to use imputation, ideally only on the technical zeros as they represent missing values. However, it is impossible to completely distinguish between biological and technical zeros based only on observed counts. Thus, there is a need to develop imputation methods that account for this.

The second major bias to be considered in scRNA-seq is batch effects (also called plate effects). scRNA-seq data is often analyzed in batches, and these batches tend to have differences between them due to technical variation. If this effect is ignored, this difference can result in “discovering” distinct clusters of cells that are in fact the same cell type from different batches. One method of handling this is to sum the counts in each batch for some stages in the analysis, such as differential expression ("pseudobulking").

Other methods include fixed effects modeling, mixed effects modeling, and the use of surrogate variables. These methods are reviewed by Chen et al. (2020), who conclude that, while methods need to be better tailored to the statistical properties of scRNA-seq data, surrogate variable analysis (Leek and Storey 2007) was the most suitable at the time when the true batches were unknown—especially when grouping data into "pseudo-cells." In general, the best method of batch correction depends on the downstream task. ComBat-seq is one of the best methods to create batch-corrected data for differential gene expression analysis when the batches are known but is less effective on various standard benchmarks (Büttner et al. 2019) compared to methods such as Liger and Harmony.

Finally, there is the cell cycle effect, large differences between cells that are in fact the same cell type at different stages in the cell cycle. The naive method of removing cell cycle-related genes is not sufficient, as the cell cycle drives the expression of other, non-cell cycle genes. The first method to account for this, scLVM, performs best only when the cell cycle effect dominates the effect of differing cell types. Newer methods, such as ccRemover, use statistical tests to detect whether the dominating effect is the cell cycle before removing it. Latent variable methods continue to be the most popular for removing the cell cycle effect, with significant work being taken to ensure they remain practical as the scale of scRNA-seq datasets continues increasing (Lalchand et al. 2022). In some cases, the cell cycle can be identified and removed merely through a linear transformation of the data (Schwabe et al. 2020).

3.3 Deconvolution

An alternative approach that can be used to obtain transcriptomic data at single-cell resolution, without having to run a costly and laborious single-cell sequencing experiment, is computational deconvolution of bulk RNA-seq datasets. Since the beginning of large-scale use of high-throughput sequencing in the mid-2000s, an increasing amount of bulk RNA-seq data has been generated, and many datasets were made publicly available. Bulk sequencing technology is still widely used and is a preferable choice in contexts where single-cell resolution is not required, mostly due to its lower cost. The approach of deconvolution allows to make use of these datasets, although it comes with several technical challenges, mainly regarding the accuracy of deconvolution algorithms in being able to correctly decompose the bulk expression and estimate the proportions of cell populations. The overall goal of a deconvolution analysis is to infer proportions of cell types from bulk sequencing data, usually based on known expression patterns of certain genes that are expressed in specific cell types. Such an approach allows to transfer the information about cell populations from one source where cell types are annotated to another that only contains information about genes and their expression levels across samples. For

Table 1 List of selected deconvolution methods developed for transcriptomic data categorized into (i) gene marker-based, (ii) reference-based, and (iii) reference-free approaches, as proposed by Jin and Liu (2021). CIBERSORTx is a web-based application, MMAD and DSection are MATLAB libraries, whereas all other tools are R packages

Gene marker-based	
DSA	Based on linear modeling approach https://github.com/zhandong/DSA
MMAD	Based on linear modeling approach http://sourceforge.net/projects/mmad/
LinDeconSeq	Based on weighted robust linear regression https://github.com/lihuamei/LinDeconSeq
MCP-counter	A multistep framework based on several approaches https://github.com/ebecht/MCPcounter
Reference-based	
CIBERSORTx	Based on nonnegative least squares regression (NNLS) https://cibersortx.stanford.edu
EPIC	Based on constrained least squares linear regression https://github.com/GfellerLab/EPIC
TIMER	Based on linear least squares (LLS) method https://github.com/hanfeisun/TIMER
MuSiC2	Based on nonnegative least squares regression https://github.com/Jiaxin-Fan/MuSiC2
Reference-free	
DSection	Based on a Markov chain Monte Carlo approach https://web.cbio.uct.ac.za/~renaud/CRAN/web/CellMix/DSection.html
debCAM	Unsupervised approach based on linear mixing modeling (it can be also used as a semi-supervised or supervised method) https://bioconductor.org/packages/debCAM
LinSeed	Based on the singular value decomposition (SVD) approach https://github.com/ctlab/LinSeed
UNDO	Unsupervised approach based on a linear latent variable model https://bioconductor.org/packages/UNDO

example, the MuSiC (multi-subject single-cell deconvolution) software uses marker genes for expression of cell type and assigns them weights based on their cross-subject variance (as a proxy for informativeness). Genes with low cross-subject variance (informative) are up-weighted and those with high cross-subject variance (uninformative) are down-weighted.

The approach of deconvolution has become increasingly popular in recent years, and numerous new algorithms have been proposed (Table 1). These algorithms can be grouped into three categories based on the type of reference data they use: (i) gene marker-based, (ii) reference-based, and (iii) reference-free (Jin and Liu 2021). Methods that fall into the two former groups make use of gene lists or entire gene expression profiles with annotated cell types, respectively. The latter group includes algorithms that do not require any reference panels for the decomposition of the cell types, but the returned output includes clusters of cell populations that need to be annotated.

One of the main challenges that deconvolution methods need to tackle is the dynamic nature of tissue's cellular composition, regardless of whether it is in a healthy normal or diseased state. Tissue samples from different organs are composed of different cell-type populations; therefore, it is crucial to make sure the reference data originate from the same or similar clinical context as the bulk dataset, namely, the cell types expected to be inferred from the bulk dataset are present and well-annotated in the single-cell dataset.

4 Applications to Immunology and Disease

4.1 Immune Cell Development and Subtyping

scRNA-seq is often used to discover novel cell types and their developmental relationships with each other. It has been used to discover distinct subpopulations of innate lymphoid cells, dendritic cells, and germinal center B cells, among many others (Papalexi and Satija 2018). scRNA-seq can also be used to investigate heterogenicity within a single population, such as by cataloguing the diversity of T cell receptor sequences in T cells.

This methodology can also be applied to other species. One study of several vertebrates found that macrophages tended to have highly diverse expression profiles across species, while B and T cells were more conserved (Jiao et al. 2023). In *Drosophila*, scRNA-seq has enabled the discovery of two additional blood cell types, as well as further subtyping of the traditional three *Drosophila* blood cell types (Cattenoz et al. 2021). The various blood cell types in invertebrates are less understood than those of vertebrates; scRNA-seq has also led to the discovery of unique blood cell types in mosquitoes (Raddi et al. 2020) and shrimp (Yang et al. 2022).

Advances in scRNA-seq have not only enabled the splitting of cell types into distinct subtypes but also the charting of the developmental pathways leading to the diversification of such subtypes. This includes the development of *i*NKT cells, B cells, and T cells. More than just cellular development, scRNA-seq has also been used to discover new pathways leading to T cell exhaustion (Sun et al. 2022b). Much of this has been enabled through pseudotime analysis.

Immune cell trajectory inference is such a common problem that it has led to the development of specialized tools, such as Pseudocell Tracer, and sciCSR. Using paired scVDJ-seq in tools such as Dandelion has been shown to improve trajectory inference (Suo et al. 2023). The development of spatial techniques such as Slide-seqV2 (Stickels et al. 2021) has enabled "spatiotemporal" trajectory inference—that is, the tracing of a cell type's lineage through space and time.

4.2 Cancer and the Tumor Microenvironment

The progression of cancer is not only affected by the cancer cells themselves but also by their interactions with the surrounding cells; this is the "seed-and-soil" hypothesis. scRNA-seq has enabled high-resolution explorations of the tumor microenvironment (TME). From an immunological perspective, this has been applied both to studying the heterogeneity of noncancerous immune cells within tumors such as breast cancer, as well as the microenvironment of immunological cancers themselves, such as in diffuse large B cell lymphoma and chronic lymphocytic leukemia. Pseudotime analysis can be performed to uncover the development of cancer, and scRNA-seq data can aid in tumor subtyping and diagnosis. This section contains an overview of the scRNA-seq-enabled advances in the knowledge of the microenvironments of three hematological cancers—in particular, for non-Hodgkin's B cell lymphomas.

There are several types of B cell lymphomas, which have been further divided by cell of origin and microenvironment composition. The composition of the microenvironment is predictive of clinical outcome and also changes as the disease progresses (Kotlov et al. 2021). Furthermore, it is known that TME affects the epigenetic makeup of surrounding immune cells and that the local interactions between cells are clinically relevant (Autio et al. 2022).

Follicular lymphoma, a type of B cell lymphoma, has had numerous scRNA-seq studies of its microenvironment. These studies have revealed new ways of subtyping follicular lymphoma; based on scRNA-seq data of B cells in follicular lymphoma, the cancer can be largely categorized into two types, one of which is associated with poor prognosis. The same study revealed that the composition of T cells within the microenvironment is related to the grade of the cancer (Wang et al. 2022).

The TME of diffuse large B cell lymphoma (DLBCL), the most common B cell lymphoma, has also been studied. TMEs of DLBCL, as well as its subtypes primary central nervous system DLBCL (PCNS-DLBCL) and primary testicular DLBCL (PT-DLBCL), have all been investigated. In PCNS-DLBCL, the mechanisms behind T cell exhaustion and its impact on prognosis have been explored; Liu et al. (2023) found that the TME consisted primarily of exhausted CD8 T cells. In PT-DLBCL, the presence of S100A4+ tumor-associated macrophages in the TME are associated with worse prognosis (Bian et al. 2024). Bulk studies have been used to divide DLBCL into two distinct subtypes, activated B cell-like DLBCL and germinal center B cell-like DLBCL; however, no significant difference in the TMEs of these two types has been found (Ye et al. 2022).

The TMEs of a few other B cell lymphomas have been studied. In relapsed mantle cell lymphoma following CAR T cell treatment, the proportion of T cells in the TME decreases, while the proportion of myeloid cells increases, compared to the pretreatment TME (Jiang et al. 2022b). The co-development of the tumor and the TME in Waldenström macroglobulinemia (WM) was studied by Sun et al. (2022a), identifying CD19+ CD3+ B cells as a precursor to WM. Most scRNA-seq studies have so far been focused on the most common B cell lymphomas, primarily follicular lymphoma and DLBCL.

4.3 Diseases and the Immune Response

scRNA-seq can be used to understand the state of disease. It has been used to identify functional signatures identifying subpopulations of Th17 cells that had differing levels of pathogenicity for multiple sclerosis (Gaublomme et al. 2015). Several rheumatic autoimmune disorders have been studied using scRNA-seq data. scRNA-seq has been used to investigate at what point autoimmune disorders begin during embryo development and can be used to predict an organ recipient's risk of chronic lung allograft dysfunction.

scRNA-seq, when coupled with B cell receptor repertoire sequencing (BCR-seq), can aid in the identification of novel antibodies, such as for SARS-CoV-2 (Li et al. 2021). scRNA-seq complements scBCR-seq due to its ability to determine the cell type (crucially, whether the cell expressing a given antibody is activated or naive). Due to the importance of posttranslational modifications to antibody function, there is a need for the capability to include more "omics" in the search for antibodies.

Transcriptomics technologies have also been applied to the study of inflammation. A scRNA-seq study of patients with ileal Crohn's disease has identified a specific microenvironment makeup in inflamed tissues that was correlated with treatment failure (Martin et al. 2019). This microenvironment was found in several independent cohorts. scRNA-seq has been combined with proteomics and epigenomics in studies investigating inflammatory bowel diseases.

Spatial transcriptomics has also found its use in the study of inflammation, being used to characterize the differences between immune cells in healthy and inflamed tissue in inflammatory bowel disease. The same study revealed the tendency of certain types of fibroblasts to cluster spatially with certain types of macrophages, hinting at interactions between the two cell types. Studies in immune infiltration of adenocarcinoma tumors have shown distinct "warmed-up" spatial regions, the presence of which greatly improves patient outcomes (Zhang et al. 2023). Noncommunicable inflammatory skin diseases have also been investigated from a spatial perspective, with differences between gene expression in lesional and non-lesional skin being charted (Schäbitz et al. 2022).

4.4 Immunotherapies

scRNA-seq plays a role in the analysis and design of several immunotherapies. They have identified T cell populations associated with positive clinical outcomes in adoptive T cell therapy (Krishna et al. 2020) and T cell populations associated with the advent of colitis due to immune checkpoint blockade therapy (Mann et al.

2023), as well as macrophage populations predictive of clinical outcomes in colorectal cancer patients receiving immunotherapy (Liu et al. 2022). Immunotherapy for treating allergies has also been studied via transcriptomics, for example to understand the effects of dosage on patient response (Starchenka et al. 2019).

Chimeric antigen receptor (CAR) T cell therapy has produced promising results in clinical practice for blood cancers such as lymphomas, leukemias, and multiple myeloma, although progress on solid tumors has been slower. Two major difficulties in CAR T cell therapy are on-target off-tumor toxicities and antigen escape; it is essential to identify an array of good targets ("combinatorial antigen recognition") to maximize precision while limiting pathways for the cancer to develop countermeasures. scRNA-seq has played a role in the identification of such targets; however, the best targets may come about as a result of posttranslational modifications, requiring a multi-omics approach. Other aspects of CAR T cell therapy research that have benefited from the use of scRNA-seq include efficacy prediction as well as understanding the effect of TME immunosuppression on CAR T cells and host-CAR T cell cross talk. As CAR T cell therapy comes with a high risk of cytokine release syndrome (CRS) and immune effector cell-associated neurotoxicity syndrome (ICANS), scRNA-seq has also been used to investigate risk factors for the development of these toxicities.

These studies are frequently multi-omic in nature. It should not come as a surprise that single-cell T cell receptor repertoire profiling is frequently used in scRNA-seq CAR T cell studies to track the changing phenotypes of CAR T cells over time (Haradhvala et al. 2022). scRNA-seq has also been augmented with scATAC-seq and CITE-seq to investigate factors of T cell persistence. Spatial transcriptomics has not yet been applied to the study of CAR T cells (Yang et al. 2023).

5 Conclusions

This chapter has charted the development of scRNA-seq and related technologies, with reference to their applications in immunology. As the cost continues to go down and the throughput and sensitivity continue to increase, transcriptomics technologies will become increasingly prevalent in immunological research, yielding insights into the tumor microenvironment, autoimmune disorders, immunotherapies, and more.

In the near term, it is expected that third-generation sequencing methods will continue closing the sensitivity gap with other methods. Isoform detection will become increasingly feasible, and throughput will no longer be limited by how much can be processed in the wet lab, but rather by the ability of the dry lab to cope with such large datasets. scRNA-seq will commonly be done in conjunction with several "omics" at a time, allowing a thorough profiling of a cell's state. Spatial technologies will achieve subcellular resolution but will have a slower adoption due to their high cost. Nevertheless, the unique perspective spatial gives will make it play an important role in the future of immunology.

References

Amarasinghe SL, Su S, Dong X et al (2020) Opportunities and challenges in long-read sequencing data analysis. Genome Biol 21:30

Angermueller C, Clark SJ, Lee HJ et al (2016) Parallel single-cell sequencing links transcriptional and epigenetic heterogeneity. Nat Methods 13:229–232

Autio M, Leivonen SK, Brück O et al (2022) Clinical impact of immune cells and their spatial interactions in diffuse large B-cell lymphoma microenvironment. Clin Cancer Res 28:781–792

Bakken TE, Hodge RD, Miller JA et al (2018) Single-nucleus and single-cell transcriptomes compared in matched cortical cell types. PLoS One 13:e0209648

Bian Z, Gu B, Shi G et al (2024) The single-cell landscape reveals unique tumor subsets and microenvironments associated with poor clinical outcomes in primary testicular diffuse large B-cell lymphoma. Genes Dis 11:80–83

Blondel V, Guillaume JL, Lambiotte R et al (2008) Fast unfolding of communities in large networks. J Stat Mech-Theory E

Bressan D, Battistoni G, Hannon GJ (2023) The dawn of spatial omics. Science 381:eabq4964

Büttner M, Miao Z, Wolf FA et al (2019) A test metric for assessing single-cell RNA-seq batch correction. Nat Methods 16:43–49

Cao J, Packer JS, Ramani V et al (2017) Comprehensive single-cell transcriptional profiling of a multicellular organism. Science 357:661–667

Cao J, Cusanovich DA, Ramani V et al (2018) Joint profiling of chromatin accessibility and gene expression in thousands of single cells. Science (New York, NY) 361:1380–1385

Cao Y, Kitanovski S, Küppers R et al (2021) UMI or not UMI, that is the question for scRNA-seq zero-inflation. Nat Biotechnol 39:158–159

Cattenoz PB, Monticelli S, Pavlidaki A et al (2021) Toward a Consensus in the Repertoire of Hemocytes Identified in Drosophila. Front Cell Dev Biol 9:9

Chen S, Lake BB, Zhang K (2019) High-throughput sequencing of the transcriptome and chromatin accessibility in the same cell. Nat Biotechnol 37:1452–1457

Chen W, Zhang S, Williams J et al (2020) A comparison of methods accounting for batch effects in differential expression analysis of UMI count based single cell RNA sequencing. Comput Struct Biotechnol J 18:861–873

Datlinger P, Rendeiro AF, Boenke T et al (2021) Ultra-high-throughput single-cell RNA sequencing and perturbation screening with combinatorial fluidic indexing. Nat Methods 18:635–642

Denisenko E, Guo BB, Jones M et al (2020) Systematic assessment of tissue dissociation and storage biases in single-cell and single-nucleus RNA-seq workflows. Genome Biol 21:130

Dey SS, Kester L, Spanjaard B et al (2015) Integrated genome and transcriptome sequencing of the same cell. Nat Biotechnol 33:285–289

Ding J, Adiconis X, Simmons SK et al (2020) Systematic comparison of single-cell and single-nucleus RNA-sequencing methods. Nat Biotechnol 38:737–746

Dobin A, Davis CA, Schlesinger F et al (2013) STAR: ultrafast universal RNA-seq aligner. Bioinformatics (Oxford, England) 29:15–21

Frangieh CJ, Melms JC, Thakore PI et al (2021) Multimodal pooled perturb-CITEseq screens in patient models define mechanisms of cancer immune evasion. Nat Genet 53:332–341

Frei AP, Bava FA, Zunder ER et al (2016) Highly multiplexed simultaneous detection of RNAs and proteins in single cells. Nat Methods 13:269–275

Gaublomme JT, Yosef N, Lee Y et al (2015) Single-cell genomics unveils critical regulators of Th17 cell pathogenicity. Cell 163:1400–1412

Gross A, Schoendube J, Zimmermann S et al (2015) Technologies for Single-Cell Isolation. Int J Mol Sci 16:16897–16919

Grosselin K, Durand A, Marsolier J et al (2019) High-throughput single-cell ChIPseq identifies heterogeneity of chromatin states in breast cancer. Nat Genet 51:1060–1066

Habib N, Avraham-Davidi I, Basu A et al (2017) Massively parallel single-nucleus RNA-seq with DroNc-seq. Nat Methods 14:955–958

Hagemann-Jensen M, Ziegenhain C, Chen P et al (2020) Single-cell RNA counting at allele and isoform resolution using Smart-seq3. Nat Biotechnol 38:708–714

Han L, Zi X, Garmire LX et al (2014) Co-detection and sequencing of genes and transcripts from the same single cells facilitated by a microfluidics platform. Sci Rep 4:6485

Haradhvala NJ, Leick MB, Maurer K et al (2022) Distinct cellular dynamics associated with response to CAR-T therapy for refractory B-cell lymphoma. Nat Med 28 (September):1848–1859

Hashimshony T, Senderovich N, Avital G et al (2016) CEL-Seq2: sensitive highlymultiplexed single-cell RNA-Seq. Genome Biol 17:77

He S, Bhatt R, Birditt B et al (2021) High-plex multiomic analysis in ffpe tissue at single-cellular and subcellular resolution by spatial molecular imaging. bioRxiv pp 2021–2011

Hedlund E, Deng Q (2018) Single-cell RNA sequencing: technical advancements and biological applications. Mol Asp Med 59:36–46

Hosic S, Murthy SK, Koppes AN (2016) Microfluidic sample preparation for single cell analysis. Anal Chem 88:354–380

Hou Y, Guo H, Cao C et al (2016) Single-cell triple omics sequencing reveals genetic, epigenetic, and transcriptomic heterogeneity in hepatocellular carcinomas. Cell Res 26:304–319

Hu T, Chitnis N, Monos D et al (2021) Next-generation sequencing technologies: an overview. Hum Immunol 82:801–811

Huang D, Ma N, Li X et al (2023) Advances in single-cell RNA sequencing and its applications in cancer research. J Hematol Oncol 16:98

Hughes TK, Wadsworth MH, Gierahn TM et al (2020) Second-Strand synthesis-based massively parallel scRNA-Seq reveals cellular states and molecular features of human inflammatory skin pathologies. Immunity 53:878–894.e7

Jiang R, Sun T, Song D et al (2022a) Statistics or biology: the zero-inflation controversy about scRNA-seq data. Genome Biol 23:31

Jiang VC, Hao D, Jain P et al (2022b) TIGIT is the central player in T-cell suppression associated with CAR T-cell relapse in mantle cell lymphoma. Mol Cancer 21:185

Jiao A, Zhang C, Wang X et al (2023) Single-cell sequencing reveals the evolution of immune molecules across multiple vertebrate species. J Adv Res 55:73. https://doi.org/10.1016/j.jare.2023.02.017

Jin H, Liu Z (2021) A benchmark for rna-seq deconvolution analysis under dynamic testing environments. Genome Biol 22:1–23

Kharchenko PV (2021) The triumphs and limitations of computational methods for scRNA-seq. Nat Methods 18:723–732

Kim TH, Zhou X, Chen M (2020) Demystifying "drop-outs" in single-cell UMI data. Genome Biol 21:196

Kotlov N, Bagaev A, Revuelta MV et al (2021) Clinical and biological subtypes of B-cell lymphoma revealed by microenvironmental signatures. Cancer Discov 11:1468–1489

Krishna S, Lowery FJ, Copeland A et al (2020) Stem-like CD8 T cells mediate response of adoptive cell immunotherapy against human cancer. Science (New York, NY) 370:1328–1334

La Manno G, Soldatov R, Zeisel A et al (2018) RNA velocity of single cells. Nature 560:494–498

Lalchand V, Ravuri A, Dann E et al (2022) Modelling technical and biological effects in scRNA-seq data with scalable GPLVMs. In: Proc Mach Learn Comput Biol, PMLR, pp. 46–60

Leek JT, Storey JD (2007) Capturing heterogeneity in gene expression studies by surrogate variable analysis. PLoS Genet 3:e161

Li F, Luo M, Zhou W et al (2021) Single cell RNA and immune repertoire profiling of COVID-19 patients reveal novel neutralizing antibody. Protein Cell 12:751–755

Li J, Comeau HY, Zhang Z et al (2020) Landscape of transcript isoforms in single T cells infiltrating in non-small-cell lung cancer. J Genet Genomics 47:373–388

Li X, Wang CY (2021) From bulk, single-cell to spatial RNA sequencing. Int J Oral Sci 13:1–6

Liao Y, Liu Z, Zhang Y et al (2023) High-throughput and high-sensitivity full-length single-cell RNA-seq analysis on third-generation sequencing platform. Cell Discov 9:1–5

Liu N, Jiang C, Yao X et al (2023) Single-cell landscape of primary central nervous system diffuse large B-cell lymphoma. Cell Discov 9:1–17
Liu X, Yan G, Xu B et al (2022) Evaluating the role of IDO1 macrophages in immunotherapy using scRNA-seq and bulk-seq in colorectal cancer. Front Immunol 13
Ma A, Xin G, Ma Q (2022) The use of single-cell multi-omics in immuno-oncology. Nat Commun 13:2728
Ma S, Zhang B, LaFave LM et al (2020) Chromatin potential identified by shared single-cell profiling of RNA and chromatin. Cell 183:1103–1116.e20
Macaulay IC, Haerty W, Kumar P et al (2015) G&T-seq: parallel sequencing of single-cell genomes and transcriptomes. Nat Methods 12:519–522
Macosko EZ, Basu A, Satija R et al (2015) Highly parallel genome-wide expression profiling of individual cells using Nanoliter droplets. Cell 161:1202–1214
Mann JE, Lucca L, Austin MR et al (2023) ScRNA-seq defines dynamic T-cell subsets in longitudinal colon and peripheral blood samples in immune checkpoint inhibitor-induced colitis. J Immunother Cancer 11:e007358
Martin JC, Chang C, Boschetti G et al (2019) Single-cell analysis of Crohn's disease lesions identifies a pathogenic cellular module associated with resistance to anti-TNF therapy. Cell 178:1493–1508.e20
Merritt CR, Ong GT, Church SE et al (2020) Multiplex digital spatial profiling of proteins and rna in fixed tissue. Nat Biotechnol 38(5):586–599
Mimitou EP, Cheng A, Montalbano A et al (2019) Expanding the CITE-seq toolkit: detection of proteins, transcriptomes, clonotypes and CRISPR perturbations with multiplexing, in a single assay. Nat Methods 16:409–412
Pan Y, Cao W, Mu Y et al (2022) Microfluidics facilitates the development of single-cell RNA sequencing. Biosensors 12:450
Papalexi E, Satija R (2018) Single-cell RNA sequencing to explore immune cell heterogeneity. Nat Rev Immunol 18:35–45
Peterson VM, Zhang KX, Kumar N et al (2017) Multiplexed quantification of proteins and transcripts in single cells. Nat Biotechnol 35:936–939
Picelli S, Björklund FOR et al (2013) Smart-seq2 for sensitive full-length transcriptome profiling in single cells. Nat Methods 10:1096–1098
Pierson E, Yau C (2015) ZIFA: dimensionality reduction for zero-inflated single-cell gene expression analysis. Genome Biol 16:241
Raddi G, Barletta ABF, Efremova M et al (2020) Mosquito cellular immunity at single-cell resolution. Science (New York, NY) 369:1128–1132
Rosenberg AB, Roco CM, Muscat RA et al (2018) Single-cell profiling of the developing mouse brain and spinal cord with split-pool barcoding. Science 360:176–182
Salmen F, De Jonghe J, Kaminski TS et al (2022) High-throughput total RNA sequencing in single cells using VASA-seq. Nat Biotechnol 40:1780–1793
Saunders K, Bert AG, Dredge BK et al (2020) Insufficiently complex unique molecular identifiers (UMIs) distort small RNA sequencing. Sci Rep 10:14593
Schäbitz A, Hillig C, Mubarak M et al (2022) Spatial transcriptomics landscape of lesions from non-communicable inflammatory skin diseases. Nat Commun 13(December):7729
Schmidt MHW, Vogel A, Denton AK et al (2017) De novo assembly of a new Solanum pennellii accession using Nanopore sequencing. Plant Cell 29:2336–2348
Schwabe D, Formichetti S, Junker JP et al (2020) The transcriptome dynamics of single cells during the cell cycle. Mol Syst Biol 16:e9946
Shembekar N, Chaipan C, Utharala R et al (2016) Droplet-based microfluidics in drug discovery, transcriptomics and high-throughput molecular genetics. Lab Chip 16:1314–1331
Sheng K, Cao W, Niu Y et al (2017) Effective detection of variation in single-cell transcriptomes using MATQ-seq. Nat Methods 14:267–270
Silverman JD, Roche K, Mukherjee S et al (2020) Naught all zeros in sequence count data are the same. Comput Struct Biotechnol J 18:2789–2798

Ståhl PL, Salmén F, Vickovic S et al (2016) Visualization and analysis of gene expression in tissue sections by spatial transcriptomics. Science 353:78–82

Starchenka S, Heath MD, Lineberry A et al (2019) Transcriptome analysis and safety profile of the early-phase clinical response to an adjuvanted grass allergoid immunotherapy. World Allergy Organization J 12:100087

Stickels RR, Murray E, Kumar P et al (2021) Highly sensitive spatial transcriptomics at near-cellular resolution with slide-seqV2. Nat Biotechnol 39:313–319

Stoeckius M, Hafemeister C, Stephenson W et al (2017) Simultaneous epitope and transcriptome measurement in single cells. Nat Methods 14:865–868

Sun H, Fang T, Wang T et al (2022a) Single-cell profiles reveal tumor cell heterogeneity and immunosuppressive microenvironment in Waldenström macroglobulinemia. J Transl Med 20: 576

Sun K, Xu R, Ma F et al (2022b) scRNA-seq of gastric tumor shows complex intercellular interaction with an alternative T cell exhaustion trajectory. Nat Commun 13:4943

Suo C, Polanski K, Dann E et al (2023) Dandelion uses the single-cell adaptive immune receptor repertoire to explore lymphocyte developmental origins. Nat Biotechnol:1–12

Svensson V, Gayoso A, Yosef N et al (2020) Interpretable factor models of singlecell RNA-seq via variational autoencoders. Bioinformatics 36:3418–3421

Traag VA, Waltman L, van Eck NJ (2019) From Louvain to Leiden: guaranteeing well-connected communities. Sci Rep 9:5233

Trapnell C, Cacchiarelli D, Grimsby J et al (2014) The dynamics and regulators of cell fate decisions are revealed by pseudotemporal ordering of single cells. Nat Biotechnol 32:381–386

Uhlen M, Quake SR (2023) Sequential sequencing by synthesis and the next-generation sequencing revolution. Trends Biotechnol 41:1565–1572

Wang X, Nissen M, Gracias D et al (2022) Single-cell profiling reveals a memory B cell-like subtype of follicular lymphoma with increased transformation risk. Nat Commun 13:6772

Xing QR, Farran CAE, Zeng YY et al (2020) Parallel bimodal single-cell sequencing of transcriptome and chromatin accessibility. Genome Res 30:1027–1039

Xiong H, Luo Y, Wang Q et al (2021) Single-cell joint detection of chromatin occupancy and transcriptome enables higher-dimensional epigenomic reconstructions. Nat Methods 18:652–660

Yang J, Chen Y, Jing Y et al (2023) Advancing CAR T cell therapy through the use of multidimensional omics data. Nat Rev Clin Oncol 20:211–228

Yang P, Chen Y, Huang Z et al (2022) Single-cell RNA sequencing analysis of shrimp immune cells identifies macrophage-like phagocytes. elife 11:e80127

Ye X, Wang L, Nie M et al (2022) A single-cell atlas of diffuse large B cell lymphoma. Cell Rep 39: 110713

Zhang L, Zhang S (2020) Comparison of computational methods for imputing single-cell RNA-sequencing data. IEEE/ACM Trans Comput Biol Bioinform 17:376–389

Zhang R, Feng Y, Ma W et al (2023) Spatial transcriptome unveils a discontinuous inflammatory pattern in proficient mismatch repair colorectal adenocarcinoma. Fundam Res 3:640–646

Ziegenhain C, Vieth B, Parekh S et al (2017) Comparative analysis of single-cell RNA sequencing methods. Mol Cell 65:631–643.e4

Systemic Communication via Extracellular Vesicles in Cancer: The Body's Wireless Internet System

Laura Patras and Doru Paul

Contents

Abstract Functional integration in eukaryotic organisms is achieved, on one hand, through a hierarchical architecture and, on the other, through the communication between the cells, tissues, and organs via secreted factors, such as cytokines, chemokines, metabolites, hormones, as well as extracellular vesicles (EVs). Communication between cells, tissues, and organs is essential both in health, when it is crucial for homeostasis maintenance, and in disease, when it becomes corrupted during the development and progression of systemic diseases, particularly in cancer. Cancer is a systemic disease even before it becomes clinically apparent. The host organism is not a passive bystander during cancer development, but it actively

---

L. Patras · D. Paul (✉)
Faculty of Biology and Geology, Department of Molecular Biology and Biotechnology and Centre for Systems Biology, Biodiversity and Bioresources (3B), Babes-Bolyai University, Cluj-Napoca, Romania

Attending Hematology/Oncology, Weill Cornell Medicine, Department of Medicine, Division of Hematology/Oncology, Weill Cornell Medical College, New York-Presbyterian Hospital, New York, USA
e-mail: dop9054@med.cornell.edu

J. Barciszewski (ed.), *Systems Biology II*, RNA Technologies 15,
https://doi.org/10.1007/978-3-031-62178-9_10

contributes to the process from its inception. Once formed, the tumor exerts its systemic effects, effectively "hacking" into the various host communication systems to support cancer growth. Here, we highlight the multifaceted roles of EVs in the tumor-host communication. This communication is especially important for the metastatic process and in orchestrating therapeutically intractable complications associated with advanced cancers, such as cachexia. Ultimately, deciphering the complex interactions between the tumor and its host will lead to novel approaches to interrupt this pathologic communication and to develop more effective anticancer therapies.

Keywords Extracellular vesicles · Systemic effects · Cancer · Homeostasis

1 Introduction

Communication is one of the key features of life. All cells belonging to one of the three life domains (Bacteria, Archaea, or Eukarya) on Earth share this feature. Among the different secreted factors used by cells for communication (i.e., cytokines, metabolites, hormones, etc.) extracellular vesicles (EVs) play kaleidoscopic roles. In the human organism, EVs are involved both in health maintenance, upholding homeostasis and integrating organism's components into a functional whole, and, also, in health deterioration, EVs inducing and perpetuating various pathologic states (Fig. 1).

In cancer, communication through EVs is essential for the metastatic process. Remarkably, EVs direct metastases to specific distal locations and have been found responsible for several systemic complications associated with the malignant state. As systemic carriers of oncogenic signals, EVs are valuable diagnostic and prognostic biomarkers, and also promising therapeutic targets to obstruct cancer progression.

2 Brief History of Extracellular Vesicles: From "Cellular Dust" to Microcosm

The field of EV biology emerged almost 75 years ago (Couch et al. 2021) with the first observations dating back to the early 1940s, when Erwin Chargaff and Randolph West isolated platelet-derived particles from human plasma upon high-speed centrifugation, describing them as "lipoproteins of a very high particle weight" with high clotting potential (Chargaff and West 1946). Later independent observations by Peter Wolf in 1967 referred to the observed subcellular coagulants by electron microscopy as "platelet dust" (Wolf 1967).

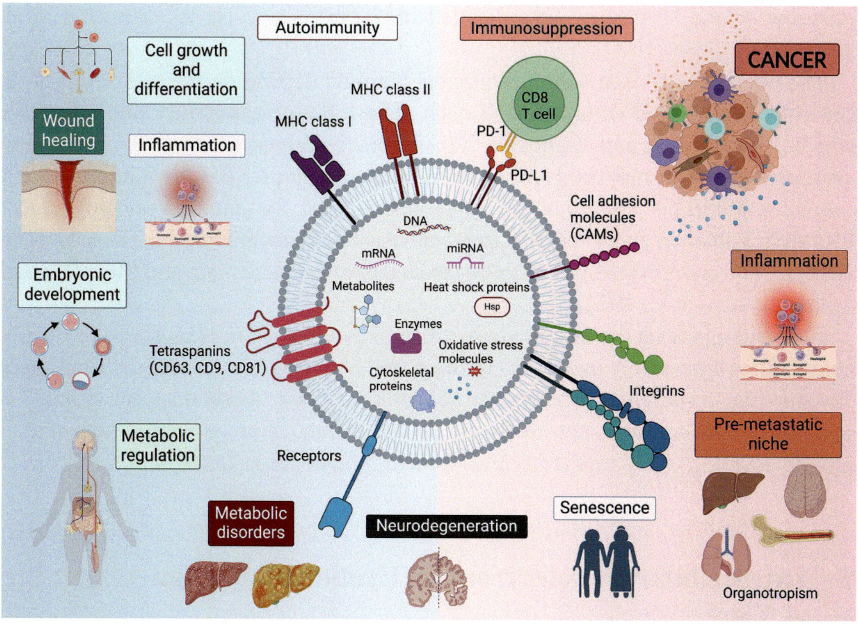

Fig. 1 Extracellular vesicles as mediators of homeostasis and systemic disease. Generalized representation of the EV architecture and cargo containing nucleic acids (DNA, RNA, miRNAs), proteins (cytoskeleton, heat shock proteins, enzymes), proteins and molecules involved in oxidative stress, and metabolites. The EV surface is commonly decorated with tetraspanins, cell adhesion molecules, integrins, receptors, and immunomodulatory proteins involved in the interaction with the immune system (e.g., MHC class I and II) and that can interfere with T cell cytotoxic function (PD-L1 is depicted here). Altogether, EVs orchestrate homeostatic processes (cell growth and differentiation, development, and metabolic regulation) or processes associated with wound healing and repair. Nevertheless, EVs are mediators of systemic disease including metabolic and neurodegenerative disorders, senescence, autoimmunity, and cancer where they don't only propagate oncogenic signals locally but also over long distances through the systemic circulation, where they reach future metastatic sites and install an inflammatory and immunosuppressive environment (pre-metastatic niches), ensuring circulating tumor cell homing and growth. Created with Biorender.com

However, it wasn't until the early 1980s that the specific EV research field was founded through two pioneering and complementary studies on the generation of intraluminal vesicles from reticulocytes—defined as exosomes—released via a specific secretory pathway (Harding et al. 1983) and that selectively externalized the transferrin receptor (Pan and Johnstone 1983). Thereon, research interest in EVs significantly increased when they were demonstrated to play a role in mediating immune responses through antigen presentation (Raposo et al. 1996) and to shuttle functional nucleic acids (mRNAs, miRNAs) between donor and recipient cells (Valadi et al. 2007). In the cancer research field, pioneering studies on EVs granted a more comprehensive, holistic view on cancer. Critically, tumor EVs were demonstrated to be a prerequisite for pre-metastatic niche (PMN) formation, and specific integrins on the surface of EVs dictated organotropic metastases (Hoshino et al. 2015).

The content of EVs and their biological function depend on the cell of origin. In general, EVs derived from immune cells have immunostimulatory and antitumor effects (Wieckowski and Whiteside 2006). Tumor-derived EVs were implicated in the immune system-mediated antitumor response as a source of tumor antigens for T cell cross-priming (Wolfers et al. 2001), relevant for the immuno-oncology field. EVs are involved in every step of carcinogenesis from cancer initiation (Kosaka et al. 2012) and promotion (Skog et al. 2008) to cancer invasion (Zhang and Ma 2012) and metastasis (Liu et al. 2022).

Key findings from the last 30 years irreversibly changed the landscape of the EV field, with progressively increasing findings and a wider geographical distribution of the scientific community. The former "cellular dust" or "waste disposal system" is now acknowledged as a key mediator of intercellular communication and that is actively shaping every aspect of development, homeostasis, and disease.

3 Extracellular Vesicles from an Evolutionary Point of View

EVs, which resemble encapsulated viruses in their structure, are widely ubiquitous both in unicellular and multicellular, evolved organisms and conserved across all life, from prokaryotes (bacteria, archaea) to eukaryotes (protists, fungi, plants, and animals) (Deatherage and Cookson 2012; Munhoz da Rocha and Amatuzzi 2020). The mechanism of release of membrane vesicles (MVs) into the environment is mostly conserved in prokaryotes. Gram-negative bacteria have two main routes for bacterial EV (bEV) formation, explosive cell lysis and membrane blebbing, whereas Gram-positive bacteria release cytoplasmic membrane vesicles via endolysins or other peptidoglycan-damaging enzymes (Roier et al. 2016; Xie et al. 2023). In most archaea, the outward protrusion of vesicles and fission of the membrane are facilitated by the cell division (Cdv) proteins that are homologous to the eukaryotic endosomal sorting complex required for transport (ESCRT-III) protein family, and vacuolar sorting protein (Vps4) homologues are also present (Deatherage and Cookson 2012; Roier et al. 2016). Moreover, archaea possess an archaeal secretion system (SecY/Sec61p) and its associated N-glycosylation machinery (Gould et al. 2016). The ESCRT-III homologues have also been found in microbial eukaryotes where they induce the formation of intraluminal vesicles (ILVs) that create multivesicular bodies (MVBs) (Deatherage and Cookson 2012).

EVs actively participate in the exchange of biological information which occurs cooperatively or offensively for survival purposes. For example, EV-mediated cargo transfer occurs between the microbial organisms residing in biofilms, in liquid environments, where the horizontal (also known as lateral) gene transfer (HGT) enables the long-distance transport of DNA or RNA protected from DNAses or

RNAses (Koeppen et al. 2016; Renelli et al. 2004) along with other biomolecules such as proteins, phospholipids, polysaccharides, and even quorum sensing signals (e.g., *Pseudomonas* quinolone signal) (Kulp and Kuehn 2010; Mashburn and Whiteley 2005; Mashburn-Warren and Whiteley 2006). This type of information transfer that is likely to have existed throughout evolution represents a driving evolutionary force that gradually contributed to prokaryotic genetic diversity, thereby emphasizing EVs as promoting evolutionary fitness and coordinating the group behavior of various unicellular life forms (Mashburn and Whiteley 2005; Sieber et al. 2017). The epitome of this effect is the rapid spread of antibiotic resistance genes between bacteria, due to selective pressure (Lee et al. 2022). However, throughout evolution, the HGT was also responsible for the transfer of genes between species or across domains of life, raising questions about the evolution of life on Earth and whether EVs might have played a critical role in this process.

Prokaryotes are unicellular life forms without a nucleus and rather distinct chemical and metabolic signatures, and strikingly different from Eukarya which are complex multicellular life forms consisting of membranous cells incorporating a nucleus and organelles with specific functions. However, all life on Earth can be traced back to about 4 billion years ago, to the last universal common ancestor (LUCA) from which it was initially believed that Eukarya, Bacteria, and Archaea have separated into the three distinct branches of the tree of life. However, due to genomic technological advances, molecular evolutionists challenged previous knowledge and proposed a new perspective on the tree of life (Lake 1988) based on the "two-domain tree" hypothesis suggesting that eukaryotes evolved later via endosymbiosis between bacteria and archaea, which led to the high genomic and cellular complexity of eukaryotes. Increasing evidence suggests that eukaryotes emerged via symbiosis with bacteria from within Asgard or Asgardarchaeota superphylum of archaea, with Heimdallarchaea being our closest ancestor to date, containing proteins with a eukaryotic signature and a more complex cell architecture consisting in actin cytoskeleton, as well as long extended protrusions (Eme et al. 2023; Zaremba-Niedzwiedzka et al. 2017). The endomembrane system is considered a key innovation in the evolutionary road of eukaryotes as it led to the compartmentalization of the cell revolutionizing energy production and use (towards entropy), as well as other cellular functions and protection from dangerous molecules (Gould et al. 2016). An intriguing hypothesis on the evolutionary origin of the eukaryotic elaborate endomembrane system (e.g., endoplasmic reticulum, nucleus, Golgi, lysosomes, peroxisomes, autophagosomes), with endosymbiosis as a prerequisite for this step, was proposed by Gould et al. (Gould et al. 2016). Gould proposed that due to lack of capacity to phagocytose—a function that would require a fully functional endomembrane system and cytosolic ATP—archaea enclosed within its membrane the bacterial endosymbiont, which in turn secreted outer membrane vesicles (OMVs). These OMVs then either accumulated and fused naturally within the host's cytosol creating the endomembrane system or fused with the host's plasma membrane, thus providing an explanation for Eukarya possessing bacterial-type cellular membranes composed of fatty acid esters rather than isoprene ethers like archaea (Gould et al. 2016).

Another intriguing aspect of EV's biology that currently remains unsolved is their resemblance to viruses (especially retroviruses) from both a structural and functional aspect, with the main differences being that EVs don't replicate and are not infectious (Nolte-'t Hoen et al. 2016). Viruses are infectious nanostructures (averaging less than 200 nm), believed to have originated via a regressive or reductive process, that contain DNA or RNA surrounded by a protein coat named capsid that co-opt the host cell biosynthetic processes for reproduction. Besides the genetic material which encodes for the viral proteins, protein-enveloped viruses exploit the host cell's protein translational modification machinery for protein glycosylation (Krishnamoorthy et al. 2009; Vigerust and Shepherd 2007). The lipid envelopes of viruses are derived from the lipids of the cells in which the viruses have replicated, either directly at the plasma membrane through the viral maturation process known as "budding" or through budding into the cellular endomembrane system and release via the secretory pathway (Lorizate and Kräusslich 2011). Thus, viruses likely hijack the cellular pathways used to make EVs to avoid being recognized by the immune system of the host (Hassan et al. 2021; Krishnamoorthy et al. 2009; Ramakrishnaiah et al. 2013), and therefore there is an overlap in EVs shed from virus-infected cells and defective viruses or virus-like particles extruded from infected cells. However, it remains unknown whether EVs appeared first as a means of intercellular communication and viruses/bacteriophages exploited them or whether EVs and viruses co-evolved (Nolte-'t Hoen et al. 2016).

Given these considerations, it appears that extracellular vesicles played a major role in shaping life on Earth and the evolution of life toward complexity and least energy expenditure. Throughout this paper we will highlight how EVs continue to shape the complex human body via their systemic effects, with an emphasis on their role in diseases such as cancer.

4 Current Consensus and Knowledge on Extracellular Vesicles

4.1 Definition, Heterogeneity, and Biogenesis Mechanisms

EVs are a heterogenous group of natural membrane-enclosed nano- and microstructures constitutively secreted by most cells into the extracellular space (Yáñez-Mó et al. 2015). Their unique role stems from the fact that EVs are a universal, evolutionarily conserved mechanism of intercellular communication, mediating both inter-kingdom and intra-kingdom communication (Chronopoulos and Kalluri 2020). EV heterogeneity is reflected by the size (micro- or nano-vesicles, small extracellular vesicles), morphology, composition, and biogenesis (exosomes and ectosomes) (van Niel et al. 2018), with different EV subpopulations named based on the cell type or tissue of origin (prostasomes, neutrophil-derived structures); the specific state of the cell: healthy or diseased, apoptotic (apoptotic body, apoptotic bleb), senescent, and stressed; or even the proposed functions of the vesicles (calcifying matrix vesicles, argosomes, tolerosomes, migrasomes) (Buzas 2023; Colombo et al. 2014).

The term "EV" is the consensus used for encompassing all types of membrane vesicles irrespective of their differences, with the main types of EVs categorized based on their biogenesis mechanism (ectosomes and exosomes with endosomal origin) or size (small and large EVs). Ectosomes (known as shedding vesicles, microparticles, or microvesicles) refer to EVs with high size range (70–1000 nm) that are released by the outward budding of the plasma membrane (Colombo et al. 2014). Exosomes are small (40–160 nm) vesicles of endosomal origin that subsequently interact with other intracellular vesicles and organelles and finally form intracellularly within the multivesicular bodies (MVBs) and are secreted via fusion of MVBs with the plasma membrane (Johnstone et al. 1987; Kalluri and LeBleu 2020). Most of the studies today focus on two EVs groups, the microvesicles and exosomes, microvesicles being the larger ones. However, small EVs appear to be the most promising vesicle group regarding to their bioactive cargo (proteins, lipids, nucleic acids, and metabolites) and the functional outcome of their uptake by target cells (Colombo et al. 2014; Kalluri and LeBleu 2020; Lucotti et al. 2022; Yáñez-Mó et al. 2015).

Another distinct type of extracellular particles (EPs) secreted by cells consists of the recently discovered exomeres and supermeres, through the efforts of dissecting the heterogeneity of EV subtypes. Exomeres (Zhang et al. 2018; Zhang et al. 2019) are a distinct population of a membranous extracellular nanoparticles (~35 nm) enriched in metabolites and proteins. Supermeres are similar in size to exomeres (~50 nm) but carry distinct protein cargo, enriched in glycolytic enzymes and oncoproteins, and the highest amount of RNA (e.g., miRNAs) among all EVPs (Zhang et al. 2019).

4.2 Extracellular Vesicle Cargo

4.2.1 Protein Composition of EVs

There is an extensive catalogue of protein cargo for different EV types both under physiological or pathological conditions, or even based on the EV isolation method. Some of the EV proteins emerged as pan-EV markers, commonly present in most EVs (Hoshino et al. 2020) with the most abundant proteins represented by cytoskeletal, cytosolic, heat shock, and plasma membrane proteins, as well as proteins involved in vesicle trafficking, with organelle-specific proteins being less abundant (Witwer et al. 2013).

EV subpopulations can often be distinguished by protein markers (although a final consensus has not been yet achieved) such as tetraspanins which are the most frequently encountered proteins (CD9, CD63, CD81, and CD82) (Witwer et al. 2013), 14-3-3 proteins, major histocompatibility complex (MHC) molecules, and cytosolic proteins such as specific stress proteins (heat shock proteins, HSPs) (Hoshino et al. 2020). For exosome characterization, proteins reflecting the exosome biogenesis mechanism are used as markers, including endosomal sorting complex required for transport (ESCRT-3), programmed cell death 6-interacting protein (PDCD6IP) or Alix (Witwer et al. 2013), and tumor susceptibility gene 101 (TSG101). RNA-binding proteins including Argonaute 1 (AGO1), RNA helicases (RHs), annexins (ANNs) such as annexin A2, and major vault protein (MVP) contribute to small RNA loading into EVs (He et al. 2021; O'Brien et al. 2020).

Tetraspanins, highly enriched in the lipid rafts of EVs, are thought to induce membrane curvature and to promote the incorporation of membrane receptors into tetraspanin-enriched microdomains which ensures their trafficking toward exosome release (Andreu and Yáñez-Mó 2014). For example, CD63 tetraspanin has been involved in EV loading with gDNA (Yokoi et al. 2019).

EV protein cargo is frequently enriched in posttranslational modifications (PTMs), including glycosylation, acetylation, phosphorylation, oxidation, methylation, and ubiquitination (Hallal et al. 2022). Of these, glycosylation is one of the most frequent PTMs of EV proteins and has garnered wide interest recently as it may hold great implications for furthering our understanding of EV biology and function, bringing new insight into the pathophysiological role of EV glycosylation. Carbohydrates that commonly occur on EVs (e.g., high mannose, polylactosamine, α-2,6 sialic acid, complex N-linked glycans, and exclusion of terminal blood group A and B antigens) or lectins (glycan-binding proteins) are involved in protein trafficking, cellular recognition, and signaling to the target recipient cells with an apparently conserved glycan and lectin signature of EVs (Batista et al. 2011; Williams et al. 2018). More intensive glycomic studies on EVs will unveil how the EV glycosylation patterns (O-linked, N-linked glycosylation, specific protein glycoforms) influence cargo recruitment in EVs, stability in systemic circulation, and their targeting potential (Liang et al. 2014; Williams et al. 2018).

4.2.2 Nucleic Acid Composition of EVs

Ribonucleic Acids

Extracellular vesicle RNA (evRNA) represents a major portion of the extracellular RNA that is enclosed in the vesicle lumen and consists in sizes smaller than 700 nucleotides (nt) compared to the cellular RNA (400–12,000 nt) (Batagov and Kurochkin 2013; Chen et al. 2010). The RNA types first described in EVs isolated from murine stem and mast cells were the messenger RNA (mRNA) and microRNA (miRNA) which also exerted functional roles (Ratajczak et al. 2006; Valadi et al. 2007). MicroRNAs (miRNAs) are 21–25 nt regulatory molecules involved in posttranscriptional gene silencing (Cheng et al. 2014). MiRNAs are transcribed as hairpin precursors (pri-miRNAs), cleaved by Dicer (into pre-miRNAs), bound by

Argonaute proteins (AGO) and loaded into the miRNA-induced silencing complex (miRISC) for mRNA target regulation (Cheng et al. 2014). Some studies suggest that miRNA loaded into EVs is rarely delivered to target cells (Albanese et al. 2021) and the roles of different EV-RNAs have not been completely elucidated yet. However, the vesicular miRNA is known to interfere with translation or/and degrade the mRNA in the target cell (O'Brien et al. 2020).

As a result of miRNA shuttle from parental to recipient cells, decreased levels of mRNA targets have been observed (Ostenfeld et al. 2014). Importantly, the miRNA cargo is specifically packaged into EVs with one described mechanism being via the ARF-Exportin-5 axis (Clancy et al. 2019). Some studies proposed that RISC proteins in EVs could process pre-miRNA maturation, thereby inducing cell-independent microRNA biogenesis (Melo et al. 2014).

The 3'-UTR mRNA fragments which contain multiple sites for miRNA regulatory binding are enriched in EVs compared to intact mRNA molecules which would normally be translated into protein in the recipient cells, highlighting that EV-RNA may compete with cellular RNA in the regulation of stability and translation by binding of miRNAs or RNA-binding proteins (Batagov and Kurochkin 2013), likely influencing both the recipient and the parental cell upon release (Ostenfeld et al. 2014).

However, with the development of the next-generation sequencing (NGS) techniques, a wide array of noncoding RNA types has been discovered in cellular transcriptomics. Thus, EVs have been also shown to contain intact mRNA, mRNA fragments, long noncoding RNA, miRNA, piwi-interacting RNA, ribosomal RNA (rRNA), and fragments of tRNA-, vault-, and Y-RNA (Yáñez-Mó et al. 2015).

Deoxyribonucleic Acids

The study of the EV-DNA which includes ssDNA, dsDNA, mtDNA, and even viral DNA has been largely overlooked until recently (Fischer et al. 2016; Lazo et al. 2021). However, some of the studies regarding cell-free DNA (cfDNA) likely disregarded that DNA could be associated with EVs (Elzanowska et al. 2021). Recent studies highlighted that EV loading with DNA is disease-specific, and this may have diagnostic relevance in the clinic. For example, mitochondrial DNA (mtDNA) which is associated with aging (Lazo et al. 2021) is loaded into EVs as a result of mtDNA leakage during mitochondrial oxidative stress (Elzanowska et al. 2021).

DNA is present both in the EV lumen (Fischer et al. 2016) and attached on the surface of EVs (Németh et al. 2017). Damaged DNA accumulated during cellular division or due to oxidative stress or radiation is released in ectosomes from the cytoplasm or shuttled via the CD63 tetraspanin which is also involved in loading exosomal cargo into MVBs via the late endosomes (Yokoi et al. 2019). The presence of dsDNA (gDNA) was demonstrated in all analyzed EVs by Lazaro-Ibanez et al. with vesicle population-specific variations likely due to selective packing of DNA (Lázaro-Ibáñez et al. 2014).

4.2.3 Lipid Composition of EVs

Lipids are essential structural and functional components of EVs, involved in EV biogenesis, trafficking, and uptake by target cells (Choi et al. 2013; Pfrieger and Vitale 2018). A distinct characteristic of the EV lipid bilayer membrane consists in the specific and cell-type dependent enrichment in cholesterol (CHOL) by two- to threefold, sphingomyelin (SM), glycosphingolipids (GSLs), and phospholipids (PL) (e.g., phosphatidyl-serine (PS), phosphatidyl-choline (PC), phosphatidyl-ethanolamine (PE)) compared to the parental cells (Record et al. 2014; Skotland et al. 2020; Wubbolts et al. 2023).

Variations in the lipid composition of EVs derived from different sources have already been found. For example, the exosomes contain high levels of PC representing 60 to 90% of their lipid content (Zhang et al. 2018). Of functional relevance in signaling and EV recognition processes is the presence of PS on the outer leaflet of the extravesicular membrane of EVs shed from the plasma membrane. This is considered an "eat-me" signal, and its functional implications have been highlighted in activated platelet-derived EVs, which contain high PS levels and exert a pro-thrombotic function, resulting in inflammatory and cardiovascular pathologies (Wei et al. 2018).

The Role of Lipids in EV Biogenesis, Cargo Sorting, and Uptake by Recipient Cells

Several specific lipids such as cholesterol, a pivotal driver of the EV formation and release, are thought to play a role in the biogenesis and function of EVs. PS is likely involved in cargo sorting due to its enrichment in lipid raft domains that have been demonstrated to participate in EV formation and structure and also function as a sorting platform for ubiquitin-based protein sorting into exosomes (de Gassart et al. 2003). Moreover, lipid rafts and proteins contained within (i.e., CD36, tetraspanins) are also critical for EV uptake, especially by receptor-mediated endocytosis (Pfrieger and Vitale 2018). Ceramide (formed by the action of neutral sphingomyelinase 2 on sphingomyelin) has also been proposed as an alternative pathway involved in the formation of ILVs within the MVBs, thereby participating in the biogenesis and cargo loading of exosome subtype of EVs (Trajkovic et al. 2008). Other lipids, such as lysobisphosphatidic acid (LBPA) and phosphatidic acid, have also been suggested to be involved or to regulate EV biogenesis (Matsuo et al. 2004; Record et al. 2014).

5 Characteristics of EVs that Impart Them with Systemic Effects (Interorgan Communication)

Extracellular vesicles represent the epitome of multidirectional, unbounded communication and information exchange, in analogy to the World Wide Web. The EV exchange takes place either in the tissue- and organ-enclosed microenvironments from complex organism or distantly throughout the body-wide network following the systemic release of EVs into bodily fluids which enables interorgan communication, as well as between different groups of organisms, such as the microbiome-host communication (Costa-Silva et al. 2015; Couch et al. 2021; Yáñez-Mó et al. 2015).

Importantly, in the context of systems biology, EV-mediated communication occurs at multiple levels and with pleiotropic effects, assigning a key role to EVs in homeostasis, metabolism, and development but also as drivers of changes associated with aging (e.g., senescence, autoimmunity) and diseases, which they reinforce. Consistent advancements in the field of EV research highlighted the ubiquitous nature of EVs and that they are not mere bystanders, but native conserved tools used for information transfer. Specifically, all mammalian cell types produce EVs (Gill et al. 2019), albeit in various quantities. The architectural properties of EVs, their structure and composition, including the small size, lipid bilayer, and decoration with specific molecules (proteins, lipid rafts), allow EVs not only to cross across physical and biological barriers (Haney et al. 2015; Yang et al. 2015) but also ensure EV biocompatibility, stealth, and low immunogenicity (Kao and Papoutsakis 2019), thereby facilitating their specific recognition and uptake potential. The study of EV organotropism, defined here as the propensity of EVs for preferential uptake by specific organs or tissues in the body, has first emerged in the context of disease, when it has been demonstrated that distinct integrins and adhesion molecules on tumor extracellular vesicles dictate to which organs the tumor vesicles go and prepare future metastatic sites thereby dictating organotropic metastases (Costa-Silva et al. 2015; Hoshino et al. 2015).

Specifically, the in vivo secretion of EVs from tissues was demonstrated by their detection in various types of body fluids, including plasma, urine, saliva, lymph, breast milk, semen, synovial fluid, cerebrospinal fluid, amniotic fluid, and ascites (Couch et al. 2021; van Niel et al. 2018; Stahl and Raposo 2019). Therefore, in the case of EVs isolated from whole tissues and screening of specific EV signatures, the enrichment of these EVs in systemic circulation could be verified, ultimately leading to the discovery of EV-based biomarkers for specific diseases (Hoshino et al. 2020; Qian et al. 2015). A completely integrated and comprehensive map of the EV-based interorgan and organ-body fluid communication trajectory along with the functional relevance has not been deciphered to date in complex organisms. However, novel findings related to the interorgan and interspecies communication via EVs in specific contexts of homeostasis and disease serve as a starting ground for the integration of EV biology into the systems biology field.

6 EV Roles in Homeostasis and Systemic Disease

EVs are emerging as a new level of physiological control in multicellular organisms holding key roles in maintaining the physiological balance and homeostasis by mediating the communication between and among tissues (Stahl and Raposo 2019) and exerting autocrine and paracrine functions. At the cellular level, one of the most physiologically relevant functions of EVs is in regulating cell guidance and sending positional cues which are important for tissue architecture under normal conditions and for cell motility and migration as response to an injury or infection (Stahl and Raposo 2019). However, the disruption of homeostasis, such as injury and stress, results in altered communication patterns in the tissue microenvironment. These consistently altered EV-mediated interactions between stromal, immune, parenchymal cells, and the extracellular environment can eventually lead to the development of disease (Stahl and Raposo 2019).

The earliest studies demonstrated that EVs activate the immune system and induce antitumor immune responses in vivo (Pitt et al. 2014), thereby instating their involvement in cellular signaling which has been observed in physiology and disease models, including cancer (Peinado et al. 2017), neurodegenerative disorders (Rajendran et al. 2014), and autoimmunity (Lu et al. 2021). Thus, EVs assist dendritic cells to acquire and process antigens (Lindenbergh and Stoorvogel 2018) and enhance surveillance under inflammatory conditions (Brown et al. 2018). During infection, EVs represent guidance queues that attract neutrophils to the site of infection (Szatmary et al. 2017).

Although only recently, it has been shown that EVs play an important function in embryogenesis and development. EVs secreted by the endometrium and embryo influence embryo implantation (Chen et al. 2022a). In the context of developmental organogenesis, EVs secreted from mesenchymal cells convey miRNAs that regulate salivary gland development (Hayashi et al. 2017).

Importantly, EVs can mediate long-distance information transfer through their release in the circulatory infrastructure of multicellular organisms, represented by biological fluids such as blood, lymph, and cerebrospinal fluid which are replete with EVs ($\sim 10^{10}$/ml of plasma of a healthy individual) (Dragovic et al. 2011), thereby highlighting the preponderance by which EVs are released by this route (Stahl and Raposo 2019). Through systemic circulation, EVs exert an epigenetic regulator function, altering the behavior of cells and whole tissues at a distance from their release site, ultimately driving epigenetic reprogramming and reinforcing systemic disease (Qian et al. 2015). The main components conveyed by EVs that drive epigenetic regulation are the miRNAs, involved in posttranscriptional gene silencing, which ultimately alter gene expression in recipient cells and regulate epigenetic processes such as DNA methylation, histone modification, and miRNA or lncRNA regulation (Qian et al. 2015).

Among the systemic diseases, EVs have been extensively studied in cancer, neurodegeneration, and cardiovascular diseases. Stem cell-derived EVs enriched in various species of miRNAs (e.g., released by mesenchymal stem cells, cardiac progenitor cells, cardiosphere cells, hematopoietic stem cells, and embryonic stem) exert cardioprotective effects (Saheera et al. 2021). A recent example of interorgan functional communication mediated by EVs showed that small EVs secreted from brown adipose tissue ensure cardioprotection during exercise by delivering miRNA species into the heart (Zhao et al. 2022).

Altogether, it remains an enigma in many contexts how EVs target a specific tissue or organ, how they transfer the signals to the proper target, and what is the conveyed message.

7 EVs as Mediators of Systemic Disease in Cancer

EVs play a key role in cancer metastasis by transferring information, at distance such as proteins, circular RNAs (circRNA), miRNA, and lncRNA. In the context of solid tumors, the EV bidirectional transfer between tumor and cells of the tumor microenvironment (TME) reinforces local tumor progression, altering the immune cell function and inducing stromal cell activation. Specifically, EV-mediated communication within the TME has been shown to orchestrate the formation of a neoplastic milieu favorable for cancer progression, facilitating immunosuppression, epithelial-to-mesenchymal transition (EMT), extracellular matrix remodeling, angiogenesis, lymphangiogenesis, vascular disruption, thrombosis, and invasiveness, inducing cancer cell resistance to therapeutic drugs and hypoxia tolerance (Chang et al. 2021; Kalluri and McAndrews 2023; Keklikoglou et al. 2019; Namee and O'Driscoll 2018; Patras et al. 2023; Zonneveld et al. 2019).

Ultimately, the chronic release of tumor-derived EVs drives the systemic effects of solid cancers along with metabolites and other soluble factors, activating platelets in the circulatory system, mobilizing bone marrow-derived cells, promoting systemic immunosuppression, and leading pre-metastatic niche formation and distant organotropic metastasis (Costa-Silva et al. 2015; Keklikoglou et al. 2019; Lucotti et al. 2022; Patras et al. 2023).

Several clinically relevant studies demonstrated that EVs dictate metastatic organotropism via differential and specific integrin or adhesion molecule expression on their surface such as laminin-binding integrins $\alpha6\beta4$ and $\alpha6\beta1$ that drive lung tropism, integrin $\beta3$ directing EVs to the brain, and cadherin 11 (CDH11) and $\alpha5$ integrin that determine bone tropism (Costa-Silva et al. 2015; Deasy and Erez 2022; Hoshino et al. 2015; Li et al. 2022; Ruoslahti 1996).

The tumor EV-mediated cross talk in the bone marrow microenvironment where stromal cells, immune cells, osteoclasts, osteoblasts, and hematopoietic cells reside induces functional changes and facilitates the EV release into systemic circulation to be further recruited to future metastatic sites, as a prerequisite for PMN formation (Peinado et al. 2012; Wang et al. 2016). For example, highly metastatic melanoma-derived EVs "educate" the bone marrow progenitor cells via MET receptor toward a pro-metastatic phenotype which determines the migration of these cells toward the PMN sites (Peinado et al. 2012).

The EV-mediated transfer from the bone marrow (BM) to other sites has just been explored in recent decades, but encouraging findings highlight role of BM-derived EVs in multiple biological processes. For example, bone marrow stem cell (BMSC)-derived EVs conveying miR-9-3p regulated the dormancy of solid tumors by suppressing cancer progression in bladder cancer, thereby highlighting that bidirectional communication occurs and determines tumor progression outcome (Costa-Silva et al. 2015; Hernández-Barranco et al. 2021).

The suppression of antitumor immunity is one of the critical steps that ensure successful tumor progression and metastatic colonization. In solid cancers, the tumor microenvironment becomes immunosuppressive due to local EV cross talk and "education" of immune cells toward pro-tumorigenic phenotypes.

At distant sites, EV homing to future metastatic organs facilitates the activation and functional alteration of resident stromal and immune cells, generating an immunosuppressive and inflammatory PMN environment. Ultimately these effects alter both innate and adaptive immunity into supporting tumor progression. For example, tumor-secreted EVs can drive macrophage differentiation toward a pro-tumor phenotype and can impair myeloid cell differentiation which generates myeloid-derived suppressor cells (MDSCs) with further orchestrate tissue and organ immunosuppression by infiltrating at distant sites (Fernández-Delgado et al. 2020; Reed et al. 2021).

Tumor EVs with immunosuppressive effects present on their surface multiple immunomodulatory molecules including inflammatory S100 proteins, antigen presentation-relevant MHC class I, immune checkpoint molecules such as programmed death ligand 1 (PDL1), cytotoxic T lymphocyte antigen 4 (CTLA4), and apoptosis-inducing ligands which affect innate and adaptive antitumor immunity at several levels: inhibition of dendritic cell maturation and function and induction of apoptosis in natural killer (NK) and $CD8^+$ T lymphocytes (Buzas 2023; Chen et al. 2018).

Last, tumor-derived EV presence and circulation in bodily fluids (blood, lymph) that represent the main highway for whole-organism-level communication ensure not only continuous immunosuppression but also interfere with immunotherapies, thereby ascribing a central role to tumor EVs in immunoregulation and immune escape. For example, EVs conveying PD-L1 contribute to the inhibition of PD1-expressing $CD8^+$ T cells, and importantly, they interfere with immune checkpoint blockade therapies (Chen et al. 2018; Dieterich 2022).

In the bodily fluids of cancer patients, tumor EVs are constantly present and exert a multitude of effects including inducing, expanding, and activating regulatory T cells (T regs), as demonstrated by the high frequency of circulating T reg cells with immunosuppressive functions (Szajnik et al. 2010). In turn, T regs secrete EVs that carry regulatory miRNAs (miR-150-5p and miR-142-3p) which when transferred to DCs induce a tolerogenic phenotype, thereby exploiting a mechanism designed to prevent autoimmunity (Tung et al. 2018). Moreover, T reg EVs containing miRNAs (miR-155, Let7b, and Let7d) can suppress $CD4^+$ T cell function or alter cytokine production by DCs (extravesicular miR-150-5p and miR-142-3p) (Buzas 2023).

Cachexia is a multifactorial metabolic syndrome associated with high morbidity and mortality in patients with advanced cancer. Cancer cachexia is defined as unintentional weight loss secondary to cancer progression of more than 5% or a BMI less than 20 and a weight loss $>2\%$ or sarcopenia and weight loss of $>2\%$ (Fearon et al. 2011). Several recent articles suggested that EV cargo may exert biological effects that contribute to cancer cachexia, but further research is needed to better define their contribution (Sun et al. 2023).

8 Integrative Overview on the Future of EVs in the Field of Fundamental Biology and Applicative Research in Systems Biology

Fundamental studies on EV biology may uncover the multidirectional trajectory and functional role of EV trafficking between cells, tissues, and organs, as well as EVs that are released into systemic circulation, and ultimately integrate these new findings with systems biology. However, the uncovering of the complex EV-mediated interactions within an organism would virtually require tracking and characterization at single-EV level and furthermore the analysis of the effect induced in the recipient cell, tissue, and organ. Current approaches include the use of multi-omics and bioinformatic tools to further help identify patterns in terms of EV cargo as well as functional changes in recipient cells to dissect these complex interactions in the context of an integrative landscape.

EVs play multiple roles in mediating the systemic effects of cancer. On one hand, cancer cell-derived EVs contribute to pre-metastatic niche formation, microenvironment remodeling, tumor cell dormancy, organotropism, metabolic reprogramming, immune escape through direct inhibition of an immune response and creating an immunosuppressive environment, angiogenesis, thrombosis, and drug resistance (reviewed in Lucotti et al. 2022); on the other hand, EVs from immune cells, such as T cells, B cells, and dendritic cells, in general, inhibit tumor growth (Von Schulze and Deng 2020). The cargo of other immune cells, such as specific neutrophils and macrophages types, may stimulate tumor growth, so it appears that the EV secretion by immune cells is context-dependent and may be a double-edge sword (Hazrati et al. 2022).

Given the crucial role played by EVs in the oncogenic process, there are multiple potential clinical applications involving EVs. Recently, the term "information therapy" (Van der Mude 2022) was used in reference to the proposal of using EVs carrying specific epigenetic active molecules (i.e., miRNA, lncRNA) to instruct the cancer cells to self-destroy. EVs are also promising reconfigurable therapeutic tools and can be used in trojan horse-like targeting strategies (i.e., using EVs loaded with drugs and chemically altered) (Chen et al. 2022b; Chen et al. 2024; Kanchanapally et al. 2019; Kar et al. 2023).

Other clinical applications include the EV-targeted therapeutic approaches focused on disrupting circulating EVs that may reinforce the progression of disease by specific targeting of biomarker molecules on the surface of EVs in systemic circulation using monoclonal antibodies. Another approach more readily applicable in clinical trials consists in the complete extracorporeal hemofiltration of EVs from the entire systemic circulation using the Hemopurifier® affinity plasmapheresis platform device known as the adaptive dialysis-like affinity platform technology (Aethlon ADAPT™) (Marleau et al. 2012). This technology has initially shown clinical efficacy in patients for hepatitis C viral particle removal and was subsequently tested as adjuvant cancer therapy consisting in the complete removal of circulating EVs smaller than 200 nm in metastatic melanoma and breast cancer patients thereby limiting the pro-metastatic role of EVs (Marleau et al. 2012).

As carriers of tumor-derived material, EVs can be used as vaccines or adjuvant to stimulate the immune system (reviewed in (Naseri et al. 2020; Xu et al. 2020). EVs are instigators of an immunosuppressive microenvironment favoring cancer development; therefore the inhibition of tumor EV secretion may be clinically beneficial as shown in several studies (reviewed in (Kim et al. 2022).

EVs circulating in plasma are a reliable mirror of both the phenotypic heterogeneity of the different cells actively involved in the cancer process and the effect of therapeutic intervention in cancer patients, and they represent promising future biomarkers for cancer detection and therapeutic monitoring (Hoshino et al. 2020).

9 Conclusions

Mammalian organism function is driven by complex, embedded, multilayered networks of cellular communication in which EVs play a key role (Fig. 2). Cancer disrupts this system of networks and induces pathologic systemic networks (CISPN) through which the whole organism is "cancerized" to support cancer development (Paul 2020). Patients with advanced cancer are not killed by cancer cells but by the changes induced by cancer at the organism level induced by CISPN. EVs, the "Hermes of modern molecular oncology" (Braicu et al. 2015), these tiny messengers of the microcosmos secreted by the cancer cells, disrupt the homeostasis of the organism's macrocosmos, contributing to its demise.

Given the non-scale-free architecture of the cancer-induced pathologic networks at the organismic level, treatments directed toward communication agents, like EVs, may be less plagued by resistance encountered by targeting intracellular or tissue networks (Paul and Komarova 2021). Expanding our treatment interventions to the organismic level and combining EVs targeting approaches with currently existing therapies may further enrich our cancer armamentarium in the immediate future.

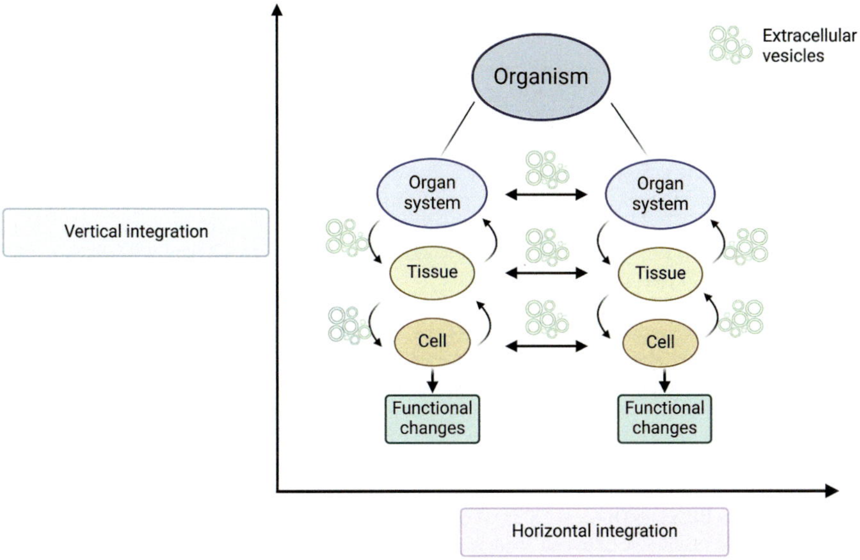

Fig. 2 Schematic overview on the body wireless Internet, highlighting that EV-mediated communication occurs at all levels, from simple-cellular to complex-whole organism. From the point of view of systems biology, a complete vertical and horizontal integration of the EV functional transfer routes should be considered, with functional changes including intracellular (epigenetic), interspecific (microbiome), or external environment-driven changes. Created with BioRender.com

References

Albanese M, Chen Y-FA, Hüls C et al (2021) MicroRNAs are minor constituents of extracellular vesicles that are rarely delivered to target cells. PLoS Genet 17(12):e1009951

Andreu Z, Yáñez-Mó M (2014) Tetraspanins in extracellular vesicle formation and function. Front Immunol 5:442

Batagov AO, Kurochkin IV (2013) Exosomes secreted by human cells transport largely mRNA fragments that are enriched in the 3′-untranslated regions. Biol Direct 8:12

Batista BS, Eng WS, Pilobello KT et al (2011) Identification of a conserved glycan signature for microvesicles. J Proteome Res 10:4624–4633

Braicu C, Tomuleasa C, Monroig P et al (2015) Exosomes as divine messengers: are they the Hermes of modern molecular oncology? Cell Death Differ 22:34–45

Brown M, Johnson LA, Leone DA et al (2018) Lymphatic exosomes promote dendritic cell migration along guidance cues. J Cell Biol 217:2205–2221

Buzas EI (2023) The roles of extracellular vesicles in the immune system. Nat Rev Immunol 23: 236–250

Chang W-H, Cerione RA, Antonyak MA (2021) Extracellular vesicles and their roles in cancer progression. Methods Mol Biol 2174:143–170

Chargaff E, West R (1946) The biological significance of the thromboplastic protein of blood. J Biol Chem 166:189–197

Chen G, Huang AC, Zhang W et al (2018) Exosomal PD-L1 contributes to immunosuppression and is associated with anti-PD-1 response. Nature 560:382–386

Chen K, Liang J, Qin T et al (2022a) The role of extracellular vesicles in embryo implantation. Front Endocrinol (Lausanne) 13:809596

Chen L, Wang L, Zhu L et al (2022b) Exosomes as drug carriers in anti-cancer therapy. Front Cell Dev Biol 10:72861

Chen TS, Lai RC, Lee MM et al (2010) Mesenchymal stem cell secretes microparticles enriched in pre-microRNAs. Nucleic Acids Res 38:215–224

Chen Z, Xiong M, Tian J et al (2024) Encapsulation and assessment of therapeutic cargo in engineered exosomes: a systematic review. J Nanobiotechnol 22:18

Cheng L, Sharples RA, Scicluna BJ et al (2014) Exosomes provide a protective and enriched source of miRNA for biomarker profiling compared to intracellular and cell-free blood. J Extracell Vesicles 3

Choi D-S, Kim D-K, Kim Y-K et al (2013) Proteomics, transcriptomics and lipidomics of exosomes and ectosomes. Proteomics 13:1554–1571

Chronopoulos A, Kalluri R (2020) Emerging role of bacterial extracellular vesicles in cancer. Oncogene 39:6951–6960

Clancy JW, Zhang Y, Sheehan C et al (2019) An ARF6-Exportin-5 Axis delivers pre-miRNA cargo to tumor microvesicles. Nat Cell Biol 21:856–866

Colombo M, Raposo G, Théry C (2014) Biogenesis, secretion, and intercellular interactions of exosomes and other extracellular vesicles. Annu Rev Cell Dev Biol 30:255–289

Costa-Silva B, Aiello NM, Ocean AJ et al (2015) Pancreatic cancer exosomes initiate pre-metastatic niche formation in the liver. Nat Cell Biol 17:816–826

Couch Y, Buzàs EI, Vizio DD et al (2021) A brief history of nearly EVerything—the rise and rise of extracellular vesicles. J Extracell Vesicles 10:e12144

Deasy SK, Erez N (2022) A glitch in the matrix: organ-specific matrisomes in metastatic niches. Trends Cell Biol 32:110–123

Deatherage BL, Cookson BT (2012) Membrane vesicle release in bacteria, eukaryotes, and archaea: a conserved yet underappreciated aspect of microbial life. Infect Immun 80:1948–1957

Dieterich LC (2022) Mechanisms of extracellular vesicle-mediated immune evasion in melanoma. Front Immunol 13:1002551

Dragovic RA, Gardiner C, Brooks AS et al (2011) Sizing and phenotyping of cellular vesicles using nanoparticle tracking analysis. Nanomedicine 7:780–788

Elzanowska J, Semira C, Costa-Silva B (2021) DNA in extracellular vesicles: biological and clinical aspects. Mol Oncol 15:1701–1714

Eme L, Tamarit D, Caceres EF et al (2023) Inference and reconstruction of the heimdallarchaeial ancestry of eukaryotes. Nature 618:992–999

Fearon K, Strasser F, Anker SD et al (2011) Definition and classification of cancer cachexia: an international consensus. Lancet Oncol 12:489–495

Fernández-Delgado I, Calzada-Fraile D, Sánchez-Madrid F (2020) Immune regulation by dendritic cell extracellular vesicles in cancer immunotherapy and vaccines. Cancers (Basel) 12:3558

Fischer S, Cornils K, Speiseder T et al (2016) Indication of horizontal DNA gene transfer by extracellular vesicles. PLoS One 11:e0163665

de Gassart A, Geminard C, Fevrier B et al (2003) Lipid raft-associated protein sorting in exosomes. Blood 102:4336–4344

Gill S, Catchpole R, Forterre P (2019) Extracellular membrane vesicles in the three domains of life and beyond. FEMS Microbiol Rev 43:273–303

Gould SB, Garg SG, Martin WF (2016) Bacterial vesicle secretion and the evolutionary origin of the eukaryotic endomembrane system. Trends Microbiol 24:525–534

Hallal S, Tűzesi Á, Grau GE et al (2022) Understanding the extracellular vesicle surface for clinical molecular biology. J Extracell Vesicles 11:e12260

Haney MJ, Klyachko NL, Zhao Y et al (2015) Exosomes as drug delivery vehicles for Parkinson's disease therapy. J Control Release 207:18–30

Harding C, Heuser J, Stahl P (1983) Receptor-mediated endocytosis of transferrin and recycling of the transferrin receptor in rat reticulocytes. J Cell Biol 97:329–339

Hassan Z, Kumar ND, Reggiori F, Khan G (2021) How viruses hijack and modify the secretory transport pathway. Cells 10:2535

Hayashi T, Lombaert IMA, Hauser B et al (2017) Exosomal MicroRNA transport from salivary mesenchyme regulates epithelial progenitor expansion during organogenesis. Dev Cell 40:95–103

Hazrati A, Soudi S, Malekpour K et al (2022) Immune cells-derived exosomes function as a double-edged sword: role in disease progression and their therapeutic applications. Biomark Res 10:30

He B, Cai Q, Qiao L et al (2021) RNA-binding proteins contribute to small RNA loading in plant extracellular vesicles. Nat Plants 7:342–352

Hernández-Barranco A, Nogués L, Peinado H (2021) Could extracellular vesicles contribute to generation or awakening of "sleepy" metastatic niches? Front Cell Dev Biol 9:625221

Hoshino A, Costa-Silva B, Shen T-L et al (2015) Tumour exosome integrins determine organotropic metastasis. Nature 527:329–335

Hoshino A, Kim HS, Bojmar L et al (2020) Extracellular vesicle and particle biomarkers define multiple human cancers. Cell 182:1044–1061.e18

Johnstone RM, Adam M, Hammond JR (1987) Vesicle formation during reticulocyte maturation. Association of plasma membrane activities with released vesicles (exosomes). J Biol Chem 262: 9412–9420

Kalluri R, LeBleu VS (2020) The biology, function, and biomedical applications of exosomes. Science 367:eaau6977

Kalluri R, McAndrews KM (2023) The role of extracellular vesicles in cancer. Cell 8:1610–1626

Kanchanapally R, Deshmukh SK, Chavva SR et al (2019) Drug-loaded exosomal preparations from different cell types exhibit distinctive loading capability, yield, and antitumor efficacies: a comparative analysis. Int J Nanomedicine 14:531–541

Kao C-Y, Papoutsakis ET (2019) Extracellular vesicles: exosomes, microparticles, their parts, and their targets to enable their biomanufacturing and clinical applications. Curr Opin Biotechnol 60:89–98

Kar R, Dhar R, Mukherjee S et al (2023) Exosome-based smart drug delivery tool for cancer Theranostics. ACS Biomater Sci Eng 9:577–594

Keklikoglou I, Cianciaruso C, Güç E et al (2019) Chemotherapy elicits pro-metastatic extracellular vesicles in breast cancer models. Nat Cell Biol 21:190–202

Kim JH, Lee C-H, Baek M-C (2022) Dissecting exosome inhibitors: therapeutic insights into small-molecule chemicals against cancer. Exp Mol Med 54:1833–1843

Koeppen K, Hampton TH, Jarek M et al (2016) A novel mechanism of host-pathogen interaction through sRNA in bacterial outer membrane vesicles. PLoS Pathog 12:e1005672

Kosaka N, Iguchi H, Yoshioka Y et al (2012) Competitive interactions of cancer cells and normal cells via secretory microRNAs. J Biol Chem 287:1397–1405

Krishnamoorthy L, Bess JW, Preston AB et al (2009) HIV-1 and microvesicles from T cells share a common glycome, arguing for a common origin. Nat Chem Biol 5:244–250

Kulp A, Kuehn MJ (2010) Biological functions and biogenesis of secreted bacterial outer membrane vesicles. Ann Rev Microbiol 64:163–184

Lake JA (1988) Origin of the eukaryotic nucleus determined by rate-invariant analysis of rRNA sequences. Nature 331(6152):184–186

Lázaro-Ibáñez E, Sanz-Garcia A, Visakorpi T et al (2014) Different gDNA content in the subpopulations of prostate cancer extracellular vesicles: apoptotic bodies, microvesicles, and exosomes. Prostate 14:1379–1390

Lazo S, Noren Hooten N et al (2021) Mitochondrial DNA in extracellular vesicles declines with age. Aging Cell e13283

Lee AR, Park SB, Kim SW et al (2022) Membrane vesicles from antibiotic-resistant Staphylococcus aureus transfer antibiotic-resistance to antibiotic-susceptible Escherichia coli. J Appl Microbiol 132:2746–2759

Li X-Q, Zhang R, Lu H et al (2022) Extracellular vesicle-packaged CDH11 and ITGA5 induce the Premetastatic niche for bone colonization of breast cancer cells. Cancer Res 82:1560–1574

Liang Y, Eng WS, Colquhoun DR et al (2014) Complex N-linked glycans serve as a determinant for exosome/microvesicle cargo recruitment. J Biol Chem 289:32526–32537

Lindenbergh MFS, Stoorvogel W (2018) Antigen presentation by extracellular vesicles from professional antigen-presenting cells. Annu Rev Immunol 36:435–459

Liu K, Gao X, Kang B et al (2022) The role of tumor stem cell exosomes in cancer invasion and metastasis. Front Oncol 12:836548

Lorizate M, Kräusslich H-G (2011) Role of lipids in virus replication. Cold Spring Harb Perspect Biol 3:a004820

Lu M, DiBernardo E, Parks E et al (2021) The role of extracellular vesicles in the pathogenesis and treatment of autoimmune disorders. Front Immunol 566299

Lucotti S, Kenific CM, Zhang H et al (2022) Extracellular vesicles and particles impact the systemic landscape of cancer. EMBO J 41:e109288

Marleau AM, Chen C-S, Joyce JA et al (2012) Exosome removal as a therapeutic adjuvant in cancer. J Transl Med 10:134

Mashburn LM, Whiteley M (2005) Membrane vesicles traffic signals and facilitate group activities in a prokaryote. Nature 437:422–425

Mashburn-Warren LM, Whiteley M (2006) Special delivery: vesicle trafficking in prokaryotes. Mol Microbiol 61:839–846

Matsuo H, Chevallier J, Mayran N et al (2004) Role of LBPA and Alix in multivesicular liposome formation and endosome organization. Science 303:531–534

Melo SA, Sugimoto H, O'Connell JT et al (2014) Cancer exosomes perform cell-independent microRNA biogenesis and promote tumorigenesis. Cancer Cell 26:707–721

Munhoz da Rocha IF, Amatuzzi RF (2020) Cross-Kingdom Extracellular Vesicles EV-RNA Communication as a Mechanism for Host–Pathogen Interaction. Front Cell Infect Microbiol 10:593160

Namee NM, O'Driscoll L (2018) Extracellular vesicles and anti-cancer drug resistance. Biochim Biophys Acta Rev Cancer 1870:123–136

Naseri M, Bozorgmehr M, Zöller M et al (2020) Tumor-derived exosomes: the next generation of promising cell-free vaccines in cancer immunotherapy. Onco Targets Ther 9:1779991

Németh A, Orgovan N, Sódar BW et al (2017) Antibiotic-induced release of small extracellular vesicles (exosomes) with surface-associated DNA. Sci Rep 7:8202

van Niel G, D'Angelo G, Raposo G (2018) Shedding light on the cell biology of extracellular vesicles. Nat Rev Mol Cell Biol 19:213–228

Nolte-'t Hoen E, Cremer T, Gallo RC et al (2016) Extracellular vesicles and viruses: are they close relatives? Proc Natl Acad Sci USA 113:9155–9161

O'Brien K, Breyne K, Ughetto S et al (2020) RNA delivery by extracellular vesicles in mammalian cells and its applications. Nat Rev Mol Cell Biol 21:585–606

Ostenfeld MS, Jeppesen DK, Laurberg JR et al (2014) Cellular disposal of miR23b by RAB27-dependent exosome release is linked to acquisition of metastatic properties. Cancer Res 74: 5758–5771

Pan BT, Johnstone RM (1983) Fate of the transferrin receptor during maturation of sheep reticulocytes in vitro: selective externalization of the receptor. Cell 33:967–978

Patras L, Shaashua L, Matei I et al (2023) Immune determinants of the pre-metastatic niche. Cancer Cell 41:546–572

Paul D (2020) The systemic hallmarks of cancer. J Cancer Metastasis Treat 6:29

Paul D, Komarova NL (2021) Multi-scale network targeting: a holistic systems-biology approach to cancer treatment. Prog Biophys Mol Biol 165:72–79

Peinado H, Alečković M, Lavotshkin S et al (2012) Melanoma exosomes educate bone marrow progenitor cells toward a pro-metastatic phenotype through MET. Nat Med 18:883–891

Peinado H, Zhang H, Matei IR et al (2017) Pre-metastatic niches: organ-specific homes for metastases. Nat Rev Cancer 17:302–317

Pfrieger FW, Vitale N (2018) Cholesterol and the journey of extracellular vesicles. J Lipid Res 59: 2255–2261

Pitt JM, Charrier M, Viaud S et al (2014) Dendritic cell-derived exosomes as immunotherapies in the fight against cancer. J Immunol 193:1006–1011

Qian Z, Shen Q, Yang X et al (2015) The role of extracellular vesicles: an epigenetic view of the cancer microenvironment. Biomed Res Int 2015:649161

Rajendran L, Bali J, Barr MM et al (2014) Emerging roles of extracellular vesicles in the nervous system. J Neurosci 34:15482–15489

Ramakrishnaiah V, Thumann C, Fofana I et al (2013) Exosome-mediated transmission of hepatitis C virus between human hepatoma Huh7.5 cells. Proc Natl Acad Sci USA 110:13109–13113

Raposo G, Nijman HW, Stoorvogel W et al (1996) B lymphocytes secrete antigen-presenting vesicles. J Exp Med 183:1161–1172

Ratajczak J, Miekus K, Kucia M et al (2006) Embryonic stem cell-derived microvesicles reprogram hematopoietic progenitors: evidence for horizontal transfer of mRNA and protein delivery. Leukemia 20:847–856

Record M, Carayon K, Poirot M et al (2014) Exosomes as new vesicular lipid transporters involved in cell-cell communication and various pathophysiologies. Biochim Biophys Acta 1841:108–120

Reed T, Schorey J, D'Souza-Schorey C (2021) Tumor-derived extracellular vesicles: a means of co-opting macrophage polarization in the tumor microenvironment. Front Cell Dev Biol 9: 746432

Renelli M, Matias V, Lo RY et al (2004) DNA-containing membrane vesicles of Pseudomonas aeruginosa PAO1 and their genetic transformation potential. Microbiology (Reading) 150: 2161–2169

Roier S, Zingl FG, Cakar F et al (2016) A novel mechanism for the biogenesis of outer membrane vesicles in gram-negative bacteria. Nat Commun 7:10515

Ruoslahti E (1996) RGD and other recognition sequences for integrins. Annu Rev Cell Dev Biol 12: 697–715

Saheera S, Jani VP, Witwer KW et al (2021) Extracellular vesicle interplay in cardiovascular pathophysiology. Am J Physiol Heart Circ Physiol 320:H1749–H1761

Sieber KB, Bromley RE, Dunning Hotopp JC (2017) Lateral gene transfer between prokaryotes and eukaryotes. Exp Cell Res 358:421–426

Skog J, Würdinger T, van Rijn S et al (2008) Glioblastoma microvesicles transport RNA and proteins that promote tumour growth and provide diagnostic biomarkers. Nat Cell Biol 10: 1470–1476

Skotland T, Sagini K, Sandvig K et al (2020) An emerging focus on lipids in extracellular vesicles. Adv Drug Deliv Rev 159:308–321

Stahl PD, Raposo G (2019) Extracellular vesicles: exosomes and microvesicles, integrators of homeostasis. Physiology (Bethesda) 34:169–177

Sun S, Wang Z, Yao F et al (2023) Breast cancer cell-derived exosome-delivered microRNA-155 targets UBQLN1 in adipocytes and facilitates cancer cachexia-related fat loss. Hum Mol Genet 32:2219–2228

Szajnik M, Czystowska M, Szczepanski MJ et al (2010) Tumor-derived microvesicles induce, expand and up-regulate biological activities of human regulatory T cells (Treg). PLoS One 7: e11469

Szatmary AC, Nossal R, Parent CA et al (2017) Modeling neutrophil migration in dynamic chemoattractant gradients: assessing the role of exosomes during signal relay. Mol Biol Cell 28:3457–3470

Trajkovic K, Hsu C, Chiantia S et al (2008) Ceramide triggers budding of exosome vesicles into multivesicular endosomes. Science 319:1244–1247

Tung SL, Boardman DA, Sen M et al (2018) Regulatory T cell-derived extracellular vesicles modify dendritic cell function. Sci Rep 8:6065

Valadi H, Ekström K, Bossios A et al (2007) Exosome-mediated transfer of mRNAs and microRNAs is a novel mechanism of genetic exchange between cells. Nat Cell Biol 9:654–659

Van der Mude A (2022) A proposed information-based modality for the treatment of cancer. Biosystems 211:104587

Vigerust DJ, Shepherd VL (2007) Virus glycosylation: role in virulence and immune interactions. Trends Microbiol 15:211–218

Von Schulze A, Deng F (2020) A review on exosome-based cancer therapy. J Cancer Metastasis Treat 6:42

Wang J, Faict S, Maes K et al (2016) Extracellular vesicle cross-talk in the bone marrow microenvironment: implications in multiple myeloma. Oncotarget 7:38927–38945

Wei H, Malcor J-DM, Harper MT (2018) Lipid rafts are essential for release of phosphatidylserine-exposing extracellular vesicles from platelets. Sci Rep Nature Publishing Group 8:9987

Wieckowski E, Whiteside TL (2006) Human tumor-derived vs dendritic cell-derived exosomes have distinct biologic roles and molecular profiles. Immunol Res 36:247–254

Williams C, Royo F, Aizpurua-Olaizola O et al (2018) Glycosylation of extracellular vesicles: current knowledge, tools and clinical perspectives. J Extracell Vesicles 7:1442985

Witwer KW, Buzás EI, Bemis LT et al (2013) Standardization of sample collection, isolation and analysis methods in extracellular vesicle research. J Extracell Vesicles 2

Wolf P (1967) The nature and significance of platelet products in human plasma. Br J Haematol 13: 269–288

Wolfers J, Lozier A, Raposo G et al (2001) Tumor-derived exosomes are a source of shared tumor rejection antigens for CTL cross-priming. Nat Med 7:297–303

Wubbolts R, Leckie RS, Veenhuizen PTM et al (2023) Proteomic and biochemical analyses of human B cell-derived exosomes. Potential implications for their function and multivesicular body formation. J Biol Chem 278:10963–10972

Xie J, Haesebrouck F, Van Hoecke L et al (2023) Bacterial extracellular vesicles: an emerging avenue to tackle diseases. Trends Microbiol 31:1206–1224

Xu Z, Zeng S, Gong Z et al (2020) Exosome-based immunotherapy: a promising approach for cancer treatment. Mol Cancer 19:160

Yáñez-Mó M, Siljander PR-M, Andreu Z et al (2015) Biological properties of extracellular vesicles and their physiological functions. J Extracell Vesicles 4:27066

Yang Y, Chen Y, Zhang F et al (2015) Increased anti-tumour activity by exosomes derived from doxorubicin-treated tumour cells via heat stress. Int J Hyperth 31:498–506

Yokoi A, Villar-Prados A, Oliphint PA et al (2019) Mechanisms of nuclear content loading to exosomes. Sci Adv 5:eaax8849

Zaremba-Niedzwiedzka K, Caceres EF, Saw JH et al (2017) Asgard archaea illuminate the origin of eukaryotic cellular complexity. Nature 541:353–358

Zhang H, Freitas D, Kim HS et al (2018) Identification of distinct nanoparticles and subsets of extracellular vesicles by asymmetric flow field-flow fractionation. Nat Cell Biol 20:332–343

Zhang J, Ma L (2012) MicroRNA control of epithelial-mesenchymal transition and metastasis. Cancer Metastasis Rev 31:653–662

Zhang Q, Higginbotham JN, Jeppesen DK et al (2019) Transfer of functional cargo in exomeres. Cell Rep 27:940–954.e6

Zhao H, Chen X, Hu G et al (2022) Small extracellular vesicles from Brown adipose tissue mediate exercise Cardioprotection. Circ Res 130:1490–1506

Zonneveld MI, Keulers TGH, Rouschop KMA (2019) Extracellular vesicles as transmitters of hypoxia tolerance in solid cancers. Cancers (Basel) 11:154

Development of Individualized Therapy for Atherosclerosis from Systems Biology

Andrew J. Buckler and Ulf Hedin

Contents

Abstract Myocardial infarction (MI) and ischemic stroke (IS), major consequences of unstable atherosclerotic lesions, are the most common causes of death worldwide. Guidance for prevention of MI and IS is currently based on treatment efficacy at the *group level*, but it is increasingly reported that distinct *individual-level* disease phenotypes exist in atherosclerotic cardiovascular disease (CVD), suggesting

The authors note that this chapter is based in part on a PhD thesis by Dr. Buckler, the full text of which may be found at: https://openarchive.ki.se/xmlui/handle/10616/48188.

A. J. Buckler
BBMSC LLC, Boston, MA, USA

Department of Molecular Medicine and Surgery, Karolinska Institute and Karolinska University Hospital, Stockholm, Sweden
e-mail: andrew.buckler@bbmsc.com

U. Hedin (✉)
Department of Molecular Medicine and Surgery, Karolinska Institute and Karolinska University Hospital, Stockholm, Sweden
e-mail: ulf.hedin@ki.se

J. Barciszewski (ed.), *Systems Biology II*, RNA Technologies 15,
https://doi.org/10.1007/978-3-031-62178-9_11

differing optimal therapies and room for improved personalized care. Whereas personalized cancer treatment strategies have been developed based on information from tissue biopsies, such approaches have previously not been possible for CVD. However, utilizing noninvasive imaging for detailed morphological phenotyping to obtain quantitative imaging biomarkers of atherosclerotic lesions, combined with large-scale molecular profiling of plaque tissue retrieved at stroke-preventive carotid surgery (endarterectomy), we have been able to create individually calibrated systems biology models capable of generating individualized guidance strategies for surgical treatment and pharmacotherapy. In this chapter, we summarize the work behind these translational research advances, which currently stand next to implementation in clinical practice for improved precision in prevention of MI and IS.

Keywords Atherosclerosis · Precision medicine · Systems biology · Proteomics · Polypharmacy

1 Motivation for Development of Individualized Therapy

1.1 The Unmet Needs in Atherosclerotic Cardiovascular Disease

Atherosclerosis lies behind most of the mortality and morbidity in CVD, where the majority is mainly related to complications such as MI and IS, accounting for a staggering more than 30% of overall global mortality (World Health Organization 2023). Atherosclerosis leads to the formation of plaques in the intima of arteries with accumulation of lipids, inflammation from recruitment of immune cells, and fibrosis by smooth muscle cells (SMCs). The disease develops over decades and can eventually impair blood flow and cause organ ischemia such as angina pectoris, and, when lesions become unstable, plaque rupture or erosions may lead to atherothrombosis, block blood flow, and cause MI or cerebral embolization and IS from atherosclerotic plaques in the carotid bifurcation (Fig. 1) (Hedin and Hansson 2016; Hafiane 2019). The unstable atheroma is characterized by a large lipid-rich necrotic core (LRNC), inflammation, intraplaque hemorrhage (IPH), cytokine and protease activity, and apoptosis of SMCs that build up the fibrous cap (FC), which shields the thrombogenic necrotic core from the vessel lumen. These processes weaken the FC, which becomes thinner, may rupture, expose thrombogenic factors such as tissue factor (TF) to the bloodstream, and initiate thrombosis and clinical manifestations (Hedin and Hansson 2016; Hafiane 2019; Mura et al. 2020).

Although therapeutic advancements for CVD have improved the prevention of IS and MI, diagnostic methods for atherosclerotic plaque phenotyping and assessment of plaque (in)stability to aid individualized therapy are lacking. Patients are today primarily identified after they have already suffered an event and, when identified, subjected to treatment established at the group rather than the individual level

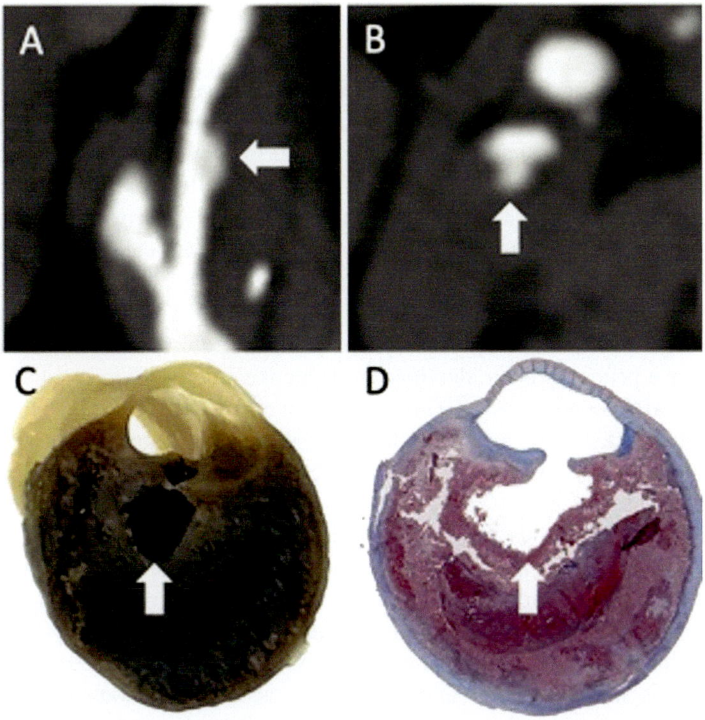

Fig. 1 CT angiography (CTA) of carotid plaque rupture (arrow) in a 77-year-old man with ischemic stroke in longitudinal (**a**) and axial view (**b**) with the corresponding endarterectomy specimen sectioned at the rupture (**c**, arrow) and stained with Masson's trichrome (D)

(Antman and Loscalzo 2016; Maiorino and Loscalzo 2023). As an example, cerebral embolism from unstable plaques in the carotid bifurcation and transient ischemic events (i.e., TIA) mandate stroke-preventive surgery, carotid endarterectomy (CEA), in patients with highly stenotic lesions above 50–70% (Naylor et al. 2023). However, since no diagnostic methods can accurately determine if the lesion is unstable and the actual source of emboli, predictive power is moderate with one prevented stroke per 6–8 procedures. The stroke-preventive efficacy for CEA in patients with asymptomatic carotid stenosis is even more modest (1/20–30), leading to the procedure being questioned for asymptomatic patients (Aday and Beckman 2017). Risk identification and prevention of MI follow the same principles. Most patients will be subjected to treatment in an emergency when they experience an MI, mainly by catheter-based methods to open up the occluded coronary artery. Intervention is then followed by secondary preventive medication, deemed suitable for many, but not all (Bergström et al. 2015; Newby et al. 2018). Nevertheless, as unstable, symptomatic atherosclerosis develops from existing, stable, and asymptomatic disease, there are strong reasons to improve diagnostics and develop methods to individualize treatment and improve effective prevention of IS and MI.

1.2 Potential for Meeting the Need

Today, risk identification and treatment for CVD largely depend on population-based scoring methods such as the Framingham risk score or secondary prevention in patients with established disease. Clinical surrogate risk markers have limited efficacy as more than 60% of CVD patients carry none (19%) or only one (43%) risk factor (Antman and Loscalzo 2016). Management of CVD would improve if biomarkers for unstable disease were available to aid the identification of individuals at risk and guide treatment. Circulating biomarkers released from unstable plaques could identify individuals at risk, whereas imaging biomarkers would specifically identify high-risk lesions. Circulating biomarkers for atherosclerosis and unstable disease have primarily been identified in deductive studies where predictive value has been correlated to future CVD (Lyngbakken et al. 2019). Once the individual at risk has been identified, it remains to detect the actual lesion in the vasculature. Plaque morphology can be imaged by several modalities (ultrasound, computed tomography angiography (CTA), magnetic resonance imaging (MRI)). However, evidence from large trials for implementing plaque imaging in clinical practice is lacking, and its role in managing individual patients is limited (Saba et al. 2019, 2021). In contrast, computer-based models that correlate imaging with biological information (qIB, quantitative imaging biomarkers) have successfully been developed to tailor and guide cancer treatment (Aerts et al. 2014; Aerts 2018). However, less progress has been seen in CVD, where accurate methods for identifying individuals and lesions at high risk for atherothrombosis in coronary or extracranial arteries and strategies to implement tailored, personalized pharmacotherapy are still lacking.

2 An Approach for Personalized CVD Treatment

Computational modeling techniques express relationships spanning multiple time- and spatial scales (Fig. 2). Outcome and cost improvements depend on improved noninvasive diagnostics to identify which patients have progressing disease despite first-line treatments. Computational modeling techniques may express relationships that span multiple temporal and spatial scales. In cancer research, the determination of molecular signatures of tumors using noninvasive imaging with biological information from tissue biopsies used as a truth basis has been met with enthusiasm (Aerts et al. 2014; Grossmann et al. 2016; Lambin et al. 2017; Lee et al. 2017; Aerts 2018; Tamez-Peña et al. 2018; Choi et al. 2020). Can this, or even something better, be achieved through vascular imaging?

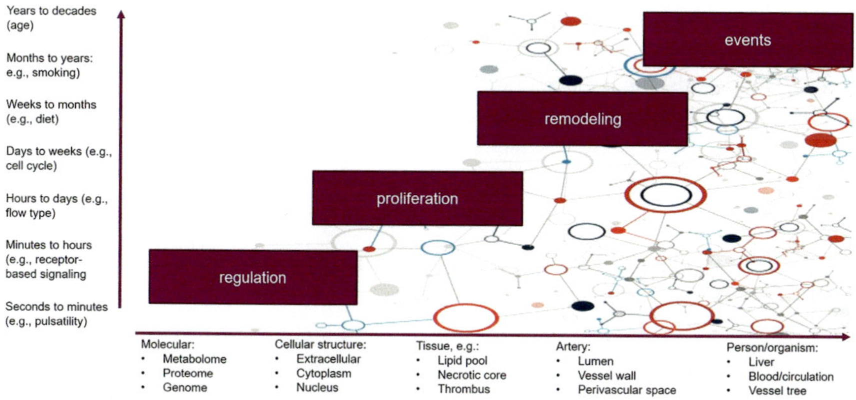

Fig. 2 Temporal and spatial scales that represent atherosclerosis

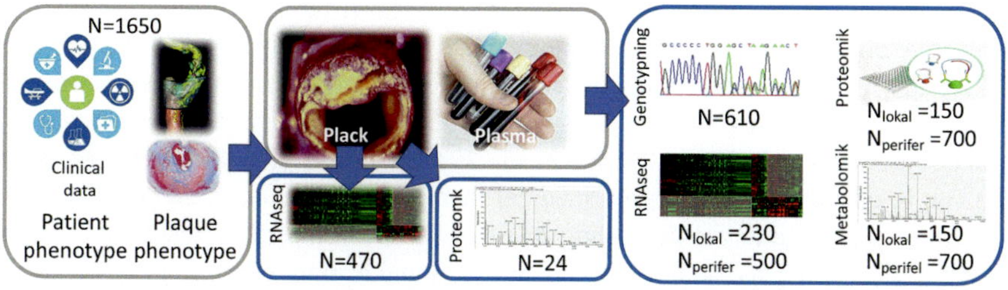

Fig. 3 Omics analyses of the BiKE cohort where clinical and imaging data stratifies patients for investigations based on genotype, plaque and PBMC transcriptomics, plasma proteomics, and metabolomics

2.1 *Carotid Atherosclerosis as a Clinical/Translational Model*

The use of noninvasive imaging to develop imaging biomarkers in coronary artery disease (CAD), established from molecular signatures of tissue biopsies reflecting the actual disease biology, is problematic as retrieval of biopsies from coronary plaque tissue cannot be obtained in living patients. In contrast, carotid atherosclerosis represents a comparable disease and a model for comprehensive studies of atherosclerotic plaque instability in humans (Ibrahimi et al. 2013). In a strategic effort to investigate atherosclerotic plaque instability in humans, we have developed a translational research platform based on a biobank of plaques and plasma from patients undergoing stroke-preventive surgery (CEA) for symptomatic (unstable) and asymptomatic (stable) carotid stenosis (BiKE, Biobank of Karolinska Endarterectomies). This unique resource contains patient clinical data, plaque imaging, molecular profiling, and outcome data. Plaques and plasma retrieved at surgery are utilized for large-scale molecular analyses of genetic, transcriptomic, and proteomic signatures which may be analyzed in combination with matching clinical and imaging data (Fig. 3) (Perisic et al. 2016; Matic et al. 2018; Karlöf et al. 2021).

Thus, carotid atherosclerosis can, in this way, serve as a clinical/translational model to build the necessary large-scale database needed to build applicable systems biology models that span from clinical baseline data, imaging data analyses, plaque morphology analyses, to large-scale molecular data from tissue and plasma (Buckler et al. 2023a, b, c).

2.2 *Noninvasive Imaging as a Quantitative Assay*

Increasing evidence shows that CTA (both carotid and coronary) may be an ideal modality to fill gaps in understanding the extent and rate of progression of atherosclerosis (Ibrahimi et al. 2013; Delfino et al. 2023). Advanced software-based techniques to extract data embedded in images, which are otherwise not readily appreciated visually or quantitatively, are being implemented to augment subjective assessment. Quantitative CTA analysis can move from subjective and qualitative to objective and quantitative to formally evaluate biomarkers (Watterson et al. 2008). The utility of CTA in detecting and managing both obstructive and nonobstructive atherosclerotic lesions is increasingly reported as documenting drug treatment effects (Sanda et al. 2017; Vaidya et al. 2018; Choi et al. 2020; Buckler et al. 2023a, b, c) and efficacy of new medical therapies (Sorger et al. 2011; Min et al. 2018). CTA utilization is increasing due to solid outcome study results that show better performance and at a cost less than functional testing (Mark et al. 2016; Moss et al. 2017; Levin et al. 2019).

Plaque morphology comprises structural anatomy and tissue composition. The tissue resources described above provide a rich source of ground truth. Ground truth data may exist at different spatial scales, such as the tissue scale (histopathology) or the molecular scale (RNA transcripts, proteomics, metabolomics, etc.). In particular, ElucidVivo (Elucid Bioimaging Inc.) was developed to exploit such truth data. ElucidVivo adopts histopathology as the performance standard of ground truth basis to allow specific tissue labeling approved for clinical use by regulatory bodies, including the FDA (Gupta et al. 2017; St. Pierre et al. 2017; Zhu et al. 2019; Abdelrahman et al. 2020; Buckler et al. 2022a, b; Delfino et al. 2023).

These capabilities are achieved using an adaptively determined, patient-specific, 3D point spread function with the effect that image intensities are restored to more closely represent the original tissues imaged, mitigating artifacts such as calcium blooming, enabling less visually conspicuous tissue types to be delineated (Pârvu and Gilbert 2016). The technique allows the software to reach an effective spatial resolution approximately 3x higher than the reconstructed voxel size with improved soft tissue plaque component differentiation relative to manual inspection. Based on this, ElucidVivo characterizes tissues using a histologically validated algorithm to accurately quantify CALC, LRNC, and IPH (Fig. 4) (Buckler et al. 2022a, b).

ElucidVivo's approach provides an ideal platform for developing multivariable and multivariate models for phenotype determination. Interpretable, rather than only black box models, are thereby enabled (Rudin 2019). Applications in carotid disease

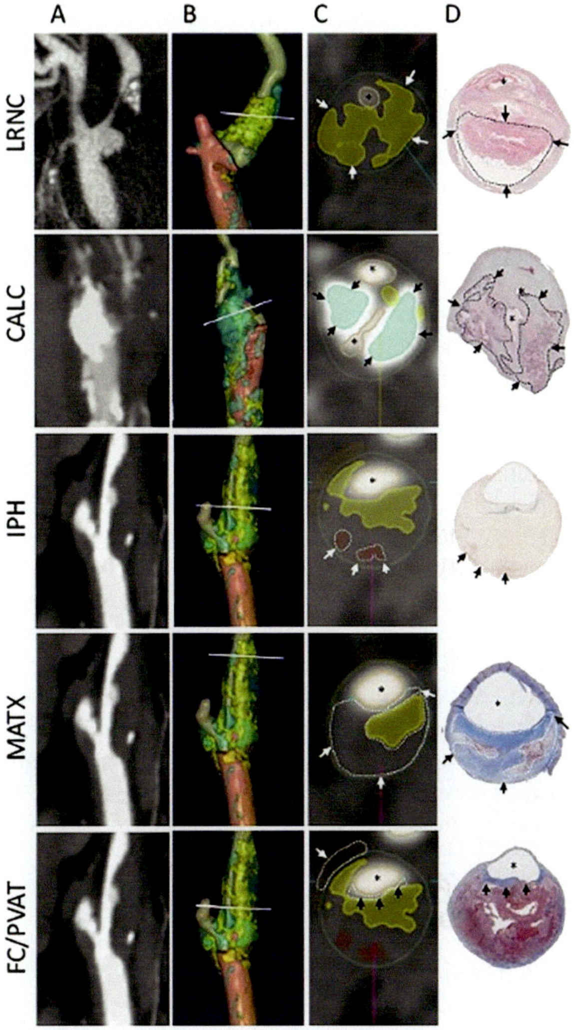

Fig. 4 ElucidVivo analyses (**b**, coronary 3Dview; **c**, axial view) of carotid plaque CTA (**a**) matched to corresponding histological sections (**d**) for identification of plaque tissue components (marked by arrows, LRNC, lipid-rich necrotic core; CALC, calcification; IPH, intraplaque hemorrhage; MATX, matrix; FC/PVAT, fibrous cap/perivascular fat; *lumen)

(Buckler et al. 2021; Saba et al. 2021; Murgia et al. 2020) and coronary disease (Abdelrahman et al. 2020; Choi et al. 2020) are increasingly reported, plaque vulnerability defined by histologic subtypes (Buckler et al. 2023a, b, c), major adverse cardiovascular event (MACE) prediction (Van Assen et al. 2019), major adverse neurovascular event (MANE) prediction (Lal et al. 2022), drug response assessment (Choi et al. 2020; Buckler et al. 2023a, b, c), and molecular profiling (Buckler et al. 2021).

2.3 *Virtual "Omics": Individualized Molecular Profiling*

Molecular-scale ground truth comprises genomics, transcriptomics, proteomics, and metabolomics. For example, gene expression may be analyzed where RNA can be extracted from tissue specimens, and gene expression may be analyzed using arrays or RNA sequencing with conventional or single-cell techniques. Various forms of mass spectrometry are used for proteomics or metabolomics. Bioinformatic processing steps may be applied, such as signal space transformation. Figure 5 shows one manifestation of a mathematical framework for multiscale analysis. The expression y(t) refers to a phenotype y at time t.

The expression x(t) designates cellular and molecular levels at time t, and z represents the patient-level outcome or state at time t. The approach is to determine systems of equations, or nonlinear models, f and g, where f goes down in scale and g goes up in scale. For example, the function f may be approached in a predictive modeling paradigm, with y being expressed as a vector of scalar values or on multidimensional data to derive expression, protein, or metabolite levels. Steps may be outlined for determining the function f (Fig. 6). Training data sets are used to optimize models of various types, and the results may be further applied to supervised or unsupervised clustering. The resulting associations, at the cohort or individual level, may be analyzed using techniques such as gene-set enrichment analysis (GSEA) to elucidate biological processes and molecular pathways at the cohort level or in individual patients. GSEA may be conducted, for example, using Enrichr (https://amp.pharm.mssm.edu/Enrichr/), further passing results from the Gene Ontology biological process, and further by analyzing data with other systems such as Revigo (http://revigo.irb.hr/) to determine, for example, non-duplicative processes. Individual patient-level inference may be applied where the degree of

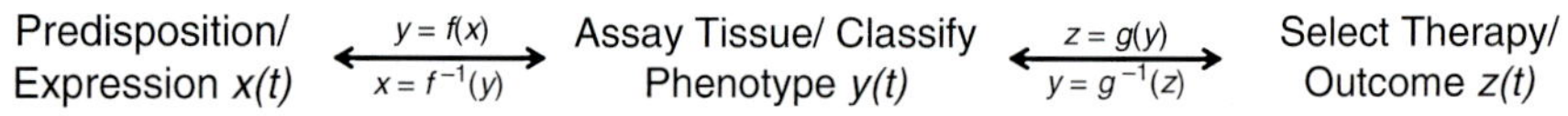

Fig. 5 Mathematical framework for multiscale analysis

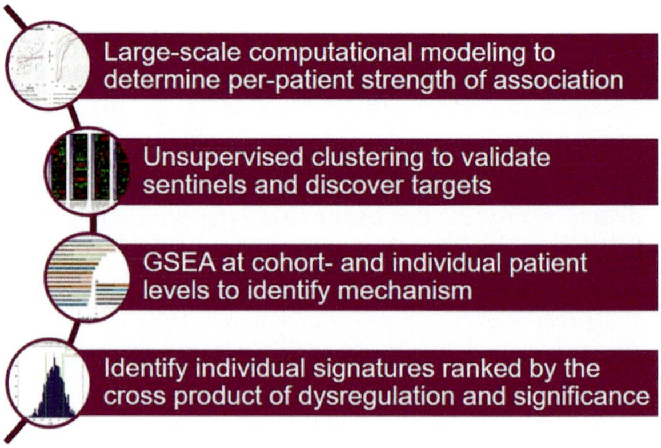

Fig. 6 Training and inference in virtual transcriptomics

dysregulation and the statistical model significance may both be considered. These techniques may be variously described as virtual transcriptomics, virtual proteomics, or virtual metabolomics.

Single- and multiple-variable regression models, covering linear and nonlinear modeling techniques, may be performed on predictor sets drawn from heterogeneous data. Predictors may be determined from plaque morphology, demographics, clinical (laboratory) values, or other variables. This is done in part to recognize that a diversity of variables could have affected the results, to inspect the added value of given variables, and to identify when variables have independent information content. One manifestation for function g may also be a predictive model, but a different model than f, one that moves up in scale. Inverse functions for f and g may also be derived.

Each model result may be output and tabulated to identify the highest-achieved performance on a species-by-species basis. Predictive performance may be determined based on the accuracy of the prediction relative to the true or reference values. Models may be built with variations, for example, differing sets of morphological measurements according to hypothesized physiological rationale; automated optimization using, for example, cross-validation while simultaneously varying tuning parameter values; or partitioning data such that a training set on which the cross-validation was performed was strictly separated from a sequestered validation data set to test performance using locked-down models. The use of histologically validated plaque features, for example, may produce interpretable models (Saba et al. 2019) and, when coupled with cross-validation, may mitigate overfitting.

The traditional models identify the differences across groups but are not generally applicable on a per-patient basis. They do not generally account for "overfitting" or consideration for accuracy on an individual patient basis, as do machine learning techniques (Kuhn and Johnson 2013). Deep learning neural networks extend this for use with higher dimensionality data sets such as 2D and 3D images. The quality of a model is only partially represented by its performance metrics; models that are likely to extrapolate well to unseen patients based on a documented rationale may be described as "interpretable" models (Rudin 2019). Model interpretability is improved by using histologically validated biomarkers as inputs and being architected according to a rigorously understood and biologically plausible mechanism. Our work has demonstrated that statistically significant prediction of plaque biology using noninvasive CTA could be accomplished for individual patients (Buckler et al. 2021).

2.4 *Creation of a Granular Systems Biology Model*

The models mentioned above capture associative but may not capture causal relationships between the predictors and the response. A formalism representing biological signaling pathways needs to be incorporated to model causality. Such models are often described as "systems biology" models. Such in silico systems biological

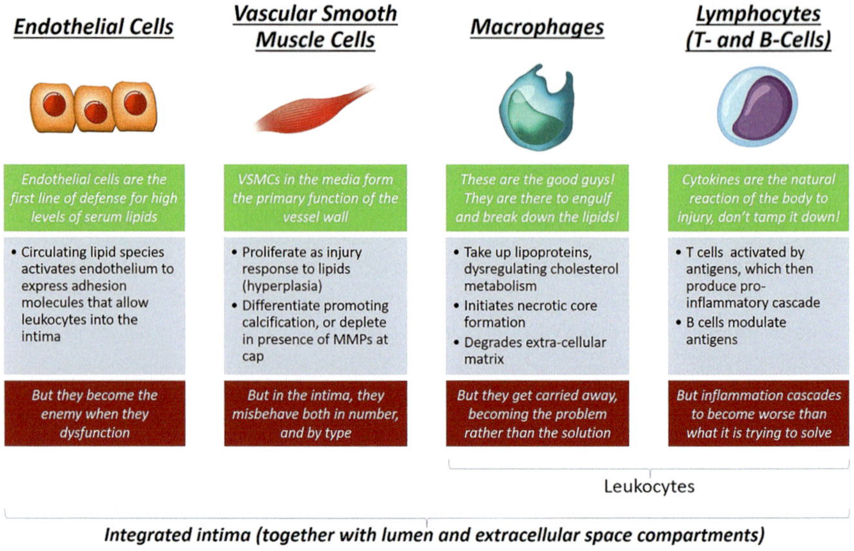

Fig. 7 Primary components necessary for a clinically relevant systems biology model of atherosclerosis

models must capture both pathology and pharmacology of atherosclerosis at a sufficient granularity (Fig. 7).

Protein-protein interactions in quantitative systems biology models may be built using "ordinary" differential equations (ODEs) interconnected with networks. The equations characterize how an independent variable changes a dependent value. This is represented by using differentiable functions:

$$a_0(x)y + a_1(x)y' + a_2(x)y'' + \ldots + a_n(x)y^{(n)} + b(x) = 0$$

where $a_0(x), \ldots, a_n(x)$ and $b(x)$ are defined functions that represent an empirically observed or hypothesized relationship in terms of change between the predictor variables x and the response variable y. The functions a do not need to be linear, but they must be differentiable. Whereas some differential equations may be solved explicitly, in general, it is necessary to use numerical methods that approximate rather than solve for the solution. Typical functions include mass action functions that express the rates of chemical or biological reactions/relations in terms of how products are produced from reactants. By way of example, Michaelis-Menten kinetics, a specialized formulation of how enzymes function, equations governing inhibition are used to express how the presence of species may cause reduced rates of production of products, as are typical of biological signaling pathways.

The term "ordinary" refers to the bulk consistency of the relationship without regard for spatial or time variation. That is, the association is the same anywhere in the postulated space, and the relationship doesn't change with time. Generally, this is applied to "compartments," localized spatial regions, or periods of time where these constraints are approximately true. These are particularly important examples of biochemical (compartment) models used in pharmacokinetic and pharmacodynamics studies. More advanced models to capture increasingly complex interdependencies that account for either spatial or temporal variant behavior are represented by systems of "partial" differential equations (PDEs) that generalize on ODEs using partial derivatives that allow for the spatial or temporal variation. If the granularity of the process is sufficiently modeled in relatively coarse "compartments," e.g., what happens in the intima vs. the liver, ODEs suffice. If, on the other hand, what happens inside of a compartment is of consequence, e.g., the endothelial surfaces vs. migration into the media, the compartment must be decomposed into smaller compartments, or if gradients or continuous variation rather than discrete variation is of interest, then PDEs are needed. In structural relationships of biomechanical forces among entities (e.g., between cells, proteins, etc.), formalisms such as CPM models may be used in conjunction with systems biology models.

Mathematical formalisms for extending multiscale models to identify causality explicitly have been explored (Eungdamrong and Iyengar 2004; Kholodenko 2006; Altafini 2007; Okuda et al. 2008; Pârvu and Gilbert 2016; Bergmann et al. 2017; Macklin 2018; Rudin 2019), such as intracellular cell cycle networks (e.g., by ODEs), regulation of the collective behavior of cellular structures (e.g., CPM), and spatially differentiated behavior (by PDEs), which in turn control intracellular behavior (returning to ODEs). Proposed model creation and validation workflows (Glass and Kauffman 1973; Eungdamrong and Iyengar 2004; Kholodenko 2006; Sorger et al. 2011; Chaouiya et al. 2012, 2013; Dejana et al. 2017) rely on high-level hypothesis generation and then the collection of data. Whereas data has been documented at individual scales, i.e., in vitro data at the cellular and molecular scale, microscopy data at the histopathology scale, and radiological data at the macroscopic scale, there is a dearth of linkages across these scales.

2.5 *Semiquantitative Simulation of Individually Calibrated Models*

Attempting clinical-scale models with complete reaction kinetics suffers from three primary challenges. First, trying to comprehensively capture all protein-to-protein interactions, much less all intermediate products in the metabolome, is generally beyond the existing scientific knowledge and, in any case, would be overwhelming. Second, even if undertaken, were it to be done by a single group, there would be a general lack of peer review or public adjudication on the biology, exacerbated by the first challenge but a problem of its own. Third, even if large swathes of the interactions were to overcome the first two challenges, even the smallest gaps would invalidate the model as the quantitative simulations do not tolerate "gaps" or inaccuracies well. A critical insight is that accomplishing the objective does not require fully quantitative models; rather, semiquantitative or even qualitative models would identify the "direction" of drug response, even if not the exact strength of that response. This observation is broadly recognized (Watterson et al. 2008; Whelan and King 2008; Gomez-Cabrero et al. 2011; Büchel et al. 2012).

Qualitative (or semiquantitative) models formulated based on Boolean algebra or other mathematics (Thomas 1973; Chaouiya et al. 2013; Aerts et al. 2014; Tian and Reed 2018; Aghamiri et al. 2020) that fall short of complete reaction kinetics still identify directionality or even levels of magnitude, to satisfy our requirements. The use of "relations" in place of "reactions" over a portion or even the whole model simultaneously solves the brittleness problem, while also allowing the use of publicly peer-reviewed pathways such as the Kyoto Encyclopedia of Genes and Genomes (KEGG) (Wrzodek et al. 2013; Du et al. 2016). Our work establishes the feasibility of using publicly curated systems of pathways with the level of completeness needed for clinical use rather than only small-scale demonstrations (Büchel et al. 2013; Chaouiya et al. 2012, 2013). Moreover, provided programming was undertaken to update from the publicly available updates through such resources continually, it would allow the model to evolve as the scholarship evolves.

In silico systems biology model(s) are initially trained with two data types. First, one uses experimentally determined data from biological specimens from development subjects. Development subjects are patients for whom actual proteomic data is available that shows differentially expressed "omic" levels that are linked to the specific characteristics and morphology of the plaques in each of those subjects. Second, one uses results from searches of public literature, experimental results, or other databases to create the initial in silico systems biology model. The in silico systems biology model is then updated with calibration data from test subjects to validate and refine the initial model. The calibration data is again based on actual biological samples that show differentially expressed protein or transcription levels linked to the specific characteristics and morphology of the plaques in each test subject. This step confirms that the model works as intended and augments and renders the model more robust, given the calibration data from many test subjects.

3 Application in Clinical Decision Support

Given the availability of the tissue, software, and systems biology resources, we can develop tools for clinical decision support. We describe a system concept with components for noninvasive, individualized molecular profiling and the use of this capability within a granular response simulation method utilizing extensive systems biology data and network representations. Clinical decision support systems (CDSS) may be developed for selecting and recommending a suitable treatment plan for a patient with CVD to analyze and process noninvasively obtained data of arteries from patients with atherosclerosis to obtain predicted proteomic and genomic information. Various alternative therapies, pharmacotherapies, or procedural interventions can be simulated based on their mechanisms of action in in silico systems biology models to provide a report for use by the healthcare provider recommending one or more specific therapy recommendations (Fig. 8).

In operation, the calibrated in silico systems biology model is again updated, but now with patient-specific personalized data based on imaging of clinical patients without the need to perform invasive procedures. The calibrated in silico systems biology model is perturbed for potential therapy mechanisms of action. The effect is simulated to provide therapy recommendations based on an automated comparison of the two or more therapies whose predicted effects are programmed into the model. This report may also be used directly with the patient to engage them in their care. Our work has developed a more comprehensive systems biology model of atherosclerotic plaque instability than has yet been reported, coupled with a practical response simulation method (Fig. 9) (Buckler et al. 2022a, b).

Our approach effectively described and predicted changes in plaque stability following individual patient treatments, differentiating outcome as a function of the drug mechanism of action. Referring to Fig. 9, recommendations generated by the software include the individual's absolute and relative radar plots, a statement on the benefit available through pharmacotherapy, and one or two heatmaps representing treated and baseline protein signatures. Patient "John Doe" exemplifies a patient with a highly unstable initial condition that may be improved with high confidence by pharmacotherapy. Patient "Bill Smith" represents a patient starting from a more stable initial condition where pharmacotherapy beyond conventional, preventive CVD medication would not be recommended. Patient "David Jones" represents a patient who would only marginally improve from pharmacotherapy and, based on the highly unstable starting point, should undergo procedural intervention.

3.1 *Use of the Simulation Capability for Clinical Care*

Given these results, a concrete roadmap of development is in view. Extrapolating the approach taken to transcriptomics for proteomics, extending the systems biology model to incorporate translation, and working with successively larger cohorts for model development and validation set directions for future work that we are now planning based on the established feasibility. Specifically, analysis outputs at various levels support clinical decision-making by informing the clinician of the current phenotype and the likely effect of different possible therapies. It can also provide tools to help discuss these options with the patient. The analyses offer recommendations based on the statistical significance of the likely improvement. They can compare potential recommendations to identify which of those considered exceeds others in the degree of improvement provided.

Such recommendations can determine a clinical action or inform decision-making that leads to a clinical action by allowing therapy to be tailored to the individual rather than based only on population statistics. Presently, clinical guidelines have not been able to use such diagnostic specificity because there has been no means to do so. Individuals have different genetic predisposition, environmental exposures, and

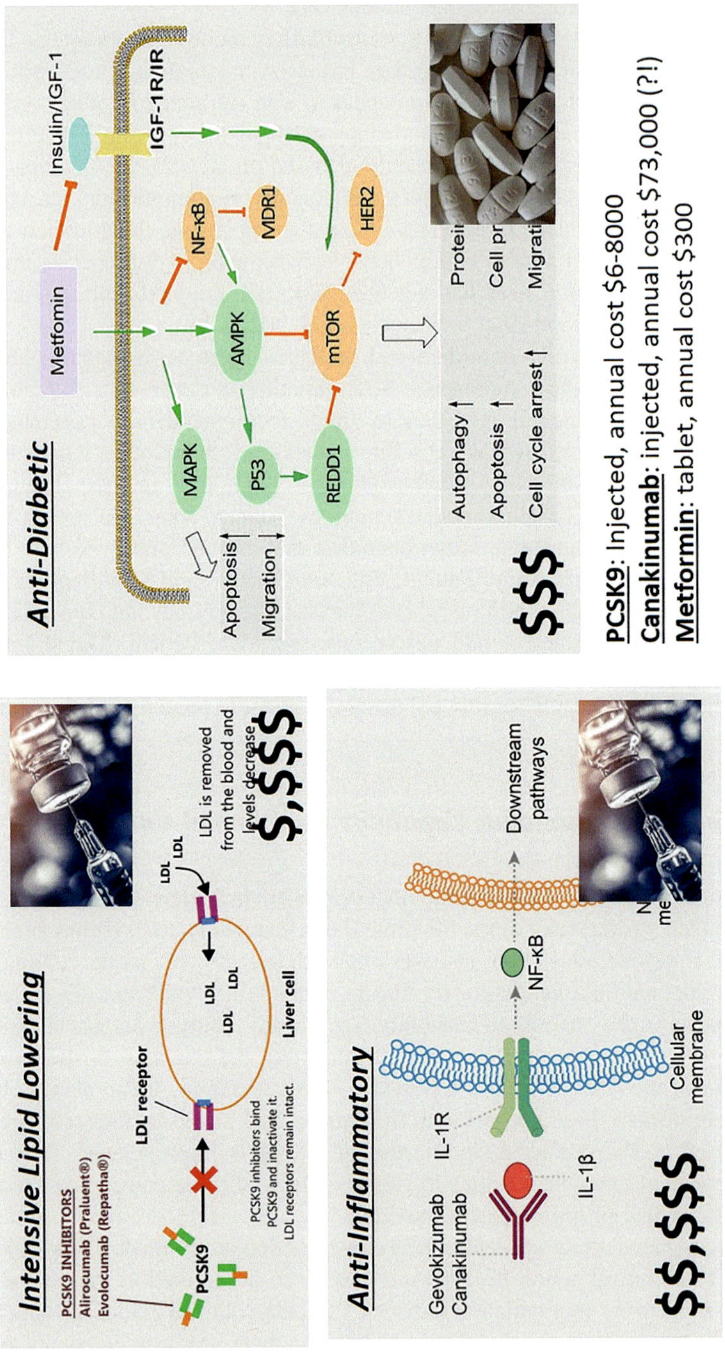

Fig. 8 Many therapeutic alternatives exist, but the differences in mechanism and cost vary widely. Here, just a few are shown to represent the range that is not currently capable of being prescribed broadly using only population-based approaches

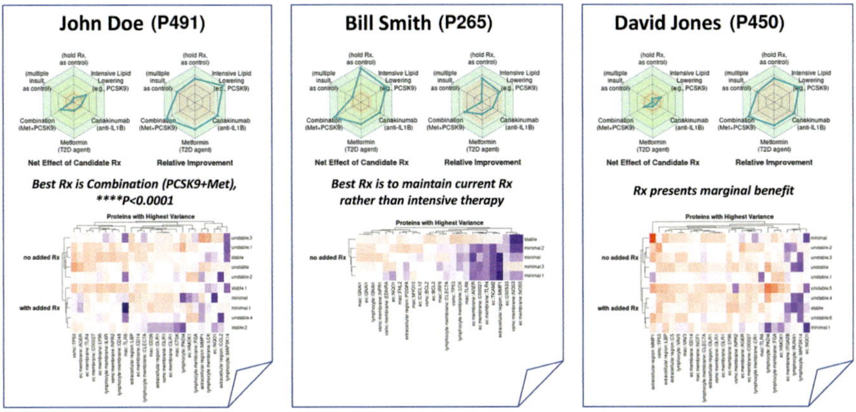

Fig. 9 Individualized patient treatment recommendations for three example patients, as might be utilized as a CDSS incorporating the technique. *Reprinted with permission from PMID: 36525832*

differing lifestyle habits. Both modifiable and non-modifiable risk factors influence what is best for that patient. The in silico systems biology model describes the disease, and this work provides a way to process and calibrate it for individual patients noninvasively. This enables the expected effect of therapies to be evaluated more precisely than previously possible. The benefit is that rather than referring to the population as a whole or, at best, subpopulations, actual molecular-level effects may be considered for the specific patient.

The utility of any diagnostic system must address what can be done with the information. Presently, numerous powerful therapies exist, both procedural, pharmaceutical, or combinations such as drug-eluting stents. By evaluating the individualized response to these therapies, diagnostics become actionable by identifying the degree of improvement and the statistical significance of the computation. More specifically, in silico systems biology model(s) may be generated, trained, and updated with patient-specific information to create the baseline condition. The in silico systems biology model representing the baseline condition is then further updated to simulate one or more potential therapies based on the mechanism of action for each treatment to arrive at simulated conditions. Based on the results, therapeutic recommendations are computed with the absolute pathology and relative improvement quantified and expressed in numeric and report forms.

After the patient has been on the recommended treatment regimen for sufficient time to elicit a therapeutic response, the simulated vs. new actual may be compared. Tangible improvement in the pathology indicates that the patient improved under the treatment. Further, suppose the actual measured changes to the protein levels were approximately as simulated. In that case, one can further determine that the treatment caused the improvement, and the method can be considered a surrogate endpoint for the treatment effect. The simulations must only be approximately correct to provide the intended utility in clinical practice.

The CDSS moves beyond patient groups to provide the facility to reach individual patient-level results. The CDSS incorporates trained virtual transcriptomic and virtual proteomic models, which, on inference, use the result vectors to initialize individual-level rate constants and concentrations, respectively. This has the effect of using a systems biology model developed with generalized data to be calibrated for an individual patient at a given point in time. This may be simulated out into the future to identify an "untreated" condition for the patient, with or without additional simulations that perturb the model according to a given candidate treatment's mechanism, thereby simulating the effect as if untreated, but also as if treated by specific pharmacologic or device interventions, thereby creating simulation for likely effect (response) of various specific treatments. Last but not least, using outcome data compiled by machine learning or other predictive models may be coupled with the mentioned per-treatment-type simulation. In this way, a family of event-free survival curves to represent the likely effect of alternative treatments for the individual patient.

3.2 Use of the Simulation Capability for Clinical Trials

This work also provides tools for improving the cost-effectiveness of clinical trials and opens the door for "virtual" clinical trials, which have been increasingly understood to provide fundamental ways to provide exploratory capability in cancer. For actual clinical trials, it is vital to improve patient selection, both to provide a way to enrich patient populations with likely responders and avoid adverse events by assessing contraindications.

Quantification of actual responses at a cohort level may be simulated, and actual responses may also be assessed at a cohort level of patients or test subjects. This can be performed in the context of an observational, randomized, or other clinical trial design.

Detecting contraindications at an individual patient level may also simulate or screen for likely contraindications. Deleterious side effects in the simulated condition are determined by evaluating how molecules are perturbed in the model, not just for evidence of improvement but also for side effects. Even if there is an apparent improvement in the pathology, there may be inadvertent other effects that are worse for the patient than the intended improvement, such as adverse reactions, toxicity, or likely future negative reactions. Likewise, assessment of actual response can be used for modifying treatments or conducting dynamic, combination, multistage, or adaptive clinical trial designs or individual patient management.

Screening tools for clinical trial enrichment to "select in" cases that increase the statistical power and to "select out" patients that decrease statistical power or who may suffer adverse events not related to the therapy may optimize clinical trials. If the likely improvement of the patient is above an inclusion criteria threshold, one could select the patient for the clinical trial provided there are no other exclusion or inclusion criteria issues. Suppose there is a simulated adverse reaction of the patient or detected likely adverse event from other causes above an exclusion criteria threshold. In that case, one could recommend not enrolling the patient in the clinical trial.

4 Conclusions

Improved prevention of CVD will ultimately require identification and treatment early enough in the disease to avoid MI and IS. Personalized therapy could increase patient access to promising therapies and maximize their intended benefits. Such an approach has been widely explored in cancer treatment based on molecular diagnostics from biopsied tumor tissue. However, as atherosclerotic plaque tissue biopsies are not available for individuals with CVD, we rather propose development of similar personalized approaches on an advanced computer-based systems biology strategy strengthened by artificial intelligence, to aid clinical decision-making.

The tissues' features are generally too complex to be readily appreciated by a human observer without aid; the analyses described herein address that problem by evaluating the data at a far more granular level. To make such a decision support system practical requires a mix of mathematical formulations, knowledge representation, and architecture in terms of user interfaces, reporting systems, and the backbone of computation. However, an engineering-only approach is insufficient, and deep biological insights are necessary to build such systems, which start with tissue characterization from clinical imaging (CT) and then proceeding to the molecular scale. Such multiscale processing also includes a temporal dimension that captures disease progression, stabilization, or regression based on spatial changes.

Utilizing this model, we demonstrated that routine radiology imaging could be analyzed to provide sufficiently granular information for selecting therapies with differing mechanisms of action. Whereas one may assume that a much higher spatial resolution to visualize cells and molecules would be needed, our studies rather indicate that a rough estimation of magnitude and directionality suffice. This is especially relevant since the number of therapy choices clinically available today is relatively low, certainly much less than could be envisioned biologically.

Sufficiently comprehensive systems biology models to provide clinically relevant analysis are necessary for application to individual patients. Until now, only relatively small population-level models have been reported, which are not useful for individual patients. In contrast, our described approach may be applied as an integrated CDSS for treatment recommendations based on simulations of how a given patient would respond to one or more therapies. On this basis, we conclude that techniques such as those developed here could enable the goal of better targeting treatments to individual patients and, through that, improve prevention of fatal events in CVD. Individual-level analysis could drive therapy recommendations based on systems biology models representing disease at a granularity that supports simulations.

References

Abdelrahman KM, Chen MY, Dey AK et al (2020) Coronary computed tomography angiography from clinical uses to emerging technologies. J Am Coll Cardiol 76:1226–1243
Aday AW, Beckman JA (2017) Medical management of asymptomatic carotid artery stenosis. Prog Cardiovasc Dis 59:585–590
Aerts HJ (2018) Data science in radiology: a path forward. Clin Cancer Res 24:532–534
Aerts HJWL, Velazquez ER, Leijenaar RTH et al (2014) Decoding tumour phenotype by noninvasive imaging using a quantitative radiomics approach. Nat Commun 5
Aghamiri SS, Singh V, Naldi A et al (2020) Automated inference of Boolean models from molecular interaction maps using CaSQ. Bioinformatics 36:4473–4482
Altafini C (2007) ODEs models in systems biology. In: International school for advanced studies
Antman EM, Loscalzo J (2016) Precision medicine in cardiology. Nat Rev Cardiol 13:591–602
Bergmann FT, Hoops S, Klahn B et al (2017) COPASI and its applications in biotechnology. J Biotechnol 261:215–220
Bergström G, Berglund G, Blomberg A et al (2015) The Swedish CArdioPulmonary BioImage study: objectives and design. J Intern Med 278:645–659
Büchel F, Rodriguez N, Swainston N et al (2013) Path2Models: large-scale generation of computational models from biochemical pathway maps. BMC Syst Biol 7:1–19
Büchel F, Wrzodek C, Mittag F et al (2012) Qualitative translation of relations from BioPAX to SBML qual. Bioinformatics 28:2648–2653
Buckler AJ, Doros G, Kinninger A et al (2023a) Quantitative imaging biomarkers of coronary plaque morphology: insights from EVAPORATE. Front Cardiovasc Med 10:1204071
Buckler AJ, Gotto AM Jr, Rajeev A et al (2023b) Atherosclerosis risk classification with computed tomography angiography: a radiologic-pathologic validation study. Atherosclerosis 366:42–48
Buckler AJ, Karlof E, Lengquist M et al (2021) Virtual transcriptomics: noninvasive phenotyping of atherosclerosis by decoding plaque biology from computed tomography angiography imaging. Arterioscler Thromb Vasc Biol 41:1738–1750
Buckler AJ, Marlevi D, Skenteris NT et al (2023c) In silico model of atherosclerosis with individual patient calibration to enable precision medicine for cardiovascular disease. Comput Biol Med 152:106364
Buckler AJ, Sakamoto A, Pierre SS et al (2022a) Virtual pathology: reaching higher standards for noninvasive CTA tissue characterization capability by using histology as a truth standard. Eur J Radiol 159:110686
Buckler AJ, Van Wanrooij M, Andersson M et al (2022b) Patient-specific biomechanical analysis of atherosclerotic plaques enabled by histologically validated tissue characterization from computed tomography angiography: a case study. J Mech Behav Biomed Mater 134:105403
Chaouiya C, Bérenguier D, Keating SM et al (2013) SBML qualitative models: a model representation format and infrastructure to foster interactions between qualitative modelling formalisms and tools. BMC Syst Biol 7:1–15
Chaouiya C, Naldi A, Thieffry D (2012) Logical modelling of gene regulatory networks with GINsim. In: Bacterial molecular networks. Springer, pp 463–479
Choi H, Uceda DE, Dey AK et al (2020) Treatment of psoriasis with biologic therapy is associated with improvement of coronary artery plaque lipid-rich necrotic core: results from a prospective, observational study. Circ Cardiovasc Imaging 13:e011199
Dejana E, Hirschi KK, Simons M (2017) The molecular basis of endothelial cell plasticity. Nat Commun 8:14361
Delfino JG, Pennello GA, Barnhart HX et al (2023) Multiparametric quantitative imaging biomarkers for phenotype classification: a framework for development and validation. Acad Radiol 30:183–195
Du J, Li M, Yuan Z et al (2016) A decision analysis model for KEGG pathway analysis. BMC Bioinformatics 17:1–12
Eungdamrong NJ, Iyengar R (2004) Modeling cell signaling networks. Biol Cell 96:355–362

Glass L, Kauffman SA (1973) The logical analysis of continuous, non-linear biochemical control networks. J Theor Biol 39:103–129
Gomez-Cabrero D, Compte A, Tegner J (2011) Workflow for generating competing hypothesis from models with parameter uncertainty. Interface Focus 1:438–449
Grossmann P, Gutman DA, Dunn WD et al (2016) Imaging-genomics reveals driving pathways of MRI derived volumetric tumor phenotype features in glioblastoma. BMC Cancer 16:611
Gupta A, Al-Dasuqi K, Kamel H et al (2017) Semi-automated detection of high-risk atherosclerotic carotid artery plaque features from computed tomography angiography. In: ESC (ed) European Stroke Conference. ESC, Berlin
Hafiane A (2019) Vulnerable plaque: characteristics, detection, and potential therapies. J Cardiovasc Dev Dis 6:26
Hedin U, Hansson G (2016) Atherosclerosis–disease mechanisms and clinical consequences. In: Oxford textbook of vascular surgery. Oxford University Press
Ibrahimi P, Jashari F, Nicoll R et al (2013) Coronary and carotid atherosclerosis: how useful is the imaging? Atherosclerosis 231:323–333
Karlöf E, Buckler A, Liljeqvist ML et al (2021) Carotid plaque phenotyping by correlating plaque morphology from computed tomography angiography with transcriptional profiling. Eur J Vasc Endovasc Surg 62:716–726
Kholodenko BN (2006) Cell-signalling dynamics in time and space. Nat Rev Mol Cell Biol 7:165–176
Kuhn M, Johnson K (2013) Applied predictive modeling. Springer, New York
Lal BK, Khan AA, Kashyap VS et al (2022) Computed tomography angiographic biomarkers help identify vulnerable carotid artery plaque. J Vasc Surg 75:1311–1322.e1313
Lambin P, Leijenaar RTH, Deist TM et al (2017) Radiomics: the bridge between medical imaging and personalized medicine. Nat Rev Clin Oncol 14:749–762
Lee G, Lee HY, Ko ES et al (2017) Radiomics and imaging genomics in precision medicine. Precis Future Med 1:10–31
Levin DC, Parker L, Halpern EJ et al (2019) Coronary CT angiography: reversal of earlier utilization trends. J Am Coll Radiol 16:147–155
Lyngbakken MN, Myhre PL, Rosjo H et al (2019) Novel biomarkers of cardiovascular disease: applications in clinical practice. Crit Rev Clin Lab Sci 56:33–60
Macklin P (2018) Key challenges facing data-driven multicellular systems biology. arXiv preprint arXiv:1806.04736
Maiorino E, Loscalzo J (2023) Phenomics and robust multiomics data for cardiovascular disease subtyping. Arterioscler Thromb Vasc Biol 43:1111–1123
Mark DB, Federspiel JJ, Cowper PA et al (2016) Economic outcomes with anatomical versus functional diagnostic testing for coronary artery disease. Ann Intern Med 165:94–102
Matic LP, Jesus Iglesias M, Vesterlund M et al (2018) Novel multiomics profiling of human carotid atherosclerotic plaques and plasma reveals biliverdin reductase B as a marker of intraplaque hemorrhage. JACC Basic Transl Sci 3:464–480
Min JK, Chandrashekhar Y, Narula J (2018) The immediate effects of statins on coronary atherosclerosis: can phenotype explain outcome? JACC Cardiovasc Imaging 11:839–841
Moss AJ, Williams MC, Newby DE et al (2017) The updated NICE guidelines: cardiac CT as the first-line test for coronary artery disease. Curr Cardiovasc Imaging Rep 10:15
Mura M, Della Schiava N, Long A et al (2020) Carotid intraplaque haemorrhage: pathogenesis, histological classification, imaging methods and clinical value. Ann Transl Med 8:1273
Murgia A, Balestrieri A, Francone M et al (2020) Plaque imaging volume analysis: technique and application. Cardiovasc Diagn Ther 10:1032–1047
Naylor R, Rantner B, Ancetti S et al (2023) Editor's choice–European Society for Vascular Surgery (ESVS) 2023 clinical practice guidelines on the Management of Atherosclerotic Carotid and Vertebral Artery Disease. Eur J Vasc Endovasc Surg 65:7–111
Newby DE, Adamson PD, Berry C et al (2018) Coronary CT angiography and 5-year risk of myocardial infarction. New Engl J Med 379:924–933

Okuda S, Yamada T, Hamajima M et al (2008) KEGG atlas mapping for global analysis of metabolic pathways. Nucleic Acids Res 36:W423–W426

Pârvu O, Gilbert D (2016) A novel method to verify multilevel computational models of biological systems using multiscale spatio-temporal meta model checking. PLoS One 11:e0154847

Perisic L, Aldi S, Sun Y et al (2016) Gene expression signatures, pathways and networks in carotid atherosclerosis. J Intern Med 279:293–308

Rudin C (2019) Stop explaining black box machine learning models for high stakes decisions and use interpretable models instead. Nat Mach Intell 1:206–215

Saba L, Brinjikji W, Spence JD et al (2021) Roadmap consensus on carotid artery plaque imaging and impact on therapy strategies and guidelines: an international, multispecialty, expert review and position statement. AJNR Am J Neuroradiol 42:1566–1575

Saba L, Saam T, Jäger HR et al (2019) Imaging biomarkers of vulnerable carotid plaques for stroke risk prediction and their potential clinical implications. Lancet Neurol 18:559–572

Sanda GE, Belur AD, Teague HL et al (2017) Emerging associations between neutrophils, atherosclerosis, and psoriasis. Curr Atheroscler Rep 19:53

Sorger PK, Allerheiligen S, Abernethy D et al (2011) Quantitative and systems pharmacology in the postgenomic era: new approaches to discovering drugs and understanding therapeutic mechanisms

St. Pierre S, Siegelman J, Obuchowski NA et al (2017) Measurement accuracy of atherosclerotic plaque structure on CT using phantoms to establish ground truth. Acad Radiol 24:1203–1215

Tamez-Peña JG, Rodriguez-Rojas JA, Gomez-Rueda H et al (2018) Radiogenomics analysis identifies correlations of digital mammography with clinical molecular signatures in breast cancer. PLoS One 13:e0193871

Thomas R (1973) Boolean formalization of genetic control circuits. J Theor Biol 42:563–585

Tian M, Reed JL (2018) Integrating proteomic or transcriptomic data into metabolic models using linear bound flux balance analysis. Bioinformatics 34:3882–3888

Vaidya K, Arnott C, Martínez GJ et al (2018) Colchicine therapy and plaque stabilization in patients with acute coronary syndrome: a CT coronary angiography study. JACC: Cardiovasc Imaging 11:305–316

Van Assen M, Varga-Szemes A, Schoepf UJ et al (2019) Automated plaque analysis for the prognostication of major adverse cardiac events. Eur J Radiol 116:76–83

Watterson S, Marshall S, Ghazal P (2008) Logic models of pathway biology. Drug Discov Today 13:447–456

Whelan KE, King RD (2008) Using a logical model to predict the growth of yeast. BMC Bioinformatics 9:1–16

World Health Organization (2023) Cardiovasc Dis. https://www.who.int/health-topics/cardiovascular-diseases

Wrzodek C, Büchel F, Ruff M et al (2013) Precise generation of systems biology models from KEGG pathways. BMC Systems Biol 7:1–12

Zhu G, Li Y, Ding V et al (2019) Semiautomated characterization of carotid artery plaque features from computed tomography angiography to predict atherosclerotic cardiovascular disease risk score. J Comput Assist Tomogr 43:452–459

Systems Biology in the Field of Vaccine Development

Sezer Okay

Contents

Abstract The conventional 3I (isolate, inactivate/attenuate, inject) strategy for vaccine development has contributed to the field of vaccinology much, and huge numbers of lives have been saved with the vaccines produced by this strategy. However, extensive knowledge on immunology and genetics was lacking during the development of such empirical vaccines. A more complete understanding of the biological processes underlying the infections such as human immunodeficiency virus needs novel systems to predict immune responses to novel vaccines in vivo. Systems biology is an approach to interrogate complex biological systems through investigation of numerous biomolecules by holistic approaches. The immune system involves more than 1500 genes/proteins in many interconnected pathways and processes, and a systems-level approach is vital in broadening our understanding of the immune response to vaccination. Critical genetic determinants of immune response found by systems biology can show the susceptibility to adverse events or reduced response to a vaccine. Systems biology also enables us to find out potent antigens to develop successful vaccines. Omics technologies including genomics, transcriptomics, and proteomics are the holistic tools that are used for systems biology. A recent field called systems vaccinology or vaccinomics has been proposed to describe the vaccine development using systems biology approaches. Thus,

S. Okay (✉)
Department of Vaccine Technology, Vaccine Institute, Hacettepe University, Ankara, Türkiye

J. Barciszewski (ed.), *Systems Biology II*, RNA Technologies 15,
https://doi.org/10.1007/978-3-031-62178-9_12

key elements for the protective immune responses and adverse events could be identified.

Keywords Immunoinformatics · Immunogenetics · Systems serology · Systems vaccinology · Vaccinomics

Abbreviations

Ab	Antibody
BCG	Bacillus Calmette-Guerin
DEG	Differentially expressed gene
HBsAg	Hepatitis B surface antigen
HIV	Human immunodeficiency virus
HLA	Human leukocyte antigen
IFN	Interferon
IL	Interleukin
OMP	Outer membrane protein

1 Introduction

Vaccines are effective tools in preventing mortality, morbidity, and disability due to infectious diseases. Therefore, they are important in terms of public health and preventive medicine. There were some applications in the history to get immunized against infections. However, the first modern vaccine was the cowpox vaccine discovered by Edward Jenner in the late eighteenth century to protect people against smallpox. Since this invention, several important diseases have been eradicated or controlled by vaccination. Various health and economic benefits have been provided by reducing the burden of infectious diseases in both developed and developing countries (Poland 2007; Blohmke et al. 2015; Brisse et al. 2020; Pollard and Bijker 2021).

Despite this remarkable success, a number of challenges remain in vaccine development and application. There are some problems such as efficacy issues in vaccine development against many infections like human immunodeficiency virus (HIV), tuberculosis, and influenza. Also, there is a need for rapid vaccine development in epidemics such as the coronavirus disease 2019 (COVID-19) pandemic (Brisse et al. 2020). A better understanding of the complex relationships within the immune system is crucial for modern vaccine development, which requires new strategies and approaches (Bragazzi et al. 2018).

Formerly, Pasteur's 3I methodology (isolate, inactivate/attenuate, and inject) was used in vaccine development. For instance, live bacillus Calmette-Guerin (BCG) vaccine against tuberculosis was obtained via 230 times passage of *Mycobacterium bovis* bacilli in culture medium to attenuate the pathogen. Later on, the poliovirus was inactivated in $<0.5\%$ formalin in development of an effective and safe polio vaccine. However, the conventional culture-based techniques in vaccinology have limitations such as the safety reasons or lack of appropriate in vitro culture conditions for pathogens, antigenic shift and drift, and molecular mimicry (Bidmos et al. 2018). Additionally, conventional empirical approaches are costly and time-consuming (Blohmke et al. 2015; Okay 2023). Figure 1 shows that vaccine development process takes a long period of time for certain diseases, and for some, this process even continues. However, it should also be noted that vaccine development paradigm accelerated and more vaccines have been commercialized since the 1980s.

As our access to modern technologies such as "omics" sciences, which provide us with a wide range of information about pathogens, has increased, it has become easier to analyze the virulence properties of disease agents, the immune responses against them, and the consequences of genetic variations associated with the protection against diseases. Advances in bioinformatics have enabled better interpretation of data obtained from omics analyses (Bowick and Barrett 2010). Increased knowledge in the fields such as genetic engineering and recombinant DNA technology, protein and carbohydrate chemistry, and combinatorial chemistry also improved vaccinology (Bragazzi et al. 2018). Such comprehensive knowledge provided rational design and reduced empiricism in vaccine development by limiting the failure and cost (Blohmke et al. 2015). Integrating modern techniques with traditional methods has led to the emergence of promising vaccine candidates against challenging pathogens.

Systems biology is a discipline that uses a holistic approach to decipher a biological system through integration of data obtained by analyses under different conditions. In this discipline, the interrelationship of biological processes is analyzed in-depth, and models are produced to predict biological responses. For example, by revealing the immune response profile against a viral infection, a model can be created for how a vaccine should be developed against this pathogen. Thus, the era of vaccinology was switched from "empirical" to "knowledge-based" by the aid of systems biology (Adams et al. 2011; Andersen-Nissen et al. 2012; Petrizzo et al. 2012).

2 Systems Vaccinology/Vaccinomics

The conventional methods used for the evaluation of immune responses after vaccination, such as the enzyme-linked immunosorbent assay (ELISA), have been used extensively in vaccinology. However, these techniques do not provide a holistic view on varying components of the immune system with their complex structures and dynamics to gain insight into the correlates of protection for a vaccine candidate (Circelli et al. 2015).

Activation or sometimes suppression of various processes such as antigen presentation and activation of a naïve T cells is required to develop protective immune

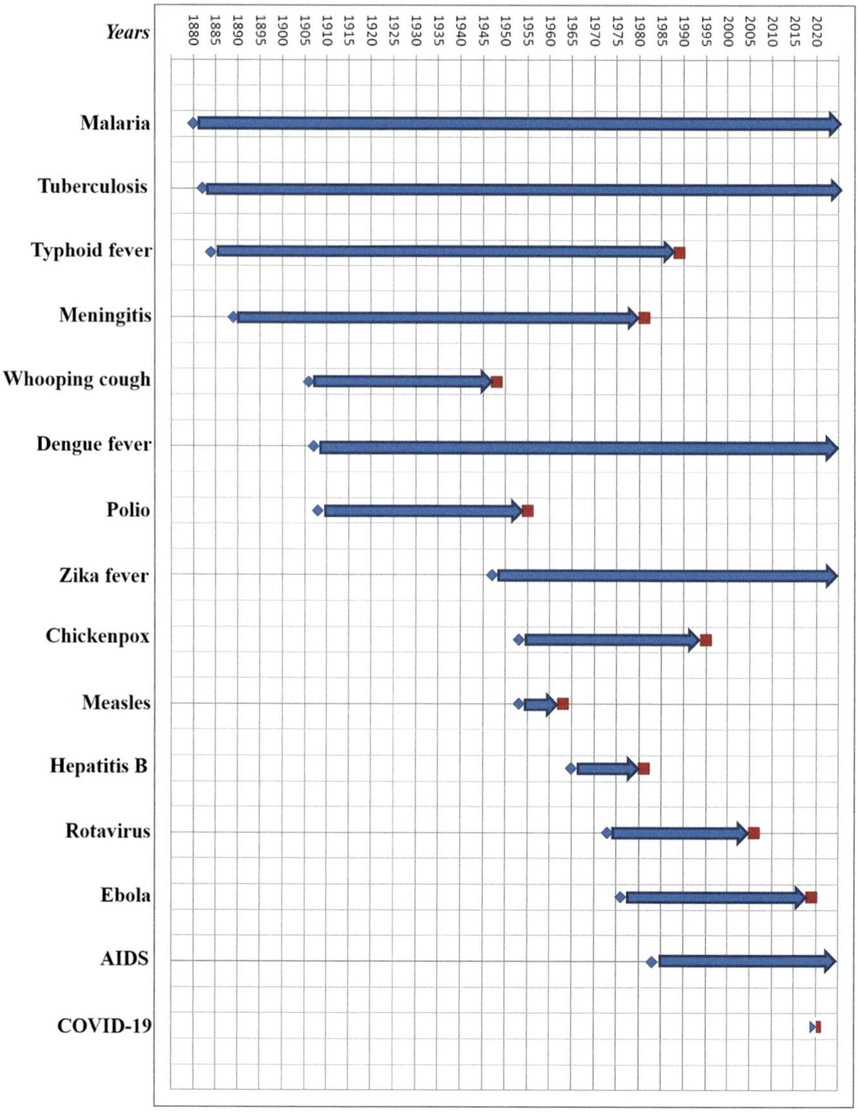

Fig. 1 Timeline of vaccine development against some diseases from 1880 to present (adapted from Mathieu et al. 2021). The blue sign shows the year that a pathogen was identified related to the disease of interest and vaccine development was started. The red sign shows the year that an efficient vaccine was approved and commercialized. For example, BCG has been used against tuberculosis; however, efforts to develop a vaccine against tuberculosis are ongoing due to some protection issues of BCG

responses using a vaccine. Upregulation and downregulation of certain genes and proteins are mediators of this control. An interactive regulation of these factors provides protective responses postvaccination (Poland et al. 2013). For example,

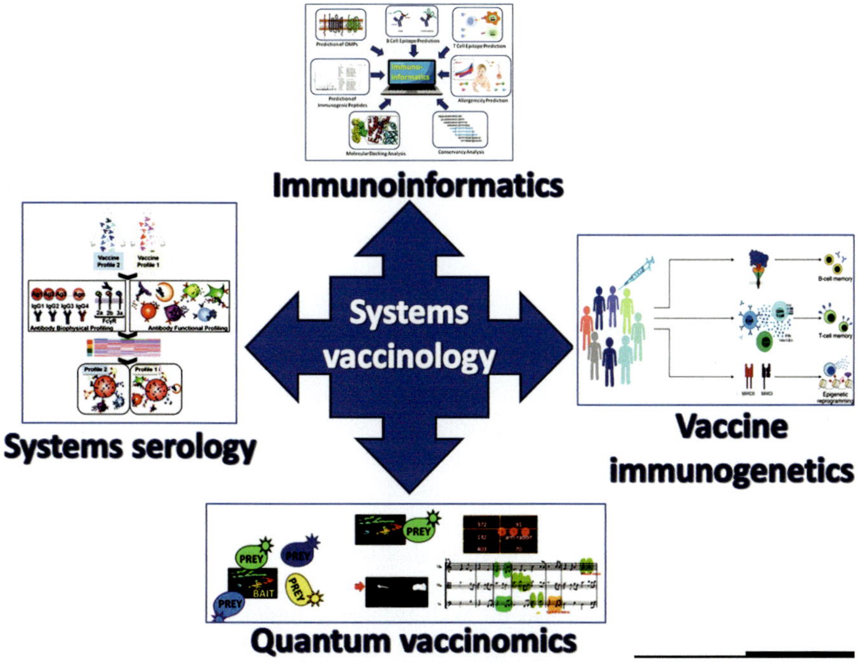

Fig. 2 Research topics which contribute to systems vaccinology. Immunoinformatics is used to identify immunoprotective epitopes with desired features. Vaccine immunogenetics investigates the effects of genetic variations on immune responses to vaccines. Quantum vaccinomics is associated with the electronic structure of molecular interactions providing immunogenicity to peptides. Systems serology methods decipher the key elements of immune responses providing protection

only the levels of certain cytokines might be monitored as an indicator of protection for a vaccine, but these cytokine levels may also show the induction of inflammation. Therefore, a holistic assessment of multiple biomarkers gives more accurate and comprehensive evaluation of immune responses for a protective vaccine (Poland et al. 2011). This kind of comprehensive analysis would also reduce the adverse events and vaccine failure (Prachi et al. 2013).

In the last two decades, important breakthrough has been achieved in omics sciences, bioinformatics, pathogenesis, immunology, and biotechnology for vaccine development and evaluation of postvaccination events (Adams et al. 2011; Goll et al. 2023). Thus, a new branch of science called systems vaccinology or vaccinomics emerged to develop safe and efficient vaccines, having a global view on the elements of immune responses following vaccination (Bernstein et al. 2011; Bragazzi et al. 2018). Four research topics, i.e., immunoinformatics, vaccine immunogenetics, quantum vaccinomics, and systems serology, have considerable contributions to the systems vaccinology (Fig. 2).

Systems vaccinology provides a global view on the innate and adaptive immune responses induced by vaccines. These two main arms of the immune system interact with each other and evolve in time. Identification of this interaction in a global perspective is useful for vaccine development because correlates or biomarkers of protective immunity can be determined in this way. Additionally, regulatory networks that need to be activated to achieve desired immune responses and those that need to be repressed to reduce adverse events can be identified (Zak and Aderem 2015).

The immune system is complex, anatomically distributed, and highly interregulated. Analysis of each component separately to interpret the immune responses is not feasible. However, holistic approaches are useful to explicate this complex system and its regulation. Various lineages and differentiation states of cells related to the immune system can be identified easily using peripheral blood samples. These cells are very sensitive to perturbation, and leukocytes found in the peripheral blood carry information about the systemic immune events after vaccination since they migrate through the circulatory system (Pulendran et al. 2010).

The network of the immune responses contains genetic elements activating or suppressing immunity. Dominance status of genetic variants, epigenetic factors, gene-gene interactions, and genes encoding for other elements of host response also affect immune responses (Oberg et al. 2011). Omics approaches are valuable tools for comprehensive screening of immune elements and their associations (Okay 2021). Large-scale datasets can be analyzed using bioinformatics tools, and the individual components and connections are translated by systems biology approaches into models. Altogether, systems vaccinology enables us to comprehend the functions of immune system and direct it to obtain protective responses for efficient vaccines (Petrizzo et al. 2014; Germain 2018).

In systems biology approaches, since a biological process is evaluated from a global perspective, the integration of data regarding the relationships between genes, transcripts, proteins, metabolites, and epigenetic regulators is very important. Current analysis and modeling tools and databases are constantly being developed for this challenging purpose. In this field, data mining, machine learning, computational biology, and bioinformatics methods are used effectively (Buonaguro et al. 2011).

A comprehensive analysis of host-pathogen interactions, called interactome, can also be performed using systems vaccinology tools, which is very important in designing effective vaccines. These complex interactions can be deciphered via the identification of dynamic/temporal activities of genes related to host responses to infections. Thereby, identified molecular pathways can shape the design of vaccine through prediction of potential immune responses to the vaccine candidate (Adams et al. 2011; Buonaguro et al. 2011; Gomez et al. 2013).

New regulators of innate and adaptive immunity can be discovered rapidly by systems vaccinology tools. More comprehensive information regarding regulatory kinetics at the transcriptional, translational, posttranscriptional, and protein-protein interaction levels can also be obtained. Given all this information, by predicting molecular markers of a vaccine's immunogenicity, we can predict vaccine efficacy and also design new vaccine antigens or formulations. Additionally, responders vs. nonresponders of a vaccine can easily be identified by these system-level approaches (Buonaguro and Pulendran 2011).

Green et al. (2023) evaluated the immune responses to adenoviral-vectored respiratory syncytial virus (PanAd3-RSV) vaccine by transcriptomic analysis using whole blood samples collected postvaccination. The routes of vaccination, i.e., intramuscular (i.m.) and intranasal (i.n.), were also evaluated. Although i.n. prime vaccination did not produce differentially expressed genes (DEGs), i.m. prime vaccination resulted in 643 DEGs. However, higher number of DEGs was obtained by i.n. priming on boosting compared to i.m. priming. Additionally, the DEGs for both prime and boost vaccination were found to be associated with lymphocytic and humoral immune responses and type-I interferon (IFN)-related pathways.

Systems vaccinology approaches have also been used to identify the biomarkers of protective immunity against HIV (Zak and Aderem 2012). The transcriptome changes after vaccination with NYVAC-expressing HIV or simian immunodeficiency virus (SIV) antigens (*env*, *gag*, *pol*, or *nef*) were evaluated using two forms of this vaccine, NYVAC-C-ΔB19R (*B19R* viral gene was deleted to prevent replication) and NYVAC-C-KC (replication capacity was restored by insertion of *K1L* and *C7L* genes). Expression of IFN and IFN-induced genes were found to be elevated by NYVAC-C-ΔB19R both in conventional and plasmacytoid dendritic cells, cDCs and pDCs, respectively. However, distinct signaling pathways were activated by NYVAC-C-KC in these cell types. Also, NYVAC-C-KC induced the pathways related to enhanced antigen processing and presentation in cDCs (Haddad and Pantaleo 2012).

A meta-analysis of transcriptomic datasets related to SIV vaccination revealed a total of 13 pathways related to the protective immunity. Among the signaling pathways, monocyte signaling, memory processes for B cells, and activation of T cells were found to be associated positively, but inflammation was associated negatively with protection. Additionally, *IL-6R* gene encoding a subunit of the interleukin 6 (IL-6) receptor was found to be highly correlated with protective immunity conferred by SIV vaccines (Kuri-Cervantes et al. 2016).

A vaccinomics approach was designed by Contreras et al. (2017a) to identify candidate antigens for a protective vaccine against *Anaplasma phagocytophilum* infection. Their pipeline included quantitative transcriptomic and proteomic analyses using *Ixodes scapularis* ticks uninfected or infected with *A. phagocytophilum*. Differential levels of mRNAs and proteins in infected ticks compared to uninfected ones, putative biological functions of identified molecules, and the antibodies related to the infection were determined. Consequently, two tick proteins, lipocalin and lectin pathway inhibitor, were identified as promising antigens to develop a protective vaccine against *A. phagocytophilum* infection. Same research group also reported the involvement of major surface protein 4 (MSP4) and heat shock protein 70 (HSP70) of *A. phagocytophilum* in the interactions with host cells during infection (Contreras et al. 2017b).

Malaria is caused by the unicellular protozoa in the genus *Plasmodium*. Systems vaccinology approaches identified the immune signatures against *P. falciparum* infection as upregulation of genes related to innate immunity and adaptive response, downregulation of genes associated with inflammation and phagocytosis, as well as DEGs involved in apoptosis and mitochondrial processes (Tomazic et al. 2022).

There are some challenges in vaccinomics research, such as integration of the datasets obtained from different omics approaches and the scarce screening platforms to identify the effective candidate antigens that potentially provide protective immunity. Intelligent big data techniques such as graph theory for network development of key molecules in host-pathogen interactions, data visualization, pattern recognition, and machine learning can be used to address these challenges (De la Fuente et al. 2018). Reif et al. (2009) used an analytical method to integrate the genetic and proteomic datasets related to the adverse events after smallpox vaccination. Random forest algorithm was used to select the most relevant attributes from categorical or continuous data, and a final decision tree model was constructed depending on the selected attributes. A comprehensive model was obtained by combining selected attributes with related data from previous studies.

3 Immunoinformatics

In the design of subunit vaccines, immunogenic B or T cell epitopes of an antigen are identified to induce protective immune responses. In the classical method, first the antigens are determined, and then the interaction between the subunits of relevant protein with B or T cells are evaluated. Traditional methods are laborious and expensive. On the other hand, developments in omics approaches, bioinformatics, molecular biology, and immunology have accelerated the immunoinformatic studies (Hegde et al. 2018).

Immunoinformatics is a recent research field using a variety of tools and databases to analyze the large amounts of immunological data obtained by experimental studies. Today, more effective vaccines can be designed through prediction of immune epitopes using immunoinformatic approaches (Bahrami et al. 2019). However, it should be noted that the data obtained by informatics are predictions. Therefore, the information resulting from in silico workflows should be integrated with in vitro and in vivo data to compose reliable design (Söllner 2011).

There are many steps in vaccine design using immunoinformatic approaches (Khalili et al. 2014; Lata et al. 2020) (Fig. 3). First, the whole protein sequences (or proteome) of the target pathogen are obtained and screened to identify the immunogenic ones using a server such as VaxiJen (http://www.ddg-pharmfac.net/vaxijen/VaxiJen/VaxiJen.html) or VIOLIN (https://violinet.org). Next, the localizations of antigens are determined because the proteins easily introduced to immune cells are more immunogenic. Therefore, surface-exposed outer membrane proteins (OMPs) are mostly preferred vaccine candidates. The CELLO server (http://cello.life.nctu.edu.tw) can be used for the identification of OMPs. Once the optimal antigenic proteins are determined, they are screened for the B and T cell epitopes. B cell epitopes are recognized by the receptors of B cells to induce humoral immune responses. There are two types of B cell epitopes: linear ones are consecutive amino

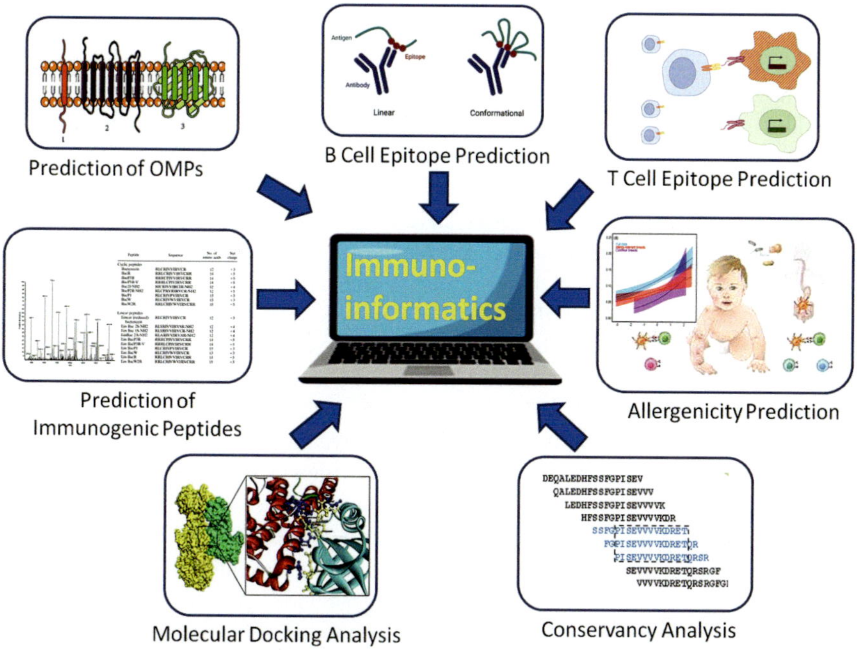

Fig. 3 Immunoinformatics includes multiple analyses to determine the most appropriate peptide epitopes for vaccine development. *OMPs* outer membrane proteins

acid sequences of an antigen, and conformational B cell epitopes include only immunogenic, discontinuous residues in three-dimensional (3D) structure of an antigen. Since T cell receptors recognize the antigenic peptides after being processed by antigen-presenting cell and introduced by major histocompatibility complex (MHC), T cell epitopes are predicted according to the MHC types (MHC class I or II) known also as human leukocyte antigen (HLA). The Immune Epitope Database (IEDB) Analysis Resource (http://tools.iedb.org/main) can be used for the prediction of B and T cell epitopes. Allergenicity score is another criterion in the selection of epitopes for vaccine design. Recently, Goto et al. (2023) proposed a novel method including machine learning and data-driven approach to predict allergen-specific patterns in peptide sequences. Conservancy analysis is also performed in immunoinformatic approach to select the epitopes conserved across the different strains or species of the pathogen, and IEDB resource can be used for this purpose. Finally, a molecular docking analysis using 3D structures of the HLA molecules and predicted epitopes can be performed using GROMACS software (https://www.gromacs.org) to simulate the molecular dynamics.

4 Vaccine Immunogenetics

The immune responses conferred by the vaccines mostly have variations among individuals depending on many factors such as age, gender, immune history (e.g., previous vaccinations or infections), mental or physical stress, and nutritional status. Genetic diversity in a population also affects the immune responses to vaccines significantly (Nakaya and Pulendran 2012). The genetic determinants related to immunity include membrane-associated receptors, toll-like receptors (TLRs), vitamin A and D receptors, cytokine receptors, as well as the genes encoding cytokines, HLAs, and immunoglobulin allotypes (Poland et al. 2008).

Vaccine immunogenetics investigates the association between diverse genetic background in a human population and varied immune responses raised after vaccination (Kennedy et al. 2020). For example, influenza vaccine induces several serum markers such as IL1RA, IL-5, and C-reactive protein (CRP) more in females compared to males, and higher antibody (Ab) titers were observed in women. Additionally, a correlation between testosterone-regulated lipid biosynthesis and decreased humoral responses to influenza vaccine was discovered in men. Efficacy of influenza vaccines is also lower in obese individuals because of reduced production of specific CD8+ T cells, IFN-γ, and granzyme. Another factor affecting immunity is immunosenescence, i.e., deterioration in the regulation of immune system by aging, due to the reasons such as chronic inflammation (inflammaging) as well as the decrease in the expression of TLRs and activity of natural killer (NK) cells (Poland et al. 2018). Similarly, efficacy of influenza vaccines is reduced in the early childhood because activation of the innate immunity is weak at this age. Moreover, the gut microbiome, chronic infections, and metabolic states also affect the immune responses (Rappuoli et al. 2018).

Genome-wide association studies are useful in the identification of genetic variants related to different human traits such as varying immune responses to a vaccine. For instance, a variant proximal to the gene encoding HLA-DPB1 was associated with both immune responses to hepatitis B vaccine and chronic hepatitis B infection because this condition results in an inability to produce antibodies against hepatitis B surface antigen (HBsAg). Omics approaches can be used to identify specific B cell repertoire patterns after vaccination (Blohmke et al. 2015).

There are some difficulties in the immunogenetic analyses. The genes associated with immune responses, such as HLA genes, are complex and extensively polymorphic. Additionally, variations can be observed due to the gender, racial, or ethnic differences. Moreover, the results obtained for gene-specific polymorphisms should be validated to determine the true-positive and false-positive associations (Poland et al. 2008).

A system-level approach includes in silico genotype/phenotype models to integrate the effects of varying markers. For instance, one gene may affect the phenotype depending on the presence of another gene, called epistatic interactions. Epistasis network can be integrated with functional interactions to obtain genotype/phenotype models. These analyses can be performed to exhibit the relations between gene variants and immune responses induced by vaccines (Kennedy et al. 2020).

Beyond the analysis of simple correlations between genetic polymorphisms and immune repertoire, sophisticated algorithms of machine learning such as discriminant analysis via mixed integer programming (DAMIP) can be used to predict the outcomes of interest. A discrete support vector machine is combined with a robust feature selection module to confer an optimization-based predictive modeling framework. Even small training sets can be classified using DAMIP with high accuracy of prediction. Thus, key elements providing protective immunity after vaccination can be predicted (Mooney et al. 2013).

5 Quantum Vaccinomics

Quantum biology investigates the topics such as dynamics maintaining the nonequilibrium state in biological systems, proton tunneling in DNA mutations, or stochastic models for gene regulation (De la Fuente and Contreras 2023). Quantum immunology was proposed for the electronic structure of molecular interactions providing immunogenicity to peptides. Since the photon is described as a quantum of light, immune protective epitopes can be defined as immunological quantum (Contreras et al. 2022), which can be combined to obtain vaccines composed of varying epitopes conferring protection against multiple pathogens (De la Fuente and Contreras 2021).

Quantum vaccinomics is a new methodological approach in vaccine research including varying in vitro, *in music*, and in silico platforms such as synthetic biology, big data, precision medicine, gene sequencing, and editing for the quantum immunology to analyze the electronic structure of molecular interactions providing immunogenicity to peptides. In this approach, *in music* analysis is very interesting that musical algorithms are used to get insights into the interactomes of proteins and their evolution (De la Fuente and Contreras 2023).

A quantum vaccinomics approach was applied to investigate the varying protection potentials of vaccines composed of *A. phagocytophilum* MSP4 antigen in sheep and in rabbits. Protective epitopes of MSP4 were characterized by microarray analysis using sera from vaccinated sheep and rabbits. Epitope mapping analysis identified the immunological quantum to design a protective antigen. A chimeric antigen including identified epitopes was used in inhibition assays, and its protection potential was shown in both sheep and rabbits (De la Fuente et al. 2022).

6 Systems Serology

Profiling the humoral immune responses has been used mostly to select suitable antigens or variants of the viruses based on Ab and neutralization titers. However, humoral response profiling is a valuable technology that can identify the correlates of protection, which are determinative in efficacy of a vaccine. Efficacy analysis for a vaccine candidate according to mechanistic correlates of protection, such as neutralization titer, is insufficient. Immune responses such as Ab-dependent cellular cytotoxicity (ADCC), Ab-dependent cellular phagocytosis (ADCP), and Ab-dependent cellular viral inhibition (ADCVI) as well as Ab-dependent complement deposition (ADCD) and Ab-dependent complement-dependent cytotoxicity (ADCDC) have critical roles in protection (Ackerman et al. 2017; Chung and Alter 2017).

Recent methodologies allow us to characterize the signatures of protective immune responses. For instance, flow cytometry is widely used to identify the immune phenotypes including varying subsets of cells. Moreover, mass cytometry, also known as cytometry by time-of-flight (CyTOF®), is a recent and updated version of flow cytometry, for high-throughput analysis and high-dimensional characterization of immune responses on single-cell basis. In this technique, heavy metal ion tags are used to label Abs or cells, and analysis is performed by ToF mass spectrometry. Thus, diverse immune elements in a single sample are identified easily (Mooney et al. 2013).

Systems serology is a systematic approach using both computational and experimental analyses to provide an analytical framework. Polyclonal Ab profiles are broadly characterized for the identification of humoral immune responses including various variants of Abs. Unique functions of Abs are determined by their structures controlled translationally and posttranslationally. Numerous different Ab structures are formed composed of fragment antigen binding (Fab) including Fv (fragment variable), which recognizes the different epitopes of antigens, and fragment crystallizable (Fc) region interacting with Fc receptors and complement system (Ackerman et al. 2017, Chung and Alter 2017).

As our knowledge of the remarkable diversity of Ab structures, modes of action, and connection between functions and features of different Abs increases, we will better understand the characteristics of humoral immune responses associated with infections. Thus, promising vaccine candidates potentially providing a high protective immunity can be designed. For instance, systems serology analysis for the RV144 HIV vaccine showed that IgG1 and IgG3 responses specific to Env antigen were interconnected with various Ab-dependent responses including ADCC, ADCP, and ADCDC (Chung and Alter 2017).

Systems serology approach was used by Loos et al. (2023) to characterize the humoral immune responses induced by hepatitis B vaccines including HBsAg and different adjuvants, i.e., $AS01_B$, $AS01_E$, AS03, AS04, or Alum. HBsAg-negative adults were vaccinated twice with adjuvanted formulations, and later they received a fractional dose of non-adjuvanted HBsAg. Higher Ab responses were measured in $AS01_{B/E}$ and AS03 groups compared to AS04 and Alum groups. Additionally, separation was observed between AS01/AS03 and AS04/Alum groups in terms of IgG1, IgA1, and IgM titers, which are associated with varying Fc-related responses.

Key elements of the protective immunity can be identified by combining the data from systems serology and machine learning methods. These markers can be generalized to design different vaccines, and the obtained knowledge can be translated for model organisms. Ab profiling by systems serology using diverse omics tools has the potential to decipher correlates of protective immune responses and cellular mechanisms resulting in development of effective vaccines against challenging infections (Pittala et al. 2019).

7 Conclusion

Vaccines have saved millions of lives for more than a century. Traditional vaccine development methods have been used successfully against many infectious agents. However, these conventional strategies have not performed well for some challenging pathogens such as HIV. Significant advances have been made in the field of life sciences in the last two decades. In particular, advances in systems biology covering omics sciences and bioinformatics have enabled us to gain a global perspective on the mechanisms underlying complex processes such as immune responses. The systems vaccinology or vaccinomics was proposed as the application of systems biology to vaccine development. This holistic approach used to unravel the key elements of protective immune responses postinfection or postvaccination leads to the design of promising vaccine candidates for particularly challenging pathogens. The future direction of systems vaccinology is to design personalized vaccines with enhanced efficacy and minimal side effects.

References

Ackerman ME, Barouch DH, Alter G (2017) Systems serology for evaluation of HIV vaccine trials. Immunol Rev 275:262–270

Adams LG, Khare S, Lawhon SD et al (2011) Enhancing the role of veterinary vaccines reducing zoonotic diseases of humans: linking systems biology with vaccine development. Vaccine 29: 7197–7206

Andersen-Nissen E, Heit A, McElrath MJ (2012) Profiling immunity to HIV vaccines with systems biology. Curr Opin HIV AIDS 7:32–37

Bahrami AA, Payandeh Z, Khalili S et al (2019) Immunoinformatics: in silico approaches and computational design of a multi-epitope, immunogenic protein. Int Rev Immunol 38:307–322

Bernstein A, Pulendran B, Rappuoli R (2011) Systems vaccinomics: the road ahead for vaccinology. OMICS 15:529–531

Bidmos FA, Siris S, Gladstone CA et al (2018) Bacterial vaccine antigen discovery in the reverse vaccinology 2.0 era: Progress and challenges. Front Immunol 9:2315

Blohmke CJ, O'Connor D, Pollard AJ (2015) The use of systems biology and immunological big data to guide vaccine development. Genome Med 7:114

Bowick GC, Barrett AD (2010) Comparative pathogenesis and systems biology for biodefense virus vaccine development. J Biomed Biotechnol 2010:236528

Bragazzi NL, Gianfredi V, Villarini M et al (2018) Vaccines meet big data: state-of-the-art and future prospects. From the classical 3Is ("isolate-inactivate-inject") vaccinology 1.0 to vaccinology 3.0, vaccinomics, and beyond: a historical overview. Front. Public Health 6:62

Brisse M, Vrba SM, Kirk N et al (2020) Emerging concepts and technologies in vaccine development. Front Immunol 11:583077

Buonaguro L, Pulendran B (2011) Immunogenomics and systems biology of vaccines. Immunol Rev 239:197–208

Buonaguro L, Wang E, Tornesello ML et al (2011) Systems biology applied to vaccine and immunotherapy development. BMC Syst Biol 5:146

Chung AW, Alter G (2017) Systems serology: profiling vaccine induced humoral immunity against HIV. Retrovirology 14:57

Circelli L, Petrizzo A, Tagliamonte M et al (2015) Systems biology approach for cancer vaccine development and evaluation. Vaccines (Basel) 3:544–555

Contreras M, Alberdi P, Fernández De Mera IG et al (2017a) Vaccinomics approach to the identification of candidate protective antigens for the control of tick vector infestations and *Anaplasma phagocytophilum* infection. Front Cell Infect Microbiol 7:360

Contreras M, Alberdi P, Mateos-Hernández L et al (2017b) *Anaplasma phagocytophilum* MSP4 and HSP70 proteins are involved in interactions with host cells during pathogen infection. Front Cell Infect Microbiol 7:307

Contreras M, Artigas-Jerónimo S, Pastor Comín JJ et al (2022) A quantum vaccinomics approach based on protein-protein interactions. Methods Mol Biol 2411:287–305

De la Fuente J, Contreras M (2023) Quantum vaccinomics platforms to advance in vaccinology. Front Immunol 14:1172734

De la Fuente J, Contreras M (2021) Vaccinomics: a future avenue for vaccine development against emerging pathogens. Expert Rev Vaccines 20:1561–1569

De la Fuente J, Moraga-Fernández A, Alberdi P et al (2022) A quantum vaccinomics approach for the design and production of MSP4 chimeric antigen for the control of *Anaplasma phagocytophilum* infections. Vaccines (Basel) 10:1995

De la Fuente J, Villar M, Estrada-Peña A et al (2018) High throughput discovery and characterization of tick and pathogen vaccine protective antigens using vaccinomics with intelligent big data analytic techniques. Expert Rev Vaccines 17:569–576

Germain RN (2018) Will systems biology deliver its promise and contribute to the development of new or improved vaccines? What really constitutes the study of "systems biology" and how might such an approach facilitate vaccine design. Cold Spring Harb Perspect Biol 10:a033308

Goll JB, Bosinger SE, Jensen TL et al (2023) The Vacc-SeqQC project: benchmarking RNA-seq for clinical vaccine studies. Front Immunol 13:1093242

Gomez G, Adams LG, Rice-Ficht A et al (2013) Host-*Brucella* interactions and the *Brucella* genome as tools for subunit antigen discovery and immunization against brucellosis. Front Cell Infect Microbiol 3:17

Goto K, Tamehiro N, Yoshida T et al (2023) Novel machine learning method allerStat identifies statistically significant allergen-specific patterns in protein sequences. J Biol Chem 299:104733

Green C, McGinley J, Sande C et al (2023) Transcriptomic response and immunological responses to chimpanzee adenovirus- and MVA viral-vectored vaccines for RSV in healthy adults. Clin Exp Immunol 211:269–279

Haddad EK, Pantaleo G (2012) Systems biology in the development of HIV vaccines. Curr Opin HIV AIDS 7:44–49

Hegde NR, Gauthami S, Sampath Kumar HM et al (2018) The use of databases, data mining and immunoinformatics in vaccinology: where are we? Expert Opin Drug Discov 13:117–130

Kennedy RB, Ovsyannikova IG, Palese P et al (2020) Current challenges in vaccinology. Front Immunol 11:1181

Khalili S, Jahangiri A, Borna H et al (2014) Computational vaccinology and epitope vaccine design by immunoinformatics. Acta Microbiol Immunol Hung 61:285–307

Kuri-Cervantes L, Fourati S, Canderan G et al (2016) Systems biology and the quest for correlates of protection to guide the development of an HIV vaccine. Curr Opin Immunol 41:91–97

Lata KS, Vaghasia V, Bhairappanvar S et al (2020) Vaccine design against leptospirosis using an immunoinformatic approach. In: Tomar N (ed) Immunoinformatics. Method Mol Biol, vol 2131. Humana, New York, USA, pp 173–184

Loos C, Coccia M, Didierlaurent AM et al (2023) Systems serology-based comparison of antibody effector functions induced by adjuvanted vaccines to guide vaccine design. NPJ Vaccines 8:34

Mathieu E, Ritchie H, Ortiz-Ospina E et al (2021) A global database of COVID-19 vaccinations. Nat Hum Behav 5:947–953

Mooney M, McWeeney S, Sékaly RP (2013) Systems immunogenetics of vaccines. Semin Immunol 25:124–129

Nakaya HI, Pulendran B (2012) Systems vaccinology: its promise and challenge for HIV vaccine development. Curr Opin HIV AIDS 7:24–31

Oberg AL, Kennedy RB, Li P et al (2011) Systems biology approaches to new vaccine development. Curr Opin Immunol 23:436–443

Okay S (2021) "Omics" technologies in vaccine research. In: Karahan M (ed) Synthetic peptide vaccine models: design, synthesis, purification and characterization, 1st edn. CRC Press, Boca Raton

Okay S (2023) Effect of coronavirus disease 2019 on the vaccine development paradigm. Explor Immunol 3:433–441

Petrizzo A, Tagliamonte M, Tornesello M et al (2014) Systems vaccinology for cancer vaccine development. Expert Rev Vaccines 13:711–719

Petrizzo A, Tornesello M, Buonaguro FM et al (2012) Immunogenomics approaches for vaccine evaluation. J Immunotoxicol 9:236–240

Pittala S, Morrison KS, Ackerman ME (2019) Systems serology for decoding infection and vaccine-induced antibody responses to HIV-1. Curr Opin HIV AIDS 14:253–264

Poland GA (2007) Pharmacology, vaccinomics, and the second golden age of vaccinology. Clin Pharmacol Ther 82(6):623–626

Poland GA, Kennedy RB, McKinney BA et al (2013) Vaccinomics, adversomics, and the immune response network theory: individualized vaccinology in the 21st century. Semin Immunol 25: 89–103

Poland GA, Ovsyannikova IG, Jacobson RM (2008) Vaccine immunogenetics: bedside to bench to population. Vaccine 26:6183–6188

Poland GA, Ovsyannikova IG, Kennedy RB et al (2011) Vaccinomics and a new paradigm for the development of preventive vaccines against viral infections. OMICS 15:625–636

Poland GA, Ovsyannikova IG, Kennedy RB (2018) Personalized vaccinology: a review. Vaccine 36:5350–5357

Pollard AJ, Bijker EM (2021) A guide to vaccinology: from basic principles to new developments. Nat Rev Immunol 21:83–100

Prachi P, Donati C, Masciopinto F et al (2013) Deep sequencing in pre- and clinical vaccine research. Public Health Genom 16:62–68

Pulendran B, Li S, Nakaya HI (2010) Systems vaccinology. Immunity 33:516–529

Rappuoli R, Siena E, Finco O (2018) Will systems biology deliver its promise and contribute to the development of new or improved vaccines? Systems biology views of vaccine innate and adaptive immunity. Cold Spring Harb Perspect Biol 10:a029256

Reif DM, Motsinger-Reif AA, McKinney BA et al (2009) Integrated analysis of genetic and proteomic data identifies biomarkers associated with adverse events following smallpox vaccination. Genes Immun 10:112–119

Söllner J (2011) Beyond epitopes: future and application of computational vaccinology. Hum Vaccin 7:795–797

Tomazic ML, Marugan-Hernandez V, Rodriguez AE (2022) Next-generation technologies and systems biology for the design of novel vaccines against apicomplexan parasites. Front Vet Sci 8:800361

Zak DE, Aderem A (2012) Overcoming limitations in the systems vaccinology approach: a pathway for accelerated HIV vaccine development. Curr Opin HIV AIDS 7:58–63

Zak DE, Aderem A (2015) Systems integration of innate and adaptive immunity. Vaccine 33:5241–5248

Proteomics in Transplant Medicine

Ira Van Matre and Timothy D. Veenstra

Contents

Abstract Solid organ transplants represent a major lifesaving procedure for tens of thousands of Americans every year. While tissue typing and immunosuppressive agents have reduced the incidence of organ rejection, the overall percentage of patients who suffer this event remains stubbornly high. Successful transplant outcomes rely on matching organ donors and recipients as well as early detection of graft injury or rejection. Biomarkers would increase the accuracy of organ matches, while decreasing the percentage of graft injuries or rejections. Modern technologies have enabled investigators to screen thousands of proteins to identify biomarkers associated with a variety of physiological conditions. However, testing these hypothetical biomarkers is laborious and challenging. In this chapter, we will describe the

---

I. Van Matre · T. D. Veenstra (✉)
School of Pharmacy, Cedarville University, Cedarville, OH, USA
e-mail: mivanmatre@cedarville.edu; tveenstra@cedarville.edu

J. Barciszewski (ed.), *Systems Biology II*, RNA Technologies 15,
https://doi.org/10.1007/978-3-031-62178-9_13

need for transplant-associated biomarkers, the challenges associated with their discovery, and several applications that have used proteomics to identify biomarkers that can increase the success of transplant procedures.

Keywords Proteomics · Organ transplant · Biomarkers · Tissue rejection · Graft injury · Differentially abundant protein

Abbreviations

ADRS	Acute respiratory distress syndrome
APC	Antigen-presenting cell
AR	Acute rejection
AUC	Area under the curve
BKVN	BK virus nephritis
CAN	Chronic allograft nephropathy
CKD	Chronic kidney disease
COVID-19	Coronavirus disease of 2019
DSA	Donor-specific antibody
ELISA	Enzyme-linked immunosorbent assay
HLA	Human leukocyte antigen
IMBAS	Immunopeptidomics by Biotinylated Antibodies and Streptavidin
iTRAQ	Isobaric tags for relative and absolute quantitation
LC/SRM	Liquid chromatography/selected reaction monitoring
MS	Mass spectrometry
MS/MS	Tandem mass spectrometry
NGS	Next-generation sequencing
NK	Natural killer
PCA	Principal component analysis
PCR	Polymerase chain reaction
STA	Stable renal allograft
ROC	Receiver operating characteristic
2D gels	Two-dimensional gel electrophoresis

1 Introduction

During the 1900s several important medical discoveries were made. These include the discovery of antibiotics, the determination of DNA's structure, the development of tissue culture, mass vaccinations against deadly diseases, the isolation of insulin for treating diabetes, and advancements in surgical procedures. Some of the most important developments were those allowing solid organ transplantation. Transferring body parts was described over 3500 years ago in the Ebers Papyrus, which describes using skin grafts to treat burns (Paul 2008). Several other documented descriptions of skin grafts were written over the next centuries, but it is unclear whether these procedures were conducted (Ehrenfried 1909). The first verifiable skin transplant occurred in 1869 when a Swiss surgeon, Jacques-Louis Reverdin, used grafted small epidermal samples taken from a healthy area to seed a location where the skin had been damaged (Hattery et al. 2015). The first successful human allograft cornea transplant was later performed in 1905. However, transplanting larger, more complex, internal organs remained elusive (Zirm 1906).

While research on methods to transplant internal organs continued, skin grafts and eye transplants remained the only types of tissues that could successfully be transplanted until the 1950s. In 1933, Dr. Yurii Voronov attempted the first human kidney allograft from a recently deceased patient, but the recipient survived only 2 days (Howard et al. 2012). In 1953, Richard Herrick, who had been diagnosed with chronic nephritis, received a kidney from his *identical* twin brother (Nordham and Ninokawa 2022). The surgery was performed by Dr. Joseph Murray, who 10 years earlier had been part of a team that performed a skin graft from a deceased soldier to a patient severely burned in an aviation accident. Murray had suspected the success of the graft was partially due to the patient's severely suppressed immune system. So, before the kidney transplant, Murray tested for possible rejection of the twin's kidney by performing small skin grafts between the brothers. With no signs of skin rejection, he successfully transplanted the kidney without complications. Richard Herrick lived another 8 years with his brother's kidney. Later discovery revealed the transplant was successful because these twins had identical human leukocyte antigen (HLA) types; allograft rejection in the recipient was impossible.

During this time, Dr. Peter Medawar's studies of tissue rejection showed that graft recipients formed antibodies against a donor graft unless the recipient and donor had been exposed to similar foreign tissue early in life (Medawar 1946; Billingham et al. 1953; Billingham et al. 1956). This discovery earned him a share of the 1960 Nobel Prize in Physiology or Medicine with Sir Macfarlane Burnet.

In 1959, the first successful kidney transplant was performed between individuals who were not genetically identical (Murray et al. 1962). The donor and recipient were fraternal twin brothers. The recipient underwent a sublethal dose of total body X-ray before surgery. Suppression of the immunological response before receiving his brother's kidney allowed the recipient to thrive for an additional 25 years. This success was particularly exciting since it expanded the population who could donate organs to recipients who were not genetically identical or closely related.

This increase in the availability of transplants led to rapid advancements. Researchers began to focus further on using immunosuppression to prevent organ rejection between unrelated donors and recipients. Though X-rays were an effective method to induce immunosuppression, several patients died from excess radiation. In 1959, certain drugs, including azathioprine, were found to suppress the immune system as effectively as X-rays (Starzl et al. 1967). A canine study in 1962 also showed that 6-mercaptopurine prolonged survival after renal transplant. Building on this, Joseph Murray showed that irradiation combined with chemical immunosuppression using azathioprine, actinomycin C, and prednisone improved short-term outcomes in humans. This study led to the first successful kidney transplant between non-related patients; it was also the first successful transplant where the donor was deceased. Tissue typing provided another leap forward. By 1962, the combination of tissue typing (Terasaki and Singal 1969) and drug-induced immune suppression was used for the first time in a successful kidney transplant. The discovery of cyclosporine by Sandozin 1972 was another breakthrough in transplant surgery (Tedesco and Haragism 2012). Its introduction as an immunosuppressant increased the 1-year survival rate of kidney transplants to more than 90% and graft survival to 80% (Shimmura et al. 2005).

2 Prevalence of Solid Organ Transplants

Data reported to the Global Observatory on Donation and Transplantation (GODT, https://www.transplant-observatory.org/) shows that ~157,526 solid organ (i.e., kidney, heart, lung, liver, pancreas, and small bowel) transplants were performed worldwide in 2022 across 90 reporting countries representing nearly 90% of the world's population. This represents an almost 65% increase from 2009 in which a similar number of countries reported their data. As of 2022, the number of kidney, liver, and heart transplants performed in the United States has almost doubled since 2000 (kidney, 26310 vs. 13,621; liver, 9528 vs. 4954; heart, 4126 vs. 2246). Lung transplants have almost tripled (2743 vs. 959). The number of pancreas transplants has slightly decreased (Fig. 1).

While the numbers above are impressive, providing safe organs for every recipient remains a challenge. Almost 16,000 kidney transplants were performed in the

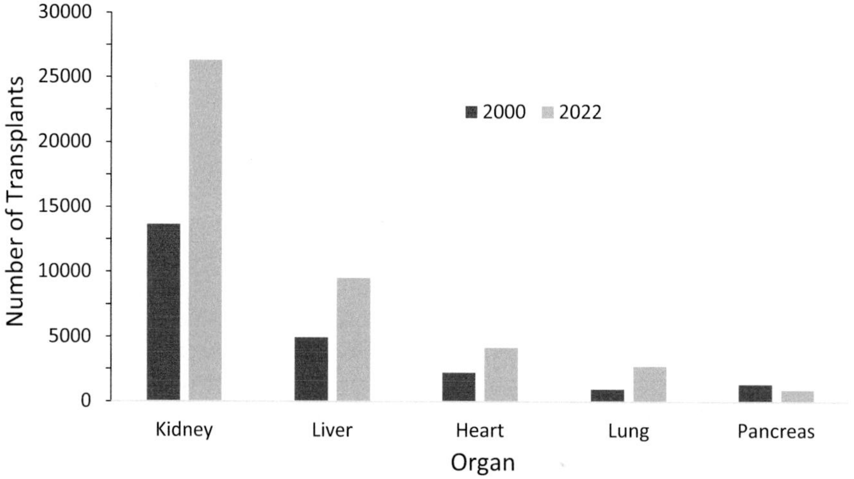

Fig. 1 Bar graph comparing the number of transplants conducted for specific solid organs in 2000 versus 2022

United States in 2023, but 90,000 patients are on the waiting list for a suitable kidney donation, and on average 17 people die each day waiting for an organ transplant (https://www.organdonor.gov/learn/organ-donation-statistics). Considering the preciousness of organs for transplant, it is imperative to maximize the chance of graft survival.

3 Tissue Typing

Tissue typing, also known as human leukocyte antigen (HLA) typing, is performed to determine if the organs of a donor are compatible with a prospective recipient. Mismatched donor and recipient tissues can lead to rejection of the tissues. In the process of tissue typing, the sequences of an individual's HLAs are determined (Thorsby 2009). In transplant medicine, tissue typing is critical since it provides a measure of the rejection risk between a donor-recipient pair by determining if any antibodies within the recipient serum are specific against donor HLA molecules. Pairing donors and recipients with closely matched tissue types (i.e., HLA profiles) lowers the risk of organ rejection and improves long-term graft survival. HLAs are the primary alloantigens—genetically encoded antigens present in some but not all individuals of a species—germane to transplant rejection (Argani 2019). HLA genes express both HLA class I and II proteins. HLA class I proteins—which are present on all nucleated cells to varying degrees—present peptides of viruses and expired or defective intracellular proteins (Bodmer 2023; Sheldon and Poulton 2006). Presentation of these peptides to CD8+ cytotoxic T cells can activate immune processes that kill the cell. HLA proteins also inhibit natural killer (NK) cell activity through their interaction with killer inhibitory receptors on the surface of NK cells. HLA class II protein expression is constitutive but is limited to immune cells such as B cells and other antigen-presenting cells (APCs) (Wernet et al. 1987). Like class I proteins, they present peptides but in this case to CD4+ helper T cells. The affinity of the peptide binding to either class of HLA proteins dictates the allogenicity response of the host.

3.1 Assessing Tissue Type

As the primary alloantigens related to transplant rejection, HLA similarities must be assessed when tissue typing. The more than 200 genes that encode HLA proteins are clustered within loci on the short arm of chromosome 6 (Zabaneh et al. 2017). Owing to their role in presenting peptides from a wide diversity of invasive organisms to immune cells, it is not surprising that HLA genes represent the most polymorphic gene cluster within the human genome (Wang et al. 2012). The IPD-IMGT/HLA Database lists 38,416 HLA-related alleles listed as of September 2023 (Barker et al. 2023).

As tissue typing technology has evolved, the focus has been on determining the antigenically important sequences within HLAs that determine the risk of recipient organ rejection. Unlike conditions such as HIV infection and cancer where malfunctions lead to disease, organ rejection occurs when the HLA system functions too well (Mosaad 2015). Advances in the use of immunosuppression agents have minimized the importance of HLA matching, but they have not eliminated the fact that the extent of HLA mismatches between recipients and donors increases the rate of organ rejection (Zachary and Leffell 2016). A study of 135,970 kidney transplants showed a significant increase in graft survival with HLA matching (Opelz and Döhler 2007). An analysis of the Scientific Registry of Transplant Recipients database between 1988 and 2007 showed that only 10% of patients with no HLA mismatches became newly sensitized after graft loss compared to almost 40% of patients with a greater extent of HLA mismatches (Meier-Kriesche et al. 2009).

Determining donor and recipient HLA types was originally performed using serological methods that used antibodies to detect HLA antigens. The three primary techniques for detecting HLA antibodies include the serological cytotoxicity method, flow cytometry, and solid-phase methods (McCaughan et al. 2019). The cytotoxicity method places lymphocytes from the patient within individual wells of a serological plate that contain specific antibodies. If the cells possess an HLA antigen that binds to the antibody within the well, the cells will die. Unfortunately, this method does not differentiate HLA and non-HLA cytotoxic antibodies but can compare HLAs from donors and recipients. Flow cytometry using fluorescently labeled anti-HLA antibodies can also detect surface HLA antigens as cells pass through the laser beam. While more delicate to perform, a solid-phase technique incubates a patient's serum with fluorochrome-dyed polystyrene beads coated with specific HLA antigens. Any reaction between antibodies in the serum and the HLA antigens is visualized using a polyethylene-conjugated Fc-domain-specific antibody. While more complicated than the other two serological techniques, this solid-phase method is more sensitive.

Advancements in molecular biology, primarily gene sequencing, have made polymerase chain reaction (PCR) and next-generation sequencing (NGS) the primary methods for HLA typing today (Erlich 2012; Profaizer and Kumánovics 2018). Both PCR and NGS provide higher resolution and greater accuracy results than serological methods. NGS methods enable concurrent sequencing of millions of DNA fragments, which allows comprehensive coverage of HLA genes. HLA typing can be performed via NGS using either targeted gene panels or whole-genome sequencing. NGS offers the greatest accuracy along with the ability to discover novel alleles with minimal labor and cost. The greatest challenge using this technique is the need for specialized bioinformatic software owing to the number of polymorphisms associated with HLA genes and the large datasets generated using NGS.

3.2 Proteomic Determination of Human Leukocyte Antigen Immunopeptides

While direct HLA typing will continue to be dominated by NGS methods for the foreseeable future, strategic advances in proteomic sample preparation methods have shown the capability of contributing to this space. The challenge for proteomics is identifying a small subset of low abundant proteins within the background of all the other proteins that comprise a proteome. While methods have been devised to identify membrane bound HLAs (mHLAs), it is a laborious process to acquire and process the necessary tissue samples. Identification of mHLAs is also tissue specific making it impossible to provide an organism-wide repertoire of HLAs.

Fortunately, HLAs also exist within the body in a soluble form (sHLAs) that circulate within the blood. These sHLAs are hypothesized to be shed from the cell, cleaved from its surface, or not possess a membrane anchoring domain due to a splice variant (Tanuwidjaya et al. 2022). Soluble HLAs are easily acquired using a minimally invasive blood withdrawal; however, several milliliters of sample are required (Shraibman et al. 2019). Since the role of HLAs is to present peptides from intracellular proteins on the surface of cells, sHLAs would continue to bind these peptides as they circulate. Considering that matching HLAs between donor and recipient patients is critical for assessing compatibility, identifying the peptides bound to the HLAs would also provide a measure of donor/recipient compatibility.

Recently a novel method termed Immunopeptidomics by Biotinylated Antibodies and Streptavidin (IMBAS) was developed that can identify thousands of sHLA-bound immunopeptides from as little as 200 microliters of plasma (Wahle et al. 2023). A key to this method is the use of a biotinylated antibody (W6/32) to immunoprecipitate sHLAs. W6/32 is an antihuman antibody that recognizes HLA-A, B, and C. By biotinylating the antibody, the captured sHLA molecules could be captured using magnetic streptavidin beads. Not only does the antibody immunoprecipitate the HLA protein, it also captures the peptides that are bound to them. To specifically capture the HLA-bound peptides (referred to as immunopeptides), the captured sHLA molecules were filtered through a 10 kDa molecular weight cutoff membrane. Only the peptides that passed through the membrane were analyzed using LC-MS.

The IMBAS method was applied to plasma samples taken from six different individuals. At least 5000 peptides were identified in each sample with an astonishing 13,000 peptides identified in one individual. The total number of unique peptides identified across all samples was almost 40,000. To evaluate the utility of using IMBAS to identify variations between immunopeptides within individuals, the data was analyzed using principal component analysis (PCA). PCA of data acquired from the same sample run in replicate showed tight clustering. PCA of data acquired from different donors, however, showed only loose correlation. These consistent and discordant PCA results demonstrate two important components of IMBAS method: (i) it is reproducible, and (ii) it can identify commonalities and variations between individuals.

Obviously the sHLA immunopeptidome identified in this study arises from several organs throughout the body. Since a vast majority of organ transplants are performed on a single organ, it would be of interest to determine the organ-specific origin of the immunopeptides. Immunopeptide origin was assigned by comparing the results of this sHLA study to a database containing the proteome of 29 healthy tissues (Wang et al. 2019). This comparison showed that the sHLA immunopeptidome was representative of almost all tissues within the body. The most notable exception was the brain, which was less represented in the sHLA immunopeptidome possibly due to their inability to cross the blood-brain barrier.

The ability to identify such large numbers of HLA-bound immunopeptides has great potential in diagnostic medicine. Since HLAs present peptides from within cells, the identification of an aberrant immunopeptide can enable the diagnosis of early-stage cancer through the identification of a neoepitope (i.e., an epitope that has not been expressed previously). Beyond neoepitopes, IMBAS could be used to identify immunopeptides that are diagnostic of known cancers in which a commonly known genetic mutation is responsible. The key is the noninvasiveness of sample collection that would enable routine (e.g., annual or biennial) screening. As it pertains to the topic of this chapter, IMBAS can provide sHLA-bound immunopeptide signatures that can be used to evaluate the match between a donor and recipient prior to solid organ transplantation.

4 Need for Proteomics

The worst-case scenario is that graft loss in kidney transplantation results in the death of the patient. A lesser, but difficult outcome, is that it requires the recipient to resume dialysis treatments. Even if graft loss is detected before patient death, the donor options become limited because anti-HLA antibodies will develop within the recipient. While remedying graft rejection via retransplantation does not alleviate the lack of available organs, early detection of graft rejection is critical for long-term patient survival. The gold standard for detecting transplant complications remains a biopsy, but serial biopsies are rarely performed owing to their invasiveness and risk of causing internal bleeding (Rainey et al. 2023). Unfortunately, early clinical biomarkers to detect failing grafts are lacking. Noninvasive markers such as proteinuria, serum creatinine, serum cystatin C, and estimated glomerular filtration rate lack the necessary sensitivity and specificity (Coll et al. 2000). Immunological graft dysfunction donor-specific antibodies (DSAs) have good diagnostic specificity for detecting graft rejection, but they are usually identified too late for effective clinical intervention (Dutra et al. 2021).

The identification of reliable, accurate, disease-specific, noninvasive methods for the early diagnosis of graft injury or rejection would continue to increase the rate of successful organ transplants. Ideally, these noninvasive methods would result in diagnostic markers that could be measured in routine follow-ups without the discomfort and risks associated with a graft biopsy. The development of advanced, high-throughput proteomic methods that can quantitate thousands of proteins in a variety of biological samples has enabled biomarker discovery to be approached through discovery-driven methods (Wollscheid et al. 2004; Veenstra et al. 2005). Discovery-driven methods survey much of the proteomic content of samples and use sophisticated software algorithms to identify species whose quantity differs in samples being compared. Owing to its ability to analyze minimally invasive samples such as urine, serum, and plasma, proteomics has the potential to identify biomarkers of graft rejection that possess higher sensitivity than current markers. Beyond more reliable diagnosis, the ability to survey minimally invasive samples (rather than analyzing tissue biopsies) can help identify graft rejection sooner, which enables earlier therapeutic intervention and improved outcomes.

For a discovery-driven proteomics to be successful, it needs to be able to identify a significant portion of the proteins present in the samples being analyzed, especially biofluids if the aim is to identify solid organ transplant biomarkers. In 2004, only 1175 proteins were identified in human plasma samples analyzed in four separate studies (Anderson et al. 2004). Today it is common for upward of 4000 proteins to be identified in serum or plasma samples (Geyer et al. 2019; Shah et al. 2024).

This increased proteomic coverage is driven almost solely by developments in analytical instrumentation and computer hardware and software. Not only have these technological improvements spurred proteomics, but basic research studies such as identifying protein interactions, posttranslationally modified proteins, and mutations have also benefited. Previously techniques used to measure changes in protein expression levels or modifications (Western blotting, ELISAs, etc.) became increasingly replaced by technologies such as MS owing to its accuracy and direct measurement.

4.1 *Quantitative Proteomic Methods*

While MS is the core technology for making the measurements to enable the quantitation of proteins within a proteome, it relies extensively on sample preparation and processing methods. Proteomics has evolved to a point where creative and innovative methods in preparing the sample prior to MS analysis have been responsible for many of the key discoveries in the field.

The two strategies used to identify proteins via MS are bottom-up and top-down. These strategies differ in the form in which the proteins are introduced into the mass spectrometer. In bottom-up proteomics (Miller and Smith 2023), proteins are digested into peptides (usually using trypsin) prior to entering the mass spectrometer. Top-down proteomics sends intact proteins into the mass spectrometer (Habeck and Lermyte 2023). Whether bottom-up or top-down is used, the mass spectrometer used MS/MS to identify the intact protein or the peptides. Bottom-up remains the dominant method for MS-based proteomics owing to several factors, including the fact that peptides can be separated with higher resolution than intact proteins.

As advances in the fields of chromatography and MS made it possible to identify thousands of proteins in a complex mixture, the focus turned to developing quantitation methods so that the abundance of proteins in different samples could be compared. The simplest method was termed subtractive proteomics (Schirmer et al. 2003). Subtractive proteomics is based on a simple hypothesis: the number of unique peptides identified for a protein is directly correlated to its abundance within a sample. A classic proof of this hypothesis is comparison of raw serum and serum in which albumin has been depleted. If performed properly, only a few albumin peptides will be identified in the depleted serum using MS; however, characterization of the raw serum will be dominated by albumin peptides.

The problem with subtractive methods was the lack of a direct peptide-to-peptide comparison, and when the mass spectrometer is operated in a data-dependent acquisition (DDA) mode, some peptides may not be selected and identified. The next methods to be developed employed stable isotope labeling. One of the earliest and most popular of these methods was SILAC (stable isotope labeling of amino acids in culture). SILAC is a metabolic labeling method in which stable isotope-labeled amino acids are added in place of normal isotopic abundance amino acids to the cell culture medium (Mann 2006; Veenstra et al. 2000). In a typical SILAC study, two different cell cultures are grown, one is cultured in normal isotopic abundance medium (light) and the other in medium containing the heavy isotopic versions of lysine or arginine. One of the cell cultures is treated with some perturbation, and then equal numbers of cells are combined. After extracting the proteins from this combined sample, they are analyzed using bottom-up LC-MS. They are then extracted from this single sample and digested into tryptic peptides. The normal and heavy isotope-labeled peptides elute into the MS at the same time, and their relative quantities are measured by comparison of their peak intensities. The biggest disadvantage is SILAC requires that cells be cultured, preventing it from being used for tissue samples.

The next big development in using stable isotopes to chemically modify proteins was iTRAQ (isobaric tag for relative and absolute quantitation). This method labels primary amines (i.e., N terminus, lysine side chains, etc.) of peptides with chemically identical, but isotopically distinct, reagents (Ross et al. 2004). Unlike SILAC, the proteins are labeled after they are extracted from the sample, which allows any type of sample to be analyzed. iTRAQ reagents come in kits of four and eight distinct reagents (i.e., isobaric tags) allowing the quantity of peptides to be compared in up to eight different proteome samples. To use iTRAQ, proteome samples are digested into tryptic peptides and labeled at primary amines with the isobaric tags. The labeled samples are combined and analyzed using LC-MS. The isobaric tags have the same overall mass resulting in peptides from the different samples producing a single peak. After cleavage of the tag using MS/MS, however, the tags produce reporter peaks with different masses ranging from 113–121 Da (depending on the iTRAQ reagents used). Quantitation is achieved by comparing the areas of these reporter peaks that arise from the different isobaric tags.

4.2 *High-Throughput Identification of Proteomes Using Mass Spectrometry*

To illustrate how peptides in a complex proteome sample are identified using a bottom-up approach, I am going to use the elements presented in Fig. 2. In the first stage, the proteins are digested into tryptic peptides. The peptides are either injected directly onto a reversed-phase LC column, which is coupled to the mass spectrometer, or they can be prefractionated into several aliquots that are individually analyzed using LC-MS. Regardless, as the peptides enter the mass spectrometer, the instrument detects them in a survey scan. At any point in time, several peptide ions enter the mass spectrometer at once. These ions are guided into the ion-trap

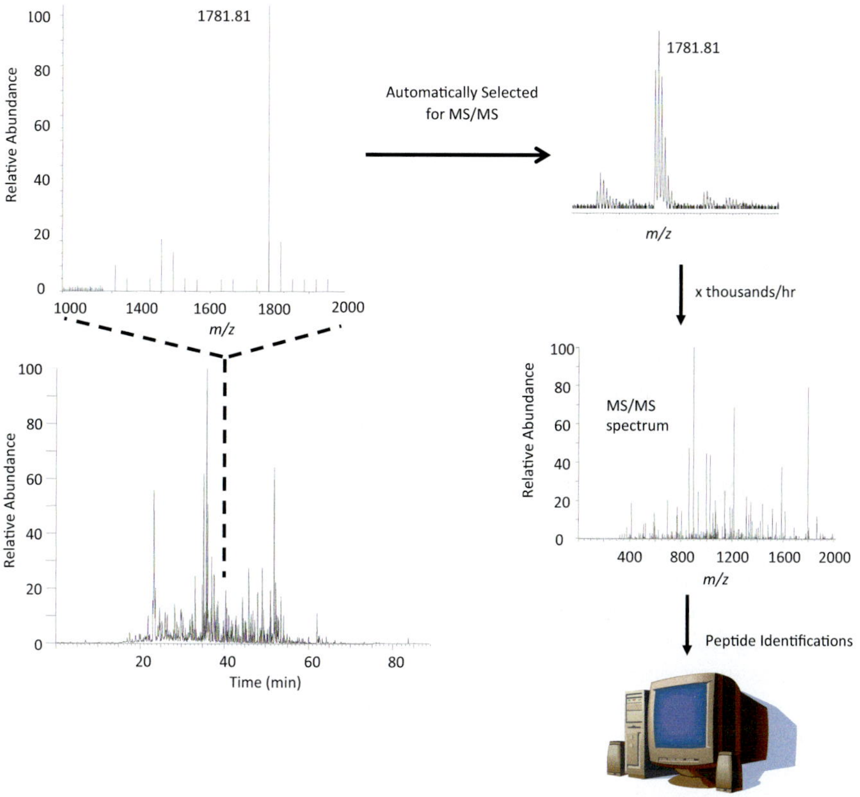

Fig. 2 High-throughput protein identification using mass spectrometry (MS). The mass spectrometer conducts a survey scan in which all the peptides entering the instrument at a specific time are detected. The mass spectrometer selects peptides based on their intensity and individually isolates them. The isolated peptide is subjected to MS/MS to obtain data on fragments created from the peptide. These fragments are analyzed along with the m/z value of the intact molecule to identify the peptide. Modern mass spectrometers can repeat this process thousands of times per hour, enabling thousands of peptides to be identified within a complex mixture

portion of the instrument, where they are temporarily stored. A peptide with a specific *m/z* value (usually the highest abundance one) is then isolated. The energy of the peptide ion is increased, and an inert gas (e.g., nitrogen, helium, or air) is added to the trap. Due to collisions with the gas, the peptide fragments into small pieces. These fragments are ejected from the trap and sent to the detector. The mass spectrometer repeats this process of isolation, fragmentation, and detection on the next most abundant peptide it detected in the survey scan. When analyzing a complex mixture, the mass spectrometer is operated in a DDA (or data-independent acquisition) mode where each full survey scan is followed by a large number (usually 10 or more) of MS/MS scans. Modern mass spectrometers repeat this cycle thousands of times per hour resulting in thousands of peptides being identified in a single LC-MS experiment. The key for the reader going forward is not to get lost in the technical details but realize that this process allows thousands of proteins in complex samples to be identified.

5 Identification of Biomarkers of Kidney Transplantation

One of the largest studies to date analyzed the urine proteome obtained from 137 individuals after kidney transplantation to identify changes related to acute rejection (AR), BK virus nephritis (BKVN), and chronic allograft nephropathy (CAN) compared to those that had a stable renal allograft (STA) (Sigdel et al. 2016). A total of 396 kidney transplant patients were enrolled in this study. The samples were divided into three groups: (i) 108 samples from the four outcomes (i.e., AR, BKVN, CAN, and STA) were analyzed using iTRAQ (isobaric tags for relative and absolute quantitation), (ii) 137 samples were analyzed using label-free LC-MS, and (iii) 151 samples were analyzed using liquid chromatography/selected reaction monitoring (LC/SRM) methods to validate select results obtained using the iTRAQ and label-free methods. A schematic detailing the number of samples (and their source) used for the discovery and validation stages is shown in Fig. 3.

The iTRAQ-labeled samples were digested using trypsin and analyzed using a two-dimensional separation strategy employing both high-pH reversed-phase chromatography followed by conventional low-pH reversed-phase chromatography leading to direct elution into and analysis of the peptides with an LTQ Orbitrap Velos mass spectrometer. The net result was the identification of 6379 unique peptides originating from 958 unique proteins. The data was analyzed using PCA. The PCA was able to segregate the sources of the samples as originating from patients with AR, BKVN, CAN, or STA. Clustering of the data revealed fibrinogen β and γ were higher in abundance in urine samples taken from AR patients. This result was confirmed using an enzyme-linked immunosorbent assay.

This iTRAQ study was then complemented with an even larger label-free LC-MS/MS study of urine samples from 137 patients. It identified almost 27,000 unique peptides originating from 2291 unique proteins, which were filtered down to 16,218 and 1574 unique peptides and proteins, respectively, after the data was

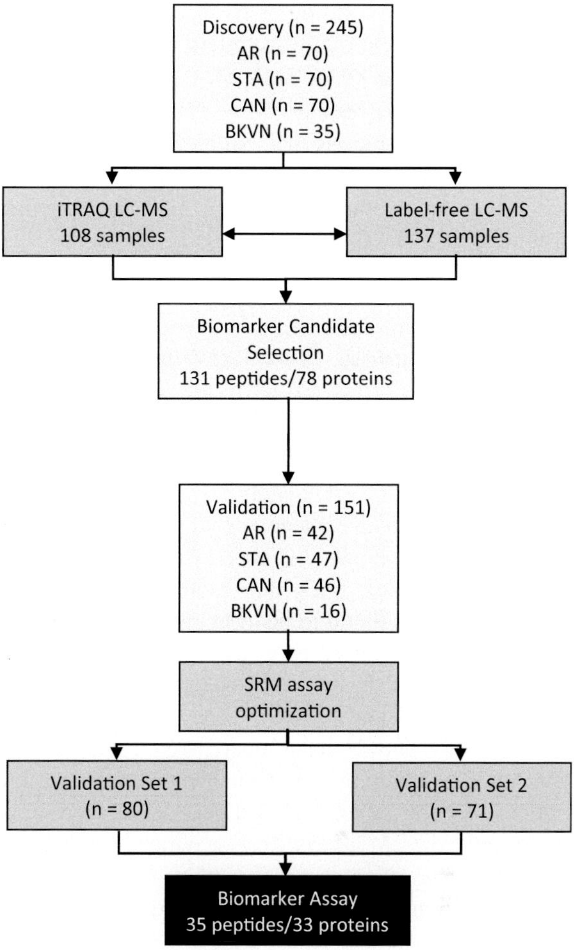

Fig. 3 Schematic showing the analysis of samples used for the discovery and validation of biomarkers to identify acute rejection (AR), BK virus nephritis (BKVN), chronic allograft nephropathy (CAN), and stable renal allograft (STA)

filtered to eliminate signals arising from outliers and contaminants. As with the iTRAQ data, PCA was able to discriminate clusters that separated cases of AR, BKVN, and CAN from STA.

Within this massive amount of data, the abundances of 811 proteins were found to be significantly affected during kidney graft injury (i.e., AR, BKVN, or CAN). The major molecular categories represented by these changes included immune response ($n = 179$ proteins), response to external stimuli ($n = 224$), and extracellular matrix organization ($n = 118$). Of this total number, 228 and 289 proteins increased and decreased, respectively, in urine obtained from patients with AR. The categories most represented within these groups were regulation of cell adhesion ($n = 51$), wound healing ($n = 48$), and regulation of body fluid levels ($n = 47$). Of the 186 proteins that were dysregulated in patients suffering from CAN, most of them were associated with negative regulation of protein metabolic processes ($n = 19$), innate immune response ($n = 20$), and multi-organism processes ($n = 19$). Just over 100 proteins were dysregulated in BKVN cases, and many of them overlapped with processes observed in AR and CAN including enrichment of protein folding ($n = 5$), regulation of stress response ($n = 19$), and xenobiotic catabolic processes ($n = 3$). Specific urinary proteins that exhibited abundance changes in BKVN compared to STA patients included lactotransferrin (TRFL), SUMO2 (SUMO2), granulins (GRNs), haptoglobin-related protein (HPTR), peptidase inhibitor 16 (PI16), alpha-1-antitrypsin (A1AT), and fibulin-1 (FBLN1).

5.1 Validation of Potential Biomarkers

A challenge with any large proteomic biomarker study is the large number of proteins that showed changes in abundance between samples being compared. This was the case in this study. To select proteins for validation, they had to be increased only in the phenotype of interest but not in other phenotypes, which included other previously studied syndromes: nephrotic syndrome (NS) and hepatocellular carcinoma (HC). For example, a protein selected for validation of AR could not also be found to be dysregulated in BVKN, CAN, STA, NS, or HC. Any protein that passed this filter also had to have been identified by at least two peptides in at least half of the samples analyzed using LC-MS. After eliminating immunoglobulins, 61, 26, and 13 AR-, CAN-, and BKVN-specific proteins (encompassing a total of 296 peptides), respectively, were selected for SRM validation in which the absolute concentration of specific peptides was measured by adding known concentrations of stable isotope-labeled peptides to each sample.

Optimized SRM assays were developed for 113 peptides encompassing 78 proteins. The 151 patient samples were split into two validation sets ($n = 80$ and 71), each containing AR, BKVN, CAN, and STA samples. Overall, the SRM analysis confirmed 35 peptides from 33 proteins that were accurately differentiated AR ($n = 11$), BKVN ($n = 12$), and CAN ($n = 12$) transplant injuries from STA. The complete list of peptides and their corresponding proteins that were measured in these SRM assays are shown in Table 1. The receiver operating characteristic (ROC) curve analysis for AR-, BKVN-, and CAN-specific panels provided areas under the curve (AUC) of 93.9%, 83.2%, and 99.5% for AR, BKVN, and CAN, respectively. Adding additional clinical parameters to the analysis did not discriminate the source of the sample better than the peptide panels measured using SRM in terms of AUC.

Overall, this study examined approximately 400 paired biopsy-urine samples from three different graft injury subtypes and, through a series of discovery and validation experiments, identified a panel of 33 protein biomarkers to distinguish the three graft subtypes. An obvious strength of this assay is the use of urine rather than tissue. Urine is not only easier on the patient but also allows for routine analysis at many time points after an organ transplant. The ease of sample acquisition permits diagnosis earlier than currently possible. This feature can detect histological injury at incipient stages when treatments are most effective. Unfortunately, the assay

Table 1 Selected reaction monitoring validated peptide panels for transplant injury-specific AR, BKVN, and CAN discovered using liquid chromatography-mass spectrometry

Protein name	Uniprot ID	Peptide
Allograft rejection (AR) specific		
Cartilage oligomeric matrix protein	COMP	ELQETNAALQDVR
Pro-epidermal growth factor	EGF	IESSLQGLGR
Brevican core protein	PGCB	ALHPEEDPEGR
Zinc alpha-2-glycoprotein	ZA2G	EIPAWVPFDPAAQITK
Fibrinogen alpha-1 chain	FIBA	NSLFEYQK
Histone 2B type 1B	H2B1B	LLLPGELAK
Keratin, type II cytoskeletal 75	K2C75	NLDLDSIIAEVK
Keratin, type II cytoskeletal 15	K1C15	ALEEANADLEVK
Dystroglycan 1	DAG1	VTIPTDLIASSGDIIK
Ig heavy variable 3–13	HV305	EVQLVESGGGLVQPGGSLR
Haptoglobin	HPT	VGYVSGWGR
BK virus nephritis (BKVN) specific		
Ig heavy chain V-III region VH26	HV303	AEDTAVYYCAK
Vesicular integral-membrane protein VIP36	LMAN2	NCIDITGVR
Alpha enolase	ENOA	GNPTVEVDLFTSK
Immunoglobulin heavy variable 1–2	HV102	DTSTSTVYMELSSLR
Complement factor B	CFAB	EAGIPEFYDYDVALIK
Beta-2 glycoprotein 1	APOH	FICPLTGLWPINTLK
Endoplasmic reticulum chaperone BiP	GRP78	ITPSYVAFTPEGER
Neutrophil defensin 1	DEF1	IPACIAGER
Cathepsin G	CATG	NVNPVALPR
Lactotransferrin	TRFL	NLLFNDNTECLAR
Probable carboxypeptidase 1	CPXM1	LLPQTWLQGGAPCLR
Annexin A2	ANXA2	GLGTDEDSLIEIICSR
Chronic allograft nephropathy (CAN) specific		
Ig heavy variable 3–13	HV305	NTLYLNMNSLR
Large ribosomal subunit protein eL18	RL18	ILTFDQLALDSPK
FAM151A	F151A	AVGPSLDLLR
TGF-beta receptor type-2	TGFR2	LTAQCVAER
L-selectin	LYAM1	AEIEYLEK
Plasminogen-like protein B	PLGB	AFQYHSK
FAM151A	F151A	TYTQAMVEK
Keratin type II, cytoskeletal 8	K2C8	LSELEAALQR
Keratin, type 1 cytoskeletal 19	K1C19	ILGATIENSR
Insulin-like growth factor-binding protein 7	IBP7	GTCEQGPSIVTPPK
DS RNA-specific adenosine deaminase	DSRAD	YLNTNPVGGLLEYAR
Ig lambda variable 1–47	LV102	WYQQLPGTAPK

developed in this study suffers from a common deficiency of most MS-based assays—the high cost and availability of mass spectrometers. While the expertise in MS is growing and the cost of MS-based assays is decreasing, making these platforms commonplace for conducting clinical assays will require several more years of development and knowledge dissemination. A solution to this quandary would be the development of a multi-panel ELISA to measure each of the targets validated in this study.

6 Proteomic Studies in Liver Transplants

While the organ may be different, the primary aim of using proteomics to study liver transplantation is similar: identify biomarkers that can predict and prevent organ dysfunction and graft rejection. The ability of proteomics to provide a non-biased global measure of the quantity (usually relative) of proteins in an organ provides a powerful tool for identifying proteins involved in a disease process. However, the bulk of data in proteomics remains challenging. Often there are too many promising proteins that could be used as biomarkers; different studies of similar cases show little overlap in the potential biomarkers they find. These factors make it difficult to move to the next step in developing assays of validated biomarkers that can identify rejection events with high sensitivity and/or specificity.

6.1 Systematic Review of Liver Transplant Proteomic Studies

Ideally, studies that analyzed similar conditions and samples would report significant overlap in the biomarkers they discover. A systematic literature review of proteomic studies of liver transplants analyzed data from 22 studies that examined graft rejection, immune tolerance, ischemia-reperfusion injury (IRI), and other transplant-derived diseases (López-López et al. 2021). The studies were conducted between 2004 and 2019 using either samples from humans (12) or rats (10). Sample types included serum (6), plasma (3), tissue biopsy (11), peripheral blood mononuclear cells (PBMCs, 1), or preservative solution (1). The condition studies included tolerance (4), IRI (9), rejection (6), liver regeneration (1), chronic kidney disease (CKD; 1), and primary graft non-function (1).

The studies used different proteomic techniques to find differentially expressed proteins. These included two-dimensional gel electrophoresis (2D gels) using silver staining for quantitation, shotgun proteomics (i.e., LC-MS/MS) primarily combined with iTRAQ, antibody arrays, and a combination of 2D gels and antibody array. Recently conducted studies were more likely to use prefractionation steps to help identify lower abundance proteins. Protein identification was performed using MS/MS in all non-antibody array studies, but only 50% validated their results, mainly using Western blot.

6.2 *Identified Proteins in Liver Transplant Dysfunctions*

A total of 497 unique proteins were identified in the 22 studies; 354 and 211 were found in humans and rats, respectively, and 68 (i.e., 14%) of these were identified in both species. The median number of proteins found to be differentially abundant in the studies was 37, from a high of 197 differentially abundant proteins identified in an LC-MS/MS study of proteins within preserving solution to a low of only 7 differentially abundant proteins found in an analysis of serum samples using 2D gels and matrix-assisted laser desorption/ionization time-of flight mass spectrometry (MALDI-TOF MS) for identification. Approximately 50% of the proteins were identified in serum and plasma, the least invasive source. Three proteins (complement C3, GC vitamin D-binding protein, and haptoglobin) were found in almost half of the serum/plasma studies. Five others (apolipoprotein A1, apolipoprotein E, apolipoprotein H, ceruloplasmin, and microtubule-associated protein 1) were found in one-third of the studies. Only 20 proteins were found in both serum and tissue samples. Four proteins (C-reactive protein, glutathione S-transferase α1, hemoglobin subunit β, and hemopexin) were found in all the different types of samples. Ten proteins were found in both tissue and serum samples, suggesting these proteins were shed or excreted from cells. Only albumin was found in all the liver biopsy samples, but seven proteins (aldehyde dehydrogenase 1 family member L1, carbonic anhydrase 1, cytochrome b5 type A, glyceraldehyde-3-phosphate dehydrogenase, ketohexokinase, protein disulfide isomerase 3, and regucalcin) were found to be differentially expressed in about one-quarter of all biopsied proteomic samples.

Of the proteins identified in all sample types, 241 and 190 were identified as upregulated and downregulated, respectively. These numbers collectively account for almost 79% of the identified proteins. Of these, 40 were found to be up- and downregulated in different studies. Of the different conditions studied, IRI had 14 proteins that overlapped between studies (estrogen sulfotransferase, ribosomal proteins SA, protein disulfide isomerase family A member 3, heat shock protein 90 alpha family class A member 1, cytochrome B5 type A, apolipoprotein A1, aldehyde dehydrogenase 1 family member L, albumin, GC vitamin D-binding protein, aldo-keto reductase, alcohol dehydrogenase 4, ATP citrate lyase, carbonic anhydrase 1, and alanine-glyoxylate aminotransferase 2), while tolerance studies had 4 (haptoglobin, microtubule-associated protein 1, GC vitamin D-binding protein, and ceruloplasmin), and graft rejection studies only showed 2 (carbonic anhydrase 1 and actin β). Only five proteins (albumin, glutathione-S-transferase A1, hemopexin, complement factor 3, and apolipoprotein A3) were found to be differentially abundant in IRI, tolerance, and rejection.

6.3 Pathway Analysis of Differentially Abundant Proteins

Grouping the proteins into pathways showed serum and liver proteins could be classified into 47 and 43 pathways, respectively. Seventy percent of the pathways were shared between studies using different types of samples. The most represented pathways were those associated with metabolism including carbon metabolism, amino acid biosynthesis, glycolysis, and gluconeogenesis. The complement/coagulation cascade and interleukin 17 signaling pathways were unique to serum studies, while the endoplasmic reticulum pathway was only observed to be differentially regulated in liver samples. In cases measuring transplant tolerance, pathways associated with cytokine receptor interaction, complement/coagulation cascade, interleukin 17, and tumor necrosis factor signaling were most prominent. For IRI, pathways associated with complement/coagulation cascades and metabolic pathways showed the greatest component of differentially expressed proteins.

7 The Biomarker Discovery Challenge

The previous collection of studies provides a lot of data concerning proteins that are differentially abundant in cases of graft rejection, IRI, and tolerance in liver transplants. The challenge, as with all large-scale proteomic studies, is to turn this data into clinically useful information. The progress in graduating potential biomarkers to clinical use has been slow, regardless of the condition being studied. Unfortunately, these liver transplant studies show significant variability and do not help to identify specific biomarkers for IRI, graft rejection, or tolerance.

The variability between studies is exemplified by the lack of overlap between proteins identified in the rat and human samples. The rat liver transplant model is well-established. However, only 14% of the proteins identified as being differentially expressed in this experimental model were also observed in human samples. Another indication of experimental variability is the fact that only 2% of the proteins identified in the studies were observed in both serum and liver tissue. This result is surprising considering that hepatocytes synthesize most serum proteins in the endoplasmic reticulum. This should result in a larger number of hepatocyte-related proteins, which reflect liver status, being found within the serum. Unfortunately, differentially abundant proteins related to endoplasmic reticulum pathways were only enriched in liver tissue samples, but not serum or plasma. Part of this issue could also have to do with the extremely large dynamic range of concentrations found for serum and plasma proteins. Tissues have a large dynamic range of concentrations. However, this is dwarfed by the range of serum and plasma that spans over ten orders of magnitude (Tirumalai et al. 2003; Paul and Veenstra 2022).

7.1 *Is the Biomarker Related to the Site of Disease?*

The challenge in using circulating biomarkers for detecting disease states has been recognized ever since blood-based proteomics studies began. It is difficult to identify biomarkers that originate from the damaged site that are expected to be in low concentration when they are mixed within the milieu of circulating proteins that are often in very high concentration (i.e., albumin, immunoglobulins, complement factors, etc.). Another challenge is correlating a potential biomarker with the site of disease. The levels of many circulating proteins could change due to a patient's illness. Ideally, the biomarker would originate from the site of damage since it would increase its sensitivity and specificity.

Finding biomarkers that originate from a damaged site and can be detected within blood (i.e., serum or plasma) can be done using two strategies. Both require analysis of blood and tissue samples. The first strategy would entail a global proteomics analysis of either blood or tissue followed by a targeted study to determine if any potential biomarkers are found in the comparative sample. For instance, global analysis of serum samples may reveal several potential biomarkers for a specific primary cancer. The next phase would develop targeted assays (i.e., LC-SRM, ELISA, etc.) to measure these candidates in tissue samples to see if the same differential abundance is observed in the tumor. The second strategy requires global analysis of both blood and tissue samples. The data would be scrutinized to identify proteins that were differentially abundant in both sample types. These potential biomarkers could then be validated using targeted methods.

This second strategy was employed in a study of acute respiratory distress syndrome (ARDS) (Gong et al. 2023). Like most conditions, ARDS suffers from a lack of easily measurable biomarkers. This study conducted a global proteomic analysis of lung tissue and blood samples from mice with lipopolysaccharide (LPS)-induced ARDS (Bannerman and Goldblum 2003). A total of 368 and 505 differentially expressed proteins were identified in LPS-induced mouse serum and lung samples, respectively. Cross comparison of the two datasets showed 50 upregulated and 10 downregulated proteins that were common to both sample types. Testing each of these potential markers using a targeted MS method showed that six of these proteins (haptoglobin, serum amyloid A1 and A2, calgranulin B, leukotriene A4 hydrolase, and α1-antichymotrypsin) provided good clinical diagnostic value for ARDS.

The value of these six potential biomarkers was then evaluated for diagnosing ADRS in serum samples obtained from 20 patients with COVID-19 and 10 healthy patients. The COVID-19 samples were taken from five non-survivors, seven severe cases, and ten mild cases of COVID-19 patients. Receiver operating characteristic curves were generated for each of the proteins individually. The AUC values for the proteins ranged from 0.660 for calgranulin to 1.000 for serum amyloid A2.

8 Conclusions

Transplantation of solid organs is a complicated procedure, so it follows that identifying biomarkers that measure their efficacy would also be complicated. Great strides have been made since the beginning of solid organ transplants including HLA typing, donor-derived cell-free DNA, and apolipoprotein L1 gene tests (Choi et al. 2019). There is a continuing need for more specific biomarkers that can identify and distinguish the various stages of transplantation efficacy and prognosis. While proteomic technology has been primarily focused on studying cancer, its analytical scale provides exciting opportunities to better understand pathologies associated with solid organ transplants. There are still significant challenges. However, proteomic technology continues to increase in speed and capability, and costs fall. What is evident from the past 20 years of using proteomics to identify protein markers of specific conditions is that a greater understanding of how the data relates to the studied condition is required. Major developments in the future will focus on how to better design studies that can help interpret the massive datasets afforded using proteomics.

Acknowledgments The author would like to thank Cedarville University School of Pharmacy for financial support during the preparation of this chapter.

References

Anderson NL, Polanski M, Pieper R et al (2004) The human plasma proteome: a nonredundant list developed by combination of four separate sources. Mol Cell Proteomics 3:311–326

Argani H (2019) Anti-HLA antibody: the role of epitopes in organ transplantation. Exp Clin Transplant 17(Suppl 1):38–42

Bannerman DD, Goldblum SE (2003) Mechanisms of bacterial lipopolysaccharide-induced endothelial apoptosis. Am J Physiol Lung Cell Mol Physiol 284:L899–L914

Barker DJ, Maccari G, Georgiou X et al (2023) The IPD-IMGT/HLA Database. Nucl Acids Res 51: D1053–D1060

Billingham RE, Brent L, Medawar PB (1953) Actively acquired tolerance of foreign cells. Nature 172:603–606

Billingham RE, Brent L, Medawar PB (1956) Quantitative studies on tissue transplantation immunity. III. Actively acquired tolerance. Phil Trans R Soc Lond B 239:357–414

Bodmer W (2023) A historical perspective on HLA. Immunother Adv 3:1–3

Choi J, Bano A, Azzi J (2019) Biomarkers in solid organ transplantation. Clin Lab Med 39:73–85

Coll E, Botey A, Alvarez L et al (2000) Serum cystatin C as a new marker for noninvasive estimation of glomerular filtration rate as a marker for renal impairment. Am J Kidney Dis 36:29–34

Dutra RS, Fabreti-Oliveira RA, Lasmar MF et al (2021) Impact of the immunotherapy induction on allograft outcome and survival in kidney transplant patients with donor-specific antibodies to HLA-DQB1. Transpl Immunol 66:101390–101396

Ehrenfried A (1909) Reverdin and other methods of skin-grafting: historical. Boston Med Surg J 161:911–917

Erlich H (2012) HLA DNA typing: past, present, and future. Tissue Antigens 80:1–11

Geyer PE, Voytik E, Treit PV et al (2019) Plasma proteome profiling to detect and avoid sample-related biases in biomarker studies. EMBO Mol Med 11:e10427
Gong R, Luo H, Long G et al (2023) Integrative proteomic profiling of lung tissues and blood in acute respiratory distress syndrome. Front Immunol 14:1158951–1158964
Habeck T, Lermyte F (2023) Seeing the complete picture: proteins in top-down mass spectrometry. Essays Biochem 67:283–300
Hattery E, Nguyen T, Baker A et al (2015) Burn care in the 1800s. J Burn Care Res 36:236–239
Howard RJ, Cornell DL, Cochran L (2012) History of deceased organ donation, transplantation, and organ procurement organizations. Prog Transplant 22:6–17
López-López V, Pérez-Sánz F, de Torre-Minguela C et al (2021) Proteomics I liver transplantation: a systematic review. Front Immunol 12:672829–672841
Mann M (2006) Functional and quantitative proteomics using SILAC. Nat Rev Mol Cell Biol 7: 952–958
McCaughan J, Xu Q, Tinckam K (2019) Detecting donor-specific antibodies: the importance of sorting the wheat from the chaff. Hepatobiliary Surg Nutr 8:37–52
Medawar PB (1946) Immunity to homologous grafted skin; the relationship between the antigens of blood and skin. Br J Exp Pathol 27:15–24
Meier-Kriesche HU, Scornik JC, Susskind B et al (2009) A lifetime versus a graft life approach redefines the importance of HLA matching in kidney transplant patients. Transplantation 88:23–29
Miller RM, Smith LM (2023) Overview and considerations in bottom-up proteomics. Analyst 148: 475–486
Mosaad YM (2015) Clinical role of human leukocyte antigen in health and disease. Scand J Immunol 82:283–306
Murray JE, Merrill JP, Dammin GJ et al (1962) Kidney transplantation in modified recipients. Ann Surg 156:337–355
Nordham KD, Ninokawa S (2022) The history of organ transplantation. Proc (Bayl Univ Med Cent) 35:124–128
Opelz G, Döhler B (2007) Effect of human leukocyte antigen compatibility on kidney graft survival: a comparative analysis of two decades. Transplantation 84:137–143
Paul C (2008) Skin grafting in burns. Wounds 20:199–202
Paul J, Veenstra TD (2022) Separation of serum and plasma proteins for in-depth proteomic analysis. Separations 9:89–109
Profaizer T, Kumánovics A (2018) Human leukocyte antigen typing by next-generation sequencing. Clin Lab Med 38:565–578
Rainey A, McKay GJ, English J et al (2023) Proteomic analysis investigating kidney transplant outcomes—a scoping review. BMC Nephrol 24:346–356
Ross PL, Huang YN, Marchese JN et al (2004) Multiplexed protein quantitation in Saccharomyces cerevisiae using amine-reactive isobaric tagging reagents. Mol Cell Proteomics 3:1154–1169
Schirmer EC, Florens L, Guan T et al (2003) Nuclear membrane proteins with potential disease links found by subtractive proteomics. Science 301:1380–1382
Shah AM, Myhre PL, Arthur V et al (2024) Large scale plasma proteomic identifies novel protein and protein networks associated with heart failure development. Nat Commun 15:528–547
Sheldon S, Poulton K (2006) HLA typing and its influence on organ transplantation. Methods Mol Biol 333:157–174
Shimmura H, Tanabe K, Ishida H et al (2005) Lack of correlation between results of ABO-incompatible living kidney transplantation and anti-ABO blood type antibody titers under our current immunosuppression. Transplantation 80:985–988
Shraibman B, Barnea E, Kadosh DM et al (2019) Identification of tumor antigens among the HLA peptidomes of glioblastoma tumors and plasma. Mol Cell Proteomics 18:1255–1268
Sigdel TK, Gao Y, He J et al (2016) Mining the human urine proteome for monitoring renal transplant injury. Kidney Int 89:1244–1252

Starzl TE, Marchioro TL, Porter KA et al (1967) The use of heterologous anti-lymphoid agents in canine renal and liver homotransplantation and in human renal homotransplantation. Surg Gynecol Obstet 124:301–318

Tanuwidjaya E, Schittenhelm RB, Faridi P (2022) Soluble HLA peptidome: a new resource for cancer biomarkers. Front Oncol 12:1069635

Tedesco D, Haragism L (2012) Cyclosporin: a review. J Transp Secur 2012:230386–230392

Terasaki PI, Singal DP (1969) Human histocompatibility antigens of leukocytes. Annu Rev Med 20:175–188. https://doi.org/10.1146/annurev.me.20.020169.001135

Thorsby E (2009) A short history of HLA. Tissue Antigens 74:101–116

Veenstra TD, Martinović S, Anderson GA et al (2000) Proteome analysis using selective incorporation of isotopically labeled amino acids. J Am Soc Mass Spectrom 11:78–82

Tirumalai RS, Chan KC, Prieto DA et al (2003) Characterization of the low molecular weight serum proteome. Mol Cell Proteomics 2:1096–1103

Veenstra TD, Conrads TP, Hood BL et al (2005) Biomarkers: mining the biofluid proteome. Mol Cell Proteomics 4:409–418

Wahle M, Thielert M, Zwiebel M et al (2023) IMBAS-MS discovers organ-specific HLA peptide patterns in plasma. Mol Cell Proteomics 23:100689–100703

Wang C, Krishnakumar S, Wilhelmy J et al (2012) High-throughput, high-fidelity HLA genotyping with deep sequencing. Proc Natl Acad Sci USA 109:8676–8681

Wang D, Eraslan B, Wieland T et al (2019) A deep proteome and transcriptome abundance atlas of 29 healthy human tissues. Mol Syst Biol 15:e8503–e8518

Wernet P, Schneider EM, Kalthoff F et al (1987) Constitutive functional heterogeneity without detectable somatic mutation of antigen receptor genes in helper T cell clones: possible regulation by novel HLA class II "DY" determinants. Immunol Rev 96:109–140

Wollscheid B, Watts JD, Aebersold R (2004) Proteomics/genomics and signaling in lymphocytes. Curr Opin Immunol 16:337–344

Zabaneh D, Krapohl E, Simpson MA et al (2017) Fine mapping genetic associations between the HLA region and extremely high intelligence. Sci Rep 7:41182–41189

Zachary AA, Leffell MS (2016) HLA mismatching strategies for solid organ transplantation—a balancing act. Front Immunol 7:575–588

Zirm E (1906) Eine erfolgreiche totale keratoplastik. Arch Ophthalmol 64:580–593

Advancing Microfluidics and Biomimetic Materials for Miniaturized Cell Screening Platforms: Unleashing Insights into Complex Disease Mechanisms and Personalized Medicine

Shraddha Chakraborty, Janaki M. Nair, and Dwaipayan Bharadwaj

Contents

S. Chakraborty (✉)
Department of Biochemical Engineering and Biotechnology, Indian Institute of Technology Delhi, New Delhi, India
e-mail: schakraborty@dbeb.iitd.ac.in

J. M. Nair · D. Bharadwaj (✉)
Systems Genomics Laboratory, School of Biotechnology, Jawaharlal Nehru University, New Delhi, India
e-mail: db@jnu.ac.in

J. Barciszewski (ed.), *Systems Biology II*, RNA Technologies 15,
https://doi.org/10.1007/978-3-031-62178-9_14

Abstract Recent developments in the field of microfluidics and fabrication of novel biomimetic materials have enabled scientists worldwide to design miniaturized cell screening platforms with a wide range of experimental applications. These miniaturized lab-on-a-chip systems provide a controlled and scalable cellular microenvironment for a large number of biological samples under different conditions. In the context of complex disease analysis, these microfabricated platforms serve as powerful tools for understanding disease mechanisms and discovering novel therapeutics through high-throughput screening of drugs and other small molecules with cells grown in miniaturized assay volumes. The integration of different experimental techniques, such as live cell monitoring, drug screening, and analysis of phenotypic and molecular outcomes, demonstrates the potential of these alternative platforms as the ultimate "sample-in-answer-out" systems. Microfluidic techniques have played a critical role in the development of portable and rapid point-of-care diagnostic devices, particularly in emergency setups. Moreover, microfluidic technologies have revolutionized the field of tissue engineering by creating sophisticated platforms that allow the formation of spheroids, human tissues, and even miniature organs on a chip, accelerating the development of personalized medicine and targeted therapeutic strategies. With the ability to recreate disease-specific microenvironments, enable real-time monitoring of cells, and integrate multiple experimental techniques with different readouts into a single device, these miniaturized platforms offer unprecedented insights into complex disease mechanisms.

Keywords Disease analysis · Microfluidic technology · Miniaturization · Lab-on-a-chip · Personalized health · Point-of-care diagnostics

1 Cellular Heterogeneity in Biological Systems

Cellular heterogeneity is a fundamental feature of complex biological systems, from multicellular organisms to organs and tissues. It refers to the presence of diverse cell types and subpopulations within a given system, with wide differences in cellular morphology, gene expression dynamics, signaling pathways, and metabolic properties (Altschuler and Wu 2010). The importance of cellular heterogeneity lies on the fact that different cell types and subpopulations are subjected to perform specific functions in a coordinated manner (Gupta and Kuznicki 2020). For example, our heart contains specialized cardiac muscle cells for synchronized beating and cardiac impulse generation, endothelial cells to form blood vessels, and fibroblasts to provide structural support. Other than enabling specialized functions, cellular heterogeneity enables the body systems to adapt to different environmental stimuli. For example, our immune system contains a diverse cellular repertoire with different cell subpopulations, in order to generate a tailor-made response to various pathogens invading the body. However, while cellular heterogeneity is indispensable for the normal functioning of biological systems, it can also contribute to difficulties in the detection and remission of various pathological conditions, such as cancer (Almendro et al. 2013). The heterogenous cell subpopulations in a developing tumor exhibit a highly diverse metabolic and gene expression profile, thus contributing to a major role in disease progression and resistance to various anticancer therapeutics (Brooks et al. 2015; Lim et al. 2019). A prominent role of cellular heterogeneity is also observed in certain complex diseases as well. For example, a decreased population of $CD63^+$ pancreatic beta cells is shown to be implicated in type 2 diabetes pathogenesis by deregulations at the rate of insulin secretion in response to glucose (Rubio-Navarro et al. 2023). Therefore, it is crucial to recognize and annotate the variations among the diverse cellular networks within diseased sites of the human body. A thorough understanding of the cellular heterogeneity would not only provide a better view in dissecting the complexity of human diseases but also aid in developing precise and articulate healthcare planning toward disease amelioration.

Recent developments in the field of high-throughput biological experimentations including single-cell omic studies have expanded the ability of researchers in decoding the cellular heterogeneity of complex diseases to a great extent (Hu et al. 2018). For example, transcriptomic studies on cancer biopsy samples at single-cell resolution could provide a deeper insight on the gene expression dynamics of different cell subpopulations in that complex tumor microenvironment (Choi and Kim 2019; Saadatpour et al. 2015). Similarly, single-cell proteomic studies based on mass spectrometry and high-throughput microscopy could provide an estimation of the abundance of various proteins (Schoof et al. 2021; Murschhauser et al. 2019). These cutting-edge technologies to detect cellular dynamics in a complex tissue environment enabled biomedical researchers to construct the "Human Cell Atlas," a comprehensive map with phenotypic and molecular features of all the diverse cellular subtypes found in a human body (Regev et al. 2017). However, analyses of large-scale data from single-cell omic studies have inherent challenges of negating amplification bias and errors (Yamada et al. 2021). Furthermore, high-throughput single-cell culture destined to drug screening studies frequently encounter problems during the process of outcome evaluation (Shen et al. 2023). Therefore, along with single-cell-based studies, miniaturized cellular arrays with low number of cells are gaining popularity among scientists. Experimentations with such low-input cellular arrays aid in increased throughput, less hands-on time, and low consumption of difficult-to-obtain biological/clinical samples.

2 High-Throughput Cell Screening (HTS) for Complex Disease Therapeutics

Complex diseases encompass a large spectrum of illnesses, including metabolic disorders, immune dysregulations, neurodegeneration, cancer, cardiovascular disorders, etc. These fairly diverse kinds of complex disease patterns have few things in common—a multifactorial origin, typically heterogenous disease manifestation pattern in different patients, and involvement of several molecular pathways. Typically, a major share of complex diseases is suspected to arise from a combination of genetic susceptibility dictated by heredity, exposure to certain environmental factors, and lifestyle choices (Mitchell 2012). Individuals affected by complex diseases exhibit a gradual development of disease symptoms over time; thus, the process of accurate disease diagnosis becomes difficult at the early stage. The involvement of multiple molecular pathways toward complex disease manifestation also creates challenges in understanding the molecular mechanisms implicated in the disease progression (Mitchell 2012). Owing to the genetic and cellular diversity in human population, the rate of disease progression and response to the same course of treatment can be fairly heterogeneous among the patients. Conventional drug development strategies often face obstacles while targeting complex diseases, due to their multifaceted nature involving a variety of intricate cellular pathways. Moreover, discovery of novel effective drugs against a particular disease via the conventional methodology is usually an elaborate and time-consuming process. Therefore, a high-throughput drug screening approach with a suitable model system is critical in order to decipher the disease pathophysiology and development of novel treatment options directed to personalized medicine for a wide spectrum of complex disorders.

In vitro cell culture has revolutionized the course of biomedical research since its conception in the mid-1900s (Earle et al. 1943). In the context of novel drug discovery, the choice of an appropriate cell line enables scientists to investigate treatment strategies for a particular disease in mind. For example, HeLa-CCL2 and Caco-2 cell lines could serve as the ideal cell culture models for cervical and colorectal neoplasms, respectively. HTS technologies refer to the establishment of novel platforms to incubate a selection of cell lines with hundreds, thousands, and even up to millions of different compounds, in order to find novel therapeutic strategies against a disease. Parallelization of experimental steps using a high-throughput system helps in curbing the experimental time and cost and reduces the number of cells needed per sample. Also, a high-throughput platform enables researchers to have multiple experimental repeats onto the same plate, thereby leading to better confidence and higher reproducibility of the data (Hertzberg and Pope 2000). Multi-well plates were the first experimental platform to be established back in the 1960s that slowly set up the stage for HTS (Ramsey et al. 1984). The traditional muti-well plates are usually made up of polystyrene, which is sterilized by irradiation and further modified by corona discharge or gas-plasma under vacuum to generate hydrophilic surface properties that are essential for cellular attachment (Barker and LaRocca 1994). Till date, there is a possibility to process a wide range of 6 to 1536 individual samples in one muti-well plate, depending on the size and the number of wells available. Potential use of these multi-well plates is usually restricted to allow cell culture, incubation of the cells with different compounds, and assessment of the outcome by phenotypic readouts.

The common first step for HTS is to prepare large libraries of compounds to be tested, based on the possible effective concentration. In some cases of unknown drugs, multiple dilutions can be prepared using a buffer or nuclease-free water or other solvents and tested onto the cells as a pilot study, so as to confirm the effective concentration or toxicity. The most frequently used type of drugs to be tested are typically small molecules that could either be laboratory-grown or derived from natural sources (Inglese et al. 2007). Other than small molecule drugs, a variety of elements to be screened using in vitro cell culture could range from molecules like small interfering RNAs (siRNA), microRNAs (miRNA), nanoparticles, and lipidoid-like particles to external stimuli like changes in pH, oxygen levels, and temperature (Echeverri and Perrimon 2006; Molla et al. 2020; Oguchi et al. 2006). The next step in HTS after compound library preparation is to incubate cells with the compounds under investigation, at least for 48–72 h. There is a wide choice of model systems to be used for HTS targeted for complex diseases, starting from various cell lines, stem cells, cells derived from tissue biopsies, spheroids, to even organoids (Chakraborty et al. 2021; Liu et al. 2021; El Harane et al. 2023). The effective concentration and incubation time of the compounds are greatly varied depending on the model system under consideration. Once the incubation time is over, the effect of the compounds on the cells could be observed in a number of ways. The most common outcome of a high-throughput drug screening with a cancerous cell line could be analyzed by measuring cellular viability post-experiment. The outcome of assay can also be determined by fluorometric and colorimetric readouts, according to the design of the assay (Flobak et al. 2019; Nierode et al. 2016). The detailed mode of outcome analysis may include targeted and global gene expression analysis and quantification of target proteins or metabolites in the cells post-HTS (Ye et al. 2018; Dueñas et al. 2023). With the cue from HTS results, compounds with a potential role in disease amelioration are shortlisted for additional experimentations to test the selectivity, potency, and pharmacokinetic properties.

The main advantage of HTS is that it is a quick and efficient method, enabling researchers to screen thousands of compounds just in a matter of weeks, whereas in traditional drug discovery methods, identification of one novel drug for a particular disease could take up to few years. Another advantage of HTS could be that it is bias-free, because a large number of potential drug candidates that are being overlooked by traditional drug discovery methods can easily be incorporated into an HTS study (Huang et al. 2021). Some of the many success stories of high-throughput drug screening against complex diseases include the discovery of novel therapeutics against various cancer types and neurodegenerative disorders (Szymański et al. 2012). Moreover, HTS opens the avenue for the establishment of combination therapies, by analyzing the synergistic effects of multiple drugs at various doses to discover the effective course of treatment for a particular disease. As per the limitations of HTS are concerned, efficient screening with large molecules, e.g., antibodies, is still difficult as per the available methodologies till date (Sink et al. 2010). Moreover, in vitro cell culture, as the most widely used model system for HTS, may not always be an ideal representative of the complexity in human physiology (Ayuso et al. 2021). However, HTS is definitely an indispensable tool in drug discovery, with scope of further developments.

3 Advantages of Miniaturization in Traditional Platforms and Research Methodologies

Traditional platforms for HTS mostly include muti-well plates, as discussed in earlier sections. However, recent advances in biomedical science have resulted in an enormous number of drug candidates to be screened for their efficacy as novel therapeutics against a wide spectrum of diseases. This provokes the need to develop platforms that can have "higher" throughput, so that larger number of reactions/samples can be assembled into one experimental setup. Moreover, primary cells or cells isolated from biopsy samples are often limited in numbers, although being crucial to develop personalized therapeutics against certain diseases. Therefore, it is important to scale down the experimental setup, so that a higher number of samples, as well as samples in limited assay volumes, can be incorporated into one HTS experiment. The key advantages of miniaturization in biomedical field are multidimensional. First of all, miniaturization allows researchers to manipulate and control the assay volumes down to nano-, pico-, even up to femtoliters (Zhang et al. 2019). Handling of such low volumes is enabled by the use of sophisticated and miniaturized liquid volume dispenser machines. Uses of such automated liquid dispenser or microfluidic devices shun the probability of manual error during most of the major events of the experimental process. Therefore, better precision is achieved using miniaturized devices in a wide array of experimental fields including genomics, proteomics, and drug discovery (Vyawahare et al. 2010). Scaling down the assay volume usually translates to reduced budget of the whole biological experiment. Not only the volume of reagents and samples but the need for repeated use of other consumables such as plasticwares is also greatly reduced with the discovery of miniaturized biological platforms. Miniaturized devices therefore lead to a sustainable solution for biological experimentations that need to be repeated multiple times within a short time frame with limited resources available (Vyawahare et al. 2010).

Miniaturization also facilitates high-throughput screening in minute sample volumes, while offering parallelization of multiple experimental steps in a large number of samples. Therefore, a miniaturized HTS experiment enables higher throughput, while saving on experiment cost and time. This experimental mode of miniaturization is particularly important in the field of drug discovery, in which a large number of drugs need to be tested with different types of in vitro samples within a short span of time. Miniaturized experimental setups enable rapid screening of such compounds in a matter of weeks, thereby increasing the chance of discovering novel disease therapeutics (Szymański et al. 2012). Figure 1 depicts the trend of increasing throughput in biological studies with the advent of miniaturization. Many of the HTS experiments require patient-derived cells rather than cultured cell lines, for better interpretation of the data in context to human physiology. The number of cells to be procured from difficult-to-obtain tissue biopsy samples or primary cell lines is greatly reduced with the use of miniaturized devices. This process of obtaining meaningful data from smaller number of cells also fits perfectly with the principles

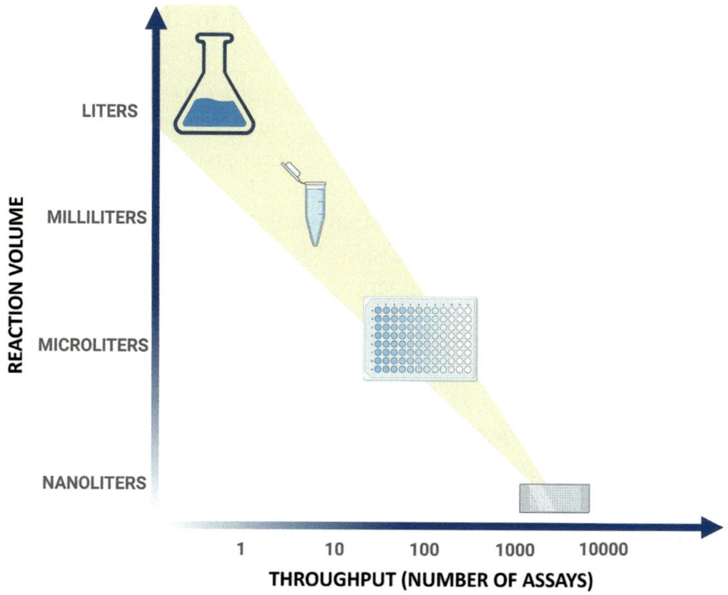

Fig. 1 Course of increased throughput in biological experimentations

of ethics in biomedical research, while providing wonderful insights onto the molecular frameworks.

One important advantage in some of the miniaturized biological devices is the fact of being much smaller in size as compared to the traditional platforms. Many of the semi-miniaturized devices can actually be carried and operated outside the laboratory setting, thus making those the perfect tools to conduct point-of-care disease testing on-site. Therefore, miniaturized devices like portable DNA sequencers or microfluidics-based diagnostic chips could be instrumental for the healthcare professionals to conduct tests and subsequent analysis as a point-of-care service in remote or resource-limited areas, without the need of special research infrastructure. Miniaturized biological devices can also be tailor-made according the requirements of the biological experiment. For example, it has been possible to culture cells, perform cell screening, and analyze the cell screening outcome both by phenotypic and transcriptomic assays, all in a nanoliter droplet-based format on a chip with down to single cell per sample sensitivity (Chakraborty et al. 2022). Such combination of multiple functionalities of a single device allows researchers to streamline various experimental workflows and increase the overall research efficiency. Moreover, integration of multiple experimental protocols onto the same platform decreases the chances of manual error and risks of contamination, while dealing with a large number of samples at a time.

4 Types of Miniaturized Systems Devised for HTS

Advances in the field of molecular diagnostics have led to rapid development of miniaturized devices aiming toward high-throughput cell screening, with subsequent analysis of the experimental outcome. The common goal of these platforms is to serve as precise and efficient biomedical platforms, by offering better experimental control, increased throughput, and reduced budget. The vast majority of these miniaturized devices can be broadly classified into three major categories based on their unique architectural features and targeted fields of application, as described in the following sections.

4.1 Microfluidic Devices

The first among the miniaturized platforms and also known as the "lab-on-a-chip" system, the world of microfluidics creates a link between the macro- and miniature world (Pattanayak et al. 2021). Most common prototype of microfluidic devices is made up of microchannels engraved on glass or different polymers [such as polymethylmethacrylate (PMMA), polydimethylsiloxane (PDMS)] (Pattanayak et al. 2021). The volume entrapped within these microchannels can be anywhere within micro- to nanoliter range. Just like our own body vasculature, the microfluidic systems are controlled by the flow of fluids inside the microchambers, using a pump, inlet, and outlet valves. The precise control on the fluid flow inside these microchannels allows researchers to trap single cells, perform sorting, and perform a broad range of functional analysis. Depending on the platform architecture, microfluidic devices can be subgrouped into a number of categories. Some of these have an open chip-based format operated by capillary forces of the microchannels lacking physical boundaries, thereby creating multiple liquid-air interfaces for several biphasic applications (Berthier et al. 2019). On the other hand, droplet-based microfluidics relies on the formation of discrete fluid spheres down to picoliter volumes, created by admixture of immiscible phases. Droplet microfluidics has revolutionized the field of single-cell omic studies, where single cell-containing droplets can be easily merged with droplets of reaction mixtures for subsequent applications (Shang et al. 2017). Continuous-flow microfluidics is gaining popularity by enabling researchers for long-term 2D cell culture, along with the formation of spheroids and organoids inside the microchambers by maintaining a steady flow of nutrient media through the inlet and outlet ports of the microchannels. It is particularly important for experimentations involving cell proliferation, migration, and communication studies (Erickson et al. 2021). Paper-based microfluidic systems have been proved to be useful in the detection of a number of disease conditions in a point-of-care setting (Campbell et al. 2018). Digital microfluidics has enabled researchers with droplet manipulations at a nanoliter level without the need for in-built microchannels (Millington et al. 2018). Novel technologies like particle detection microfluidics and magnetophoresis-assisted hydrophoretic devices are examples of advancements in the field of microfluidic systems in experimental research (Howell et al. 2020; Yan et al. 2016).

4.2 *Microarray Platforms*

A microarray can be defined by a high-density distribution of microscopic spots or droplets onto a platform, each containing a particular sample or single cells or some reagents. The foremost benefit of microarray-based miniaturized systems is that these are high-throughput assay devices without any physical partitioning among different samples. This “wall-less” structural feature prevents the loss of material from the microscopic spots, by avoiding nonspecific adsorption of the samples onto the surrounding walls. Based on the chip material, the microarray platforms for molecular diagnostics are suitable for several applications. For example, a glass substrate-based microarray can be formed by coating the same with polylysine or some other biocompatible chemicals (Fernandes et al. 2009). One such example is the “droplet microarray” system, which enables the formation of hundreds of nanoliter droplets on a standard microscope glass slide, coated with either some polymer or nanostructured nonpolymer or dendrimer compounds. It has been shown that under sterile conditions, high-throughput screening with cultured cells can be performed in those nanoliter reservoirs, without any need for external medium, e.g., mineral oil to prevent evaporation (Chakraborty et al. 2021; Jogia et al. 2016). Moreover, recent experimental developments demonstrated the applicability of droplet microarray as a “sample-in-answer-out” platform that enables cell screening, along with the assessment of the outcome by both phenotypic and transcriptomic analysis, with down to single cell per droplet sensitivity (Chakraborty et al. 2022). Microarrays can also be produced by in situ synthesis of the biomaterial directly onto the substrate, or by dispensing pre-synthesized biological materials, i.e., nucleic acids, proteins, and other biomolecules directly onto the platform (Venkatasubbarao 2004). Self-assembled microarrays are another popular choice for high-throughput biological screenings in miniaturized volumes that are frequently made by the creation of fiber-optic arrays arranged by various biological materials coated onto polystyrene beads (Bumgarner 2013). There are living cell microarrays that enable HTS on cultured cells, both in 2D and 3D formats, with the help of hydrogel scaffolds (Jonczyk et al. 2016). Henceforth, microarrays are highly customizable platforms enabling a broad spectrum of assays for a wide variety of biological samples including cells, tissues, DNA, cDNA, protein, antibody, metabolites, etc. and can be combined with other technologies to serve broader experimental goals, such chromatin immunoprecipitation (MMChips) and interferometric reflectance image sensing (IRIS) to analyze the cross talk between various biological molecules (Chen et al. 2011; Avci et al. 2015). Detailed discussion on the principle of analysis of different biological molecules by microarray has been provided in the subsequent section.

4.3 High-Content Screening (HCS) Systems

High-content screening (HCS) systems have emerged as powerful tools in the field of biomedical experimental research. The main idea of HCS is to enable simultaneous analysis of several cellular parameters at a single-cell level, providing thorough insights on both qualitative and quantitative information about various cellular processes (Taylor 2007). It is a wonderful tool for molecular dissection of various pathways implicated in disease manifestation and analyzing the response of cells to potential drug candidates (Zanella et al. 2010). HCS methodology was first developed in 1996, to enable fluorescent imaging of cellular arrays at a subcellular resolution (Taylor 2007). A large share of HCS systems are miniaturized assays that aim to combine high-throughput cell screening results with advanced whole cell quantitative imaging techniques, i.e., "cellomics." With the help of cutting-edge bioimaging systems, HCS can provide simultaneous information on cellular morphology, quantitative estimation of different proteins of interest, and subcellular localization of the target proteins (Nierode et al. 2016). In the context of drug screening with cells in culture, HCS provides a thorough evaluation of the ADME (absorption, distribution, metabolism, and excretion) of the drugs under question, along with the toxicity evaluation (Zanella et al. 2010). Moreover, since HCS allows studies involving interactions between several fluorochromes, it can be customized to be appropriate for various experimental readouts including RNAi analysis, stem cell research, and protein binding assays (Mattiazzi Usaj et al. 2016). A deeper insight on the cellular processes via bridging the gap between the depth and throughput of biological data with HCS facilitates the course of biomedical research on the development of novel therapeutics against a vast number of complex diseases.

5 Analysis of Different Biological Components in a Miniaturized System

Advancements in the field of miniaturized chip and microfluidic technologies enabled researchers for high-throughput and cost-effective analysis of DNA, RNA, protein, and metabolite repertoire in a wide variety of samples, e.g., cultured cells, tissue fluids (blood, urine, saliva, etc.) microbial samples from clinical biopsies, etc. The tailor-made miniaturized systems for various samples utilize specific combinations of basic experimental principles and methodologies for the isolation of different biological components and provide valuable information regarding the molecular state of the cells. By integrating multiple techniques onto a single chip or by accommodating multiple samples in miniaturized volumes, these platforms would further lead to a drastic reduction in experimental cost and rapid analysis of the system with higher accuracy. This section of the book chapter would deal with mentioning a few miniaturized systems to isolate and analyze the "omic" profiles for different biological entities, i.e., genome, transcriptome, proteome, and metabolome.

5.1 Genome

Miniaturized systems aimed to study the entire genome of a biological sample rely on the basic principles of some of the established methodologies, e.g., polymerase chain reaction (PCR), isothermal amplification, capillary electrophoresis, laser-induced fluorescence detection, and next-generation sequencing. Scientists have developed integrated microfluidic systems to enable short tandem repeat (STR) profiling of the DNA samples isolated directly from biological samples (Yang et al. 2019). Impressive developments have been made in building ultrafast Sanger DNA sequencing devices based on microchip electrophoresis to sequence up to 100 bases per minute (Fredlake et al. 2008). DNA microarray technologies designed by a number of biomedical companies, e.g., Affymetrix, Illumina, Agilent Technologies, etc., have enabled researchers to determine the polymorphism status of individuals across the whole genome (Bumgarner 2013). Microfluidics has particularly revolutionized the field of single-cell genomics, with an automated pipeline for sorting single cells, DNA extraction, and sequencing at high coverage (Thompson et al. 2014). Other than disease diagnostics and development of personalized medicine, these microdevices are instrumental in the field of forensic science, enabling rapid analysis of sample DNA often from a tiny sample volume (Divne and Allen 2005).

5.2 Transcriptome

Global transcriptomic analysis provides key information on the molecular status of the cells by in-depth evaluation of gene expression levels and regulatory mechanisms. A number of miniaturized systems have been devised by scientists across the globe, to evaluate the RNA expression patterns of the cells in a rapid and high-throughput manner. The basic methodologies employed in these devices are predominantly but not limited to reverse transcription PCR (RT-PCR), real-time PCR, microarrays, and RNA sequencing. Sanfilippo et al. have shown the use of microfluidics-based transcriptomic approach to decode cellular mechanosensing (Sanfilippo et al. 2019). Scientists have demonstrated the use of a droplet microarray platform to isolate cDNA for transcriptomic analysis from down to single cell per sample in a 200 nL volume (Chakraborty et al. 2022). Single-cell global transcriptomic studies have particularly gained popularity to decode the heterogeneity in complex tissues (Zilionis et al. 2017). Further development of novel microfluidics-based single-cell RNA sequencing technologies using unique molecular barcodes would lead to a paradigm shift in the understanding of the molecular response of cells to different stimuli (Macosko et al. 2015).

5.3 Proteome

Study of the complete set of proteins inside a cell provides a thorough insight into the cellular processes and functioning machineries. Techniques like immunoassays [e.g., enzyme-linked immunosorbent assays (ELISA) and radioimmunoassays (RIA)], protein microarrays, mass spectrometry, and quantification methods are employed into tailor-made microfluidic devices to analyze the proteomic status of the cells in miniaturized volumes at a high-throughput scale. Scientists have shown a novel microfluidic approach to achieve up to 100 folds of downscaling in the sample requirement for regular mass spectrometry (Furlan et al. 2019). Single-cell proteomic studies have gained new momentum with the discovery of novel miniaturized systems to streamline the entire experimental workflow in miniaturized volumes. Gebreyesus et al. have shown a streamlined technology for single-cell proteomic sample preparation and data acquisition by mass spectrometry (Gebreyesus et al. 2022). Automated microfluidic chips have been devised for cell culture along with real-time monitoring for the levels of various inflammatory cytokines secreted by the cells (Kaestli et al. 2017). Digital microfluidics has enabled researchers to do quantitative proteomic analysis even at the whole organism level (Steinbach et al. 2022).

5.4 Metabolome

Metabolites represent the repertoire of small molecules in a cell at a particular point of time that provide crucial insights into cellular metabolism, signaling pathways, and molecular response to several types of stimuli, e.g., drugs, foreign nucleic acids, etc. The list of primary metabolites in a body includes carbohydrates, proteins, fat, amino acids, enzymes, organic acids, nucleotides, and vitamins. Miniaturized systems aimed toward high-resolution metabolite screenings in a large number of samples rely on techniques like mass spectrometry, nuclear magnetic resonance (NMR) spectroscopy, and various chromatographic methods. In a recent study, scientists have discovered a microfluidic system aided by NMR spectroscopy for live metabolite screening of cells in culture (Rogers et al. 2023). Assessment of metabolites by mass spectroscopy within a nanoliter-range volume has been made possible by manipulating the cells on the droplet microarray platform (RamalloGuevara et al. 2021). Using a microfluidic device aided by mass spectroscopy, scientists were even successful in real-time metabolite profiling for a developing blastocyst, providing critical information on the molecular pathways implicated in embryo development and implantation process (Mancini et al. 2021).

6 Point-of-Care Disease Testing Using Miniaturized Devices

Point-of-care disease testing refers to the wide variety of technologies that guide medical professionals toward rapid and accurate disease diagnosis in patients at the site of sample collection. These disease testing methods eliminate the time required for transportation of the samples to laboratories and include rapid disease testing techniques, remote testing, and satellite testing (Vashist 2017). Development of miniaturized devices has revolutionized the field of disease diagnostics in a point-of-care setup, since these small devices can be easily carried to the site of sample collection. Moreover, many of these devices work with a "sample-in-answer-out" strategy and can be indispensable for remote areas with limited access to laboratory facilities. The point-of-care disease testing devices usually have minimal sample requirements, and the processing time is very fast (<10 min). Furthermore, miniaturization of disease diagnosis methods has provided better accuracy by simplifying the analyte detection process, use of samples with lower complexity (e.g., blood, saliva, etc.), and eliminating chances of manual error during sample processing. Figure 2 is a representation of the common point-of-care rapid diagnostic devices in use, as discussed in detail in the subsequent sections.

6.1 Lateral Flow Assays

Simple and portable devices like membrane-based lateral flow tests (LFT) share a large proportion of rapid point-of-care testing kits. These devices have three major advantages to be used as on-site disease diagnostics—ease of use, cost-effectiveness,

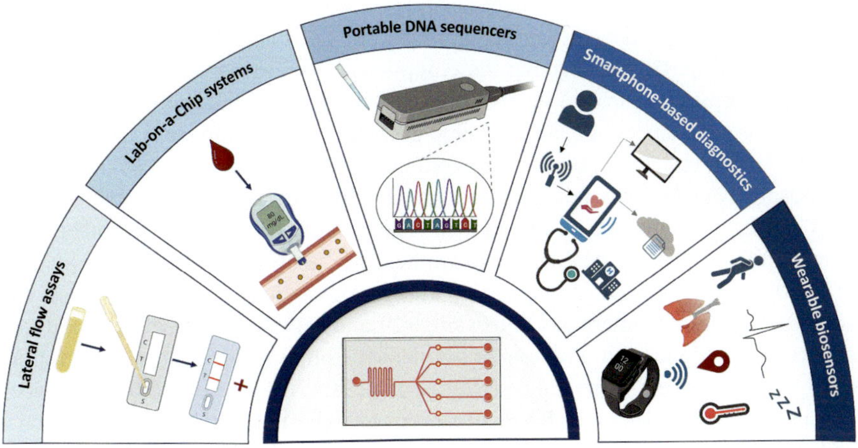

Fig. 2 Biomedical applications of point-of-care diagnostic devices

and quick (1–2 min) turnaround time. LFTs can be employed for the detection of a variety of analytes including blood gases, electrolytes, antigens, antibodies, and various pathogenic organisms (Koczula and Gallotta 2016). Most LFT devices contain a membrane-based test strip with specific reagents that lead to the appearance of visible colored lines upon finding the target analyte in the provided sample that migrates through the strip by capillary force. Due to the simplicity of these devices, handling these strips require minimal to no guidance. For example, LFT technology is used abundantly in home-based pregnancy kits and blood glucose monitors. Furthermore, the simplicity of LFT devices enables point-of-care testing in remote areas with limited healthcare resources. Depending on the sample type and the nature of the disease marker, LFT strips can be manufactured to perform a wide variety of immunological, chemical, and nucleic acid detection methods. For example, LFT strips are widely used for the detection of pathogens like SARS-CoV-2 and HIV in patient samples (Azhar et al. 2021; Martiskainen et al. 2021). The use of LFT devices in diagnosing infectious diseases is crucial for early detection, containment measures, prompt treatment, and effective disease control. Certain LFT devices have shown to yield satisfactory results in drug abuse testing at a point-of-care setting (Li and Wang 2020). Despite having a large number of advantages, some LFT strips might fail to ensure the acceptable range of sensitivity and specificity for disease testing, and therefore a thorough quality control is important before introducing them in the public health domain. Nevertheless, the versatility of LFT devices is unarguably beneficial to reduce the burden of healthcare facilities and to empower patients to have better control on monitoring their health situations.

6.2 Lab-on-a-Chip Systems

Predominantly manufactured using modern microfluidic technologies, lab-on-a-chip systems help in point-of-care analysis of small volumes of tissue fluids collected from patients. These devices can integrate a number of laboratory functions onto a single microchip, thereby making these applicable for a wide variety of diagnostic tests with a quick turnaround time (Arshavsky-Graham and Segal 2022). One of the most important features of a lab-on-a-chip device is the miniaturization, aiding in the portability to a variety of point-of-care settings. Moreover, these devices can be used to measure a number of parameters from the same patient sample. A typical lab-on-a-chip device is made up of an intricate network of microfluidic chambers and microcapillaries that are designed to manipulate the samples by counting the number of various cell types and enable a large number of chemical reactions to detect the target analytes or biomarkers. For example, rapid estimation of whole blood cell counts can be done with a lab-on-a-chip device from a single drop of blood at an emergency setup, e.g., ambulance (Hassan et al. 2015). These microfluidic devices can also be used to detect the levels of various analytes in blood, urine, saliva, tears, and other bodily exudates (Srinivasan et al. 2004). One important attribute of lab-on-a-chip devices is the possibility to connect these with digital platforms, allowing telemedicine and remote patient monitoring. Therefore, results obtained from lab-on-a-chip devices can be remotely accessed by healthcare professionals, allowing them to make immediate clinical decisions in an emergency setup. However, point-of-care microfluidic platforms also have certain limitations. For example, the development of these devices might be cumbersome, as it requires experts from a variety of scientific fields including clinicians, biomedical scientists, material scientists, chemists, and engineers. Once the devices are made to work, thorough quality control checks are advisable to maintain the regulatory standards. Despite aforementioned concerns, the integration of multiple laboratory features into these portable lab-on-a-chip systems has definitely revolutionized the way of disease diagnosis in remote and disaster-stricken areas with little access to healthcare facilities.

6.3 Portable DNA Sequencers

In a time of unprecedented development in biological research, the medicine of today should be guided by the genetic makeup of each patient, so as to provide precise medical help. Using genetic information, it would be easier to know whether a patient would be cured by a drug or develop serious side effects. Portable DNA sequencers are primarily based on nanopore technology and semiconductor-based platforms generating fluorescent signals, enabling analysis of the DNA sequences from patient samples right at the hospital bedside (Pembaur et al. 2021). Measuring at the size of a USB flash drive and having the potential to decode the patient genome within a matter of few hours, these portable DNA sequencers can be carried to a point-of-care setting, distant healthcare center, or remote biomedical research station with a great ease. Real-time analysis of the patient DNA sequences by portable sequencers could be invaluable in a multitude of clinical scenarios. In case of newborns with undiagnosed health issues or congenital disorders, DNA sequencing using portable sequencer machines could be lifesaving by accurate diagnosis of the condition and providing immediate clinical attention. Rapid analysis of DNA mutations in tumor samples of known malignancies can guide clinicians toward the most effective therapeutic approach in patients. In case of communicable diseases, sequencing of DNA from the body fluids of patients can accurately identify the disease-causing pathogen, along with the analysis of any antibiotic resistance pattern in the same. Portable DNA sequencers can be crucial in the field of forensic science, enabling rapid DNA analysis from biological samples found at the crime site. Portable sequencers also reduce the overall cost of DNA sequencing by eliminating the need for a centralized sequencing facility. However, these point-of-care DNA sequencers must undergo regular quality checks to ensure the accuracy of the results produced. In conclusion, portable DNA sequencers represent a transformative technology at emergency healthcare setups with the potential to revolutionize a multitude of experimental frameworks for mutation analysis, cancer genomics, infectious disease diagnostics, forensic science, and personalized medicine (Xu et al. 2020).

6.4 Smartphone-Based Disease Diagnostics

"Mobile health," i.e., smartphone-based software applications that enable healthcare services, is soaring up in popularity during recent times. These healthcare technologies are predominantly based on artificial intelligence and can use smartphone hardwares, e.g., camera, microphones, processing power, etc., to track the vital signs of a person. There are large number of such mobile health applications that exploit the ubiquity and ability of smartphones to provide real-time, accurate, and cost-effective health monitoring, along with providing appropriate diagnostic solutions. Ranging from counting the footsteps to monitoring the heart rate, daily intake of calories, and pulse activity, these smartphone-based disease diagnosis platforms can report the early signs of a metabolic disease, e.g., diabetes and cardiovascular disorders (Hernández-Neuta et al. 2019). The transformative potential of mobile health applications stems from the widespread availability of smartphones, making this technology useful everywhere in the world without the need for any special equipment or personal training. The use of smartphones to diagnose medical conditions further allows remote monitoring of health reports and real-time communication with healthcare professionals or central health research databases. Some smartphone-based applications have also proven to be extremely useful in women with fertility issues, by tracking conditions like ovulation and possible time window for a successful conception (Potluri et al. 2018). Using the built-in camera of smartphones, certain mobile health applications also claim to accurately predict the occurrence of diseases with a visible cue, e.g., skin cancer and retinal disorders. Moreover, mobile health applications can be used for biosensing and imaging analysis of LFT strips for diagnosis of pathogen-borne infectious diseases (Hernández-Neuta et al. 2019). However, the use of mobile health applications for disease diagnosis can sometimes yield erroneous outputs in terms of false-positive or false-negative reports. Confidentiality and security of personal health data are also important considerations for these applications, in order to follow the ethical guidelines. Nevertheless, mobile health has transformed smartphones into portable diagnostics devices, and this technology has immense potential to revolutionize healthcare systems by improving patient outcomes and democratizing healthcare services across the globe.

6.5 Wearable Biosensors

Wearable biosensors have emerged as powerful tools to revolutionize the field of personalized healthcare with real-time monitoring of health conditions. These are miniaturized biosensors integrated into wearable devices, e.g., a smartwatch, wristband, or skin patch, and can perform continuous monitoring of a large number of physiological parameters including body temperature, blood glucose and electrolyte levels, heart rate, blood pressure, sleep pattern, physical activity, oxygen saturation, etc. (Kim et al. 2019). The most important feature of wearable biosensors is the noninvasiveness of these healthcare platforms—thus providing reliable healthcare analysis reports with minimal discomfort to the individuals. Therefore, wearable biosensors are mostly epidermal, and operate on different working principles including electrochemistry, fluorescence detection, electrocardiography, photoplethysmography (PPG), and various other biochemical detection methodologies (Kim et al. 2019). With the virtue of being able to collect real-time health reports, these wearable biosensors empower people to be proactive and be in control of their physical state. In case of supervised weight reduction, wearable biosensors provide essential information on energy expenditure and calorie intake on a daily basis, enabling individuals to achieve personal fitness goals. These devices can help in chronic disease management, e.g., diabetes, by providing real-time data leading to tailoring of appropriate treatment plans. There are multiple incidences of accurate prediction of undiagnosed heart conditions by healthcare biosensors integrated in smartwatch devices. These innovative technologies can also detect the early signs of cancer in patients by analyzing the alterations in the metabolic activities of the body (Smith et al. 2023). Furthermore, the data from wearable biosensors can be remotely tracked by clinicians, leading to precise healthcare management and disease prevention. The use of wearable biosensors can thereby lead to a dramatic reduction in the healthcare cost by early detection of health conditions and avoiding hospitalization. However, the integrated biosensors must undergo a thorough quality check to ensure accuracy and reliability of the produced health reports. Nevertheless, wearable biosensors are poised to have a transformative potential in providing a holistic view of personal health and disease tracking with minimal expenses.

7 Past, Present, and Future of "Organ-on-a-Chip"

Microfluidic platforms with miniaturization have led to remarkable advancement in the development of customizable techniques for tissue engineering. "Organ-on-a-chip" is perhaps the most impressive advancement in the field, which claims to have the transformative potential of culturing living cells in miniaturized systems, with a goal of conserving the structure and function of a particular human organ. The concept of organ-on-a-chip originated in early 2000 and soon gained popularity as a more relevant model system for drug testing and toxicity studies—thereby replacing the need for traditional in vitro cell culture and in vivo disease models (Ma et al. 2021). Figure 3 is a schematic representation of the common tissue types cultured via organ-on-a-chip technology. These microphysiological platforms were previously focused on creating simple tissue constructs with cultured cells in a controlled environment such as lung, liver, colon etc., so as to mimic the response of a particular organ to different stimuli or medical conditions in the human. However, recent developments in the field enabled researchers to fabricate complex multi-organ systems by co-culturing different cell types within the microchips, with

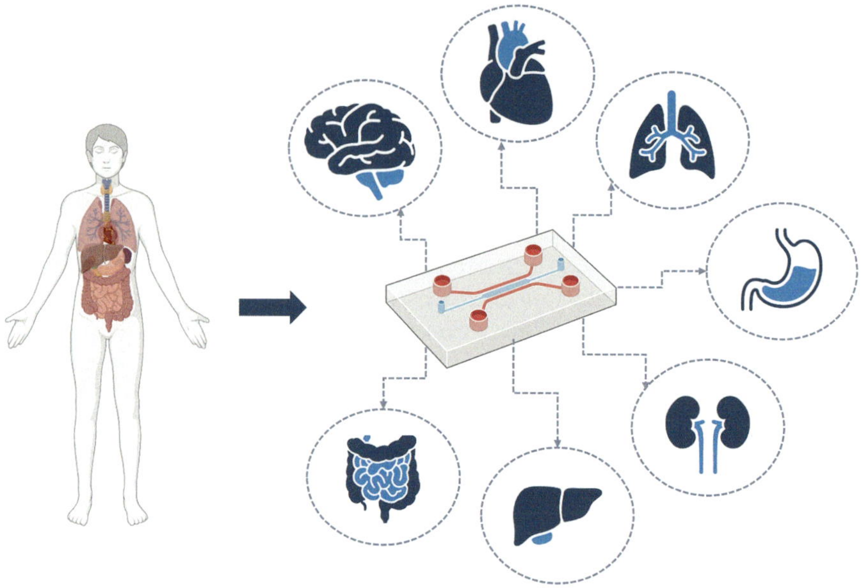

Fig. 3 "Organ-on-a-chip" technique for development of personalized precision medicine

microcapillaries serving as the vasculature. These multi-organ systems are crucial to understand the organ-to-organ cross talk at different physiological conditions, study the course of drug metabolism with the body, understand the process of disease progression, and develop personalized treatment strategies for various conditions (Wu et al. 2020). A number of intricacies are involved in manufacturing a sophisticated microphysiological system like organ-on-a-chip. First and foremost, the chip architecture has to be conceptualized and manufactured according to the need of the particular organ to grow. Next, a biocompatible inner layer of the microfluidic system needs to be fabricated (could be a polymer, peptide, or nanoparticle in chemical composition) and coated to act as the basement membrane for cellular attachment. Thereafter, the cells under consideration are seeded onto relevant microchambers of the chip. Lastly, life-supporting systems (e.g., sensor, mechanical stimuli, incubator, perfusion chambers, etc.) should be active onto or surrounding the microchip, as the need may be (Leung et al. 2022). Organ-on-a-chip platforms are often miniaturized, making these useful for high-throughput analysis of drugs for efficacy and toxicity analysis. Moreover, the use of these microchips eliminates the need for model organisms to probe a human disease, with a promise of being a robust and validated system of disease testing and therapeutic advancements (Sosa-Hernández et al. 2018). These platforms can also be prepared directly with patient-derived cells for drug screening, thereby revolutionizing the field of personalized health and drug development, while providing a better understanding of the disease. For example, human liver-on-a-chip models are proven to be useful in developing novel therapeutic strategies against fatty liver disease (Hassan et al. 2020). Modern microfluidic technologies have enabled researchers to create breathing lung organoids onto miniaturized devices, thereby serving it for the purpose of screening the environmental particulates for toxicity (Zamprogno et al. 2021). The creation of miniaturized guts in vitro with novel microfluidic technologies has been instrumental to learn gut-microbiota cross talk and study drug pharmacokinetics (Xiang et al. 2020). Even for the most complex organs like the brain, scientists have been able to create tiny organoids to model various neurovascular malfunctions like Parkinson's and Alzheimer's disease (Venkataraman et al. 2022).

8 Future Directions on Establishment of Novel Microfluidic/Miniaturized Chips for Disease Analysis

The mighty universe of miniaturized systems and microfluidics devices has definitely broadened the horizon of disease analysis and development of novel therapeutics, by providing innovative strategies toward rapid, sensitive, and cost-effective diagnostic testing. There are a number of advantages to implement lab-on-a-chip devices in personalized health and disease diagnostics outside the laboratory setup, owing to the portability and potential to be used as point-of-care diagnostic devices. Integration of multiple functionalities onto a single chip or microfluidic device has given rise to organ-on-a-chip model systems, allowing researchers for a comprehensive disease analysis and thorough understanding drug metabolism inside a body system. As we look to the future, endless opportunities are lying ahead, for the advancements of the miniature laboratory world created by inquisitive minds all across the globe. Futuristic designs of microfluidic chips and devices would potentially witness advancements in the design and fabrication techniques, along with the use of novel biomimetic surface coatings to enhance the chip performance with better biocompatibility and increased sensitivity. Additionally, the development of novel 3D scaffolds to support complex human tissues and vasculature would give rise to a better understanding of human physiology and personalized medicine development (Sun et al. 2020). The establishment of scalable chip manufacturing processes would further increase the throughput in supporting healthcare facilities with miniaturized diagnostic platforms for point-of-care applications. A vast majority of the researchers are keen to develop a universal "sample-in-answer-out" miniaturized microfluidic chip or microarrays, where the entire process of cellular screening and evaluation (i.e., cell culture, incubation with drugs or siRNA or small molecule or miRNA libraries, phenotypic microscopy-based toxicity analysis, analysis of protein biomarkers, and global genomic/transcriptomic analysis by DNA/RNA sequencing direct from the cells) can be automatized inside a sterile chamber operated by robotic handlers. The integration of several experimental workflows into one chip with automation would be instrumental in achieving higher accuracy of disease analysis by complete avoidance of manual error, reducing the experimental budget, and drastic increase in the experimental throughput. To decode the enormous amount of data generated from multiple experimentations done on a chip with a large number of samples, futuristic miniaturized diagnostic devices are likely to be highly dependent on artificial intelligence and machine learning algorithms. Development of these cutting-edge technologies would also allow real-time data analysis directly from cells on the chip or microfluidic device, enabling researchers to classify the samples according to responsiveness to different drugs, identify early signs of disease, formulate preventive measures by a predictive modeling approach, and develop precise and personalized therapeutic measures. Furthermore, development of novel microfluidic platforms could be instrumental in the fabrication of new tools for assisted reproduction in couples with fertility issues, e.g., quality analysis, sorting, manipulation, and culture of gamete cells (sperm and oocyte), and development of novel cryopreservation machineries and aid in the fertilized embryo implantation process (Smith and Takayama 2017). Encashing on the impressive amount of research done on organ-on-a-chip platforms, biomedical researchers and material scientists across the globe are now on the way to create a "human-on-a-chip" model. This ultimate on-chip human model would be instrumental in understanding the organ-organ cross talk, studying drug pharmacokinetics, and developing a critical understanding on disease development inside the body, within a controlled system in a microphysiological environment (Marx et al. 2021). The transformative potential of novel miniaturized devices and microfluidics shall therefore continue to pave its journey as the next-generation platform to provide a thorough understanding of human systems biology, complex disease analysis, and development of precision medicine based on personalized health.

References

Almendro V, Marusyk A, Polyak K (2013) Cellular heterogeneity and molecular evolution in cancer. Annu Rev Pathol 8:277–302

Altschuler SJ, Wu LF (2010) Cellular heterogeneity: do differences make a difference? Cell 141: 559–563

Arshavsky-Graham S, Segal E (2022) Lab-on-a-Chip devices for point-of-care medical diagnostics. Adv Biochem Eng Biotechnol 179:247–265

Avci O, Ünlü NL, Özkumur AY, Ünlü MS (2015) Interferometric reflectance imaging sensor (IRIS)–a platform technology for multiplexed diagnostics and digital detection. Sensors (Basel) 15:17649–17665

Ayuso JM, Park KY, Virumbrales-Muñoz M et al (2021) Toward improved *in vitro* models of human cancer. APL Bioeng 5:010902

Azhar M, Phutela R, Kumar M et al (2021) Rapid and accurate nucleobase detection using FnCas9 and its application in COVID-19 diagnosis. Biosens Bioelectron 183:113207

Barker SL, LaRocca PJ (1994) Method of production and control of a commercial tissue culture surface. J Tissue Culture Methods 16:151–153

Berthier E, Dostie AM, Lee UN et al (2019) Open microfluidic capillary systems. Anal Chem 91: 8739–8750

Brooks MD, Burness ML, Wicha MS (2015) Therapeutic implications of cellular heterogeneity and plasticity in breast cancer. Cell Stem Cell 17:260–271

Bumgarner R (2013) Overview of DNA microarrays: types, applications, and their future. Curr Protoc Mol Biol 22:Unit 22-1

Campbell JM, Balhoff JB, Landwehr GM et al (2018) Microfluidic and paper-based devices for disease detection and diagnostic research. Int J Mol Sci 19:2731

Chakraborty S, Gourain V, Benz M et al (2021) Droplet microarrays for cell culture: effect of surface properties and nanoliter culture volume on global transcriptomic landscape. Mater Today Bio 11:100112

Chakraborty S, Luchena C, Elton JJ et al (2022) "Cells-to-cDNA on Chip": phenotypic assessment and gene expression analysis from live cells in Nanoliter volumes using droplet microarrays. Adv Healthc Mater 11:e2102493

Chen Y, Meyer CA, Liu T et al (2011) MM-ChIP enables integrative analysis of cross-platform and between-laboratory ChIP-chip or ChIP-seq data. Genome Biol 12:R11

Choi YH, Kim JK (2019) Dissecting cellular heterogeneity using single-cell RNA sequencing. Mol Cells 42:189–199

Divne AM, Allen M (2005) A DNA microarray system for forensic SNP analysis. Forensic Sci Int 154:111–121

Dueñas ME, Peltier-Heap RE, Leveridge M et al (2023) Advances in high-throughput mass spectrometry in drug discovery. EMBO Mol Med 15:e14850

Earle WR, Schilling EL, Stark TH et al (1943) Production of malignancy in vitro. IV. The mouse fibroblast cultures and changes seen in the living cells. J Natl Cancer Inst 4:165–212

Echeverri CJ, Perrimon N (2006) High-throughput RNAi screening in cultured cells: a user's guide. Nat Rev Genet 7:373–384

El Harane S, Zidi B, El Harane N et al (2023) Cancer spheroids and organoids as novel tools for research and therapy: state of the art and challenges to guide precision medicine. Cells 12:1001

Erickson P, Houwayek T, Burr A et al (2021) A continuous flow cell culture system for precision cell stimulation and time-resolved profiling of cell secretion. Anal Biochem 625:114213

Fernandes TG, Diogo MM, Clark DS et al (2009) High-throughput cellular microarray platforms: applications in drug discovery, toxicology and stem cell research. Trends Biotechnol 27:342–349

Flobak Å, Niederdorfer B, Nakstad VT et al (2019) A high-throughput drug combination screen of targeted small molecule inhibitors in cancer cell lines. Sci Data 6:237

Fredlake CP, Hert DG, Kan CW et al (2008) Ultrafast DNA sequencing on a microchip by a hybrid separation mechanism that gives 600 bases in 6.5 minutes. Proc Natl Acad Sci USA 105:476–481

Furlan C, Dirks RAM, Thomas PC et al (2019) Miniaturised interaction proteomics on a microfluidic platform with ultra-low input requirements. Nat Commun 10:1525

Gebreyesus ST, Siyal AA, Kitata RB et al (2022) Streamlined single-cell proteomics by an integrated microfluidic chip and data-independent acquisition mass spectrometry. Nat Commun 13:37

Gupta RK, Kuznicki J (2020) Biological and medical importance of cellular heterogeneity deciphered by single-cell RNA sequencing. Cells 9:1751

Hassan S, Sebastian S, Maharjan S et al (2020) Liver-on-a-Chip models of fatty liver disease. Hepatology 71:733–740

Hassan U, Reddy B Jr, Damhorst G et al (2015) A microfluidic biochip for complete blood cell counts at the point-of-care. Technology (Singap World Sci) 3:201–213

Hernández-Neuta I, Neumann F, Brightmeyer J et al (2019) Smartphone-based clinical diagnostics: towards democratization of evidence-based health care. J Intern Med 285:19–39

Hertzberg RP, Pope AJ (2000) High-throughput screening: new technology for the 21st century. Curr Opin Chem Biol 4:445–451

Howell J, Hammarton TC, Altmann Y et al (2020) High-speed particle detection and tracking in microfluidic devices using event-based sensing. Lab Chip 20:3024–3035

Hu Y, An Q, Sheu K et al (2018) Single cell multi-omics technology: methodology and application. Front Cell Dev Biol 6:28

Huang L, Yi X, Yu X et al (2021) High-throughput strategies for the discovery of anticancer drugs by targeting transcriptional reprogramming. Front Oncol 11:762023

Inglese J, Shamu CE, Guy RK (2007) Reporting data from high-throughput screening of small-molecule libraries. Nat Chem Biol 3:438–441

Jogia GE, Tronser T, Popova AA et al (2016) Droplet microarray based on Superhydrophobic-Superhydrophilic patterns for single cell analysis. Microarrays (Basel) 5:28

Jonczyk R, Kurth T, Lavrentieva A et al (2016) Living cell microarrays: an overview of concepts. Microarrays (Basel) 5:11

Kaestli AJ, Junkin M, Tay S (2017) Integrated platform for cell culture and dynamic quantification of cell secretion. Lab Chip 17:4124–4133

Kim J, Campbell AS, de Ávila BE et al (2019) Wearable biosensors for healthcare monitoring. Nat Biotechnol 37:389–406

Koczula KM, Gallotta A (2016) Lateral flow assays. Essays Biochem 60:111–120

Leung CM, de Haan P, Ronaldson-Bouchard K et al (2022) A guide to the organ-on-a-chip. Nat Rev Methods Primers 2:1–29

Li Z, Wang P (2020) Point-of-care drug of abuse testing in the opioid epidemic. Arch Pathol Lab Med 144:1325–1334

Lim SB, Yeo T, Lee WD et al (2019) Addressing cellular heterogeneity in tumor and circulation for refined prognostication. Proc Natl Acad Sci USA 116:17957–17962

Liu Y, Chakraborty S, Direksilp C et al (2021) Miniaturized droplet microarray platform enables maintenance of human induced pluripotent stem cell pluripotency. Mater Today Bio 12:100153

Ma C, Peng Y, Li H et al (2021) Organ-on-a-Chip: a new paradigm for drug development. Trends Pharmacol Sci 42:119–133

Macosko EZ, Basu A, Satija R et al (2015) Highly parallel genome-wide expression profiling of individual cells using Nanoliter droplets. Cell 161:1202–1214

Mancini V, Schrimpe-Rutledge AC, Codreanu SG et al (2021) Metabolomic analysis evidences that uterine epithelial cells enhance blastocyst development in a microfluidic device. Cells 10:1194

Martiskainen I, Juntunen E, Salminen T et al (2021) Double-antigen lateral flow immunoassay for the detection of anti-HIV-1 and -2 antibodies using Upconverting nanoparticle reporters. Sensors (Basel) 21:330

Marx U, Accastelli E, David R et al (2021) An individual patient's "body" on chips-how organismoid theory can translate into your personal precision therapy approach. Front Med (Lausanne) 8:728866

Mattiazzi Usaj M, Styles EB, Verster AJ (2016) High-content screening for quantitative cell biology. Trends Cell Biol 26:598–611

Millington D, Norton S, Singh R et al (2018) Digital microfluidics comes of age: high-throughput screening to bedside diagnostic testing for genetic disorders in newborns. Expert Rev Mol Diagn 18:701–712

Mitchell KJ (2012) What is complex about complex disorders? Genome Biol 13:237

Molla MR, Chakraborty S, Munoz Sagredo L et al (2020) Combinatorial synthesis of a lipidoid library by thiolactone chemistry: in vitro screening and in vivo validation for siRNA delivery. Bioconjug Chem 31:852–860

Murschhauser A, Röttgermann PJF, Woschée D et al (2019) A high-throughput microscopy method for single-cell analysis of event-time correlations in nanoparticle-induced cell death. Commun Biol 2:35

Nierode G, Kwon PS, Dordick JS et al (2016) Cell-based assay design for high-content screening of drug candidates. J Microbiol Biotechnol 26:213–225

Oguchi S, Saito H, Tsukahara M et al (2006) pH condition in temperature shift cultivation enhances cell longevity and specific hMab productivity in CHO culture. Cytotechnology 52:199–207

Pattanayak P, Singh SK, Gulati M et al (2021) Microfluidic chips: recent advances, critical strategies in design, applications and future perspectives. Microfluid Nanofluidics 25:99

Pembaur A, Sallard E, Weil PP et al (2021) Simplified point-of-care full SARS-CoV-2 genome sequencing using Nanopore technology. Microorganisms 9:2598

Potluri V, Kathiresan PS, Kandula H et al (2018) An inexpensive smartphone-based device for point-of-care ovulation testing. Lab Chip 19:59–67
RamalloGuevara C, Paulssen D, Popova AA et al (2021) Fast Nanoliter-scale cell assays using droplet microarray-mass spectrometry imaging. Adv Biol (Weinh) 5:e2000279
Ramsey WS, Hertl W, Nowlan ED et al (1984) Surface treatments and cell attachment. In Vitro 20: 802–808
Regev A, Teichmann SA, Lander ES et al (2017) The human cell atlas. Elife 6:e27041
Rogers G, Barker S, Sharma M et al (2023) Operando NMR metabolomics of a microfluidic cell culture. J Magn Reson 349:107405
Rubio-Navarro A, Gómez-Banoy N, Stoll L et al (2023) A beta cell subset with enhanced insulin secretion and glucose metabolism is reduced in type 2 diabetes. Nat Cell Biol 25:565–578
Saadatpour A, Lai S, Guo G et al (2015) Single-cell analysis in cancer genomics. Trends Genet 31: 576–586
Sanfilippo JE, Lorestani A, Koch MD et al (2019) Microfluidic-based transcriptomics reveal force-independent bacterial rheosensing. Nat Microbiol 4:1274–1281
Schoof EM, Furtwängler B, Üresin N et al (2021) Quantitative single-cell proteomics as a tool to characterize cellular hierarchies. Nat Commun 12:3341
Shang L, Cheng Y, Zhao Y (2017) Emerging droplet microfluidics. Chem Rev 117:7964–8040
Shen S, Zhang F, Zhang Y et al (2023) Construction of multiple concentration gradients for single-cell level drug screening. Microsyst Nanoeng 9:46
Smith AA, Li R, Tse ZTH (2023) Reshaping healthcare with wearable biosensors. Sci Rep 13:4998
Smith GD, Takayama S (2017) Application of microfluidic technologies to human assisted reproduction. Mol Hum Reprod 23:257–268
Sink R, Gobec S, Pečar S, Zega A (2010) False positives in the early stages of drug discovery. Curr Med Chem 17:4231–4255
Sosa-Hernández JE, Villalba-Rodríguez AM, Romero-Castillo KD et al (2018) Organs-on-a-Chip module: a review from the development and applications perspective. Micromachines (Basel) 9: 536
Srinivasan V, Pamula VK, Fair RB (2004) An integrated digital microfluidic lab-on-a-chip for clinical diagnostics on human physiological fluids. Lab Chip 4:310–315
Steinbach MK, Leipert J, Blurton C et al (2022) Digital microfluidics supported microproteomics for quantitative proteome analysis of single Caenorhabditis elegans Nematodes. J Proteome Res 21:1986–1996
Sun H, Jia Y, Dong H et al (2020) Combining additive manufacturing with microfluidics: an emerging method for developing novel organs-on-chips. Curr Opin Chem Eng 28:1–9
Szymański P, Markowicz M, Mikiciuk-Olasik E (2012) Adaptation of high-throughput screening in drug discovery-toxicological screening tests. Int J Mol Sci 13:427–452
Taylor DL (2007) Past, present, and future of high content screening and the field of cellomics. Methods Mol Biol 356:3–18
Thompson AM, Paguirigan AL, Kreutz JE et al (2014) Microfluidics for single-cell genetic analysis. Lab Chip 14:3135–3142
Vashist SK (2017) Point-of-care diagnostics: recent advances and trends. Biosensors (Basel) 7:62
Venkataraman L, Fair SR, McElroy CA et al (2022) Modeling neurodegenerative diseases with cerebral organoids and other three-dimensional culture systems: focus on Alzheimer's disease. Stem Cell Rev Rep 18:696–717
Venkatasubbarao S (2004) Microarrays–status and prospects. Trends Biotechnol 22:630–637
Vyawahare S, Griffiths AD, Merten CA (2010) Miniaturization and parallelization of biological and chemical assays in microfluidic devices. Chem Biol 17:1052–1065
Wu Q, Liu J, Wang X et al (2020) Organ-on-a-chip: recent breakthroughs and future prospects. Biomed Eng Online 19:9
Xiang Y, Wen H, Yu Y et al (2020) Gut-on-chip: recreating human intestine in vitro. J Tissue Eng 11:2041731420965318

Xu H, Xia A, Wang D et al (2020) An ultraportable and versatile point-of-care DNA testing platform. Sci Adv 6:eaaz7445

Yamada R, Okada D, Wang J et al (2021) Interpretation of omics data analyses. J Hum Genet 66: 93–102

Yan S, Zhang J, Chen H et al (2016) Development of a novel magnetophoresis-assisted hydrophoresis microdevice for rapid particle ordering. Biomed Microdevices 18:54

Yang J, Hurth C, Nordquist A et al (2019) Integrated microfluidic system for rapid DNA fingerprint analysis: a miniaturized integrated DNA analysis system (MiDAS)-swab sample-in to DNA profile-out. Methods Mol Biol 1906:207–224

Ye C, Ho DJ, Neri M et al (2018) DRUG-seq for miniaturized high-throughput transcriptome profiling in drug discovery. Nat Commun 9:4307

Zamprogno P, Wüthrich S, Achenbach S et al (2021) Second-generation lung-on-a-chip with an array of stretchable alveoli made with a biological membrane. CommunBiol 4:168

Zanella F, Lorens JB, Link W (2010) High content screening: seeing is believing. Trends Biotechnol 28:237–245

Zhang Y, Minagawa Y, Kizoe H et al (2019) Accurate high-throughput screening based on digital protein synthesis in a massively parallel femtoliter droplet array. Sci Adv 5:eaav8185

Zilionis R, Nainys J, Veres A et al (2017) Single-cell barcoding and sequencing using droplet microfluidics. Nat Protoc 12:44–73

Systems-Wide Analyses of Immune Programming of Regulatory T Cells in Early Life Providing Lifelong Protection from Diseases

Rasmus Ibsen Dehli and Susanne Brix

Contents

Abstract Regulatory T cells (Tregs) are fundamental for lifelong protection against a wide range of noncommunicable diseases in which the immune system contributes to undesirable host-destructive processes, such as autoimmune diseases, asthma, and allergies. Based on epidemiological and experimental studies, it has become evident

R. I. Dehli · S. Brix (✉)
Department of Biotechnology and Biomedicine, Technical University of Denmark, Kgs. Lyngby, Denmark
e-mail: rasdeh@dtu.dk; sbrix@dtu.dk

J. Barciszewski (ed.), *Systems Biology II*, RNA Technologies 15,
https://doi.org/10.1007/978-3-031-62178-9_15

that genetic, cultural, social, pregnancy-, and birth-related predisposition influence the induction of tolerance in early life to self-antigens and harmless non-self-antigens, but specific mechanisms driving the influence have remained elusive. We here review and discuss recently generated knowledge from systems-wide studies combined with omics technologies which are starting to untangle the network of environmental factors (including birth mode, breastfeeding, bacteria, metabolites, and time window) influencing generation of Tregs. Additionally, we present data from studies demonstrating how multi-omics and epigenetic studies may bridge environmental factors and autoimmune diseases and address how these findings point at new therapeutic avenues for treatment or ways to prevent disease development via modulation of Tregs.

Keywords Systems immunology · Systems biology · Host-microbe interactions · Early-life immune imprinting · Regulatory T cells · Tolerance · Window of opportunity · Epigenetics · Noncommunicable diseases · Multi-omics technologies

Abbreviations

ATF	Activating transcription factor
ALDH	Aromatic lactate dehydrogenase
AhR	Aryl hydrocarbon receptor
ATAC-seq	Assay for transposase-accessible chromatin using sequencing
Bcl6	B cell lymphoma 6
B. longum ssp. *infantis*	*Bifidobacterium longum* subsp. *infantis*
Blimp1	B lymphocyte-induced maturation protein-1
Bach2	BTB domain and CNC homolog 2
CREB	cAMP response element-binding protein
CNS	Central nervous system
CXCR	Chemokine receptor
CHIP-seq	Chromatin immunoprecipitation followed by sequencing
CNS2	Conserved noncoding sequence 2
CTLA4	Cytotoxic T-lymphocyte associated protein 4
Dnmt3a	DNA methyltransferase 3 alpha
FAO	Fatty acid β-oxidation
Foxp3	Forkhead box p3
Foxo1	Forkhead box o1
GATA3	GATA-binding protein 3
GITR	Glucocorticoid-induced TNFR-related gene
HDAC7	Histone deacetylase 7
HMOs	Human milk oligosaccharides
HCAR3	Hydroxycarboxylic acid receptor 3
IgE	Immunoglobulin E

ILA	Indole-3-lactic acid
iTreg	Induced Treg
IBD	Inflammatory bowel disease
IFN	Interferon
IL	Interleukin
Mafk	MAF BZIP transcription factor K
MS	Multiple sclerosis
MEF2D	Myocyte enhancer factor 2D
Nfil3	Nuclear factor interleukin-3
RelA	NF-κB p65
Nrf2	Nuclear factor erythroid 2-related factor 2
NFAT	Nuclear factor of activated T cells
OXPHOS	Oxidative phosphorylation
pTregs	Peripheral-derived Tregs
PTGER2	Prostaglandin E receptor 2
Tregs	Regulatory T cells
RAR	Retinoid acid receptor
RXR	Retinoid X receptor
Runx1	Runt-related transcription factor 1
Smad	Small mothers against decapentaplegic
Satb1	Special AT-rich sequence-binding protein 1
STAT	Signal transducer and activator of transcription
TNBS	2,4,6-trinitrobenzenesulfonic acid
T-bet	T-box transcription factor
tTregs	Thymic-derived Tregs
TFs	Transcription factors
TGF-β	Transforming growth factor-β
TNF	Tumor necrosis factor
VAT	Visceral adipose tissue

1 Introduction

Immunology is not only the study of how our immune system protects us from potential pathogenic environmental dangers, like bacterial, viral, fungal, and parasitic infections, but is as importantly the study of how our immune system protects us from itself and how it allows for beneficial and harmless interactions with environmental factors, like food antigens, aero-antigens, and commensal microorganisms.

While induction of inflammatory processes is a vital protective mechanism, it also constitutes an inherent risk for the host itself and must accordingly be regulated by several mechanisms and checkpoints. Without proper regulation, inflammation may lead to tissue damage, impaired healing, long-term tissue remodeling, and reduced function (Suvas et al. 2003; Nikolova et al. 2009; D'Alessio et al. 2009; Barros et al.

2021). Without proper recognition and tolerance of self, the inflammatory processes may even target our own cells and lead to severe chronic, autoimmune disorders (Sakaguchi et al. 2020). The regulation of inflammation and recognition of self are especially governed by CD4+ CD25+ Foxp3+ regulatory T cells (Tregs), a heterogenous population of helper T cells which home to nonlymphoid tissues for regulation of several different cell types, including other T cells (Sakaguchi et al. 2010; Ferreira et al. 2020). They are vital modulators in resolution of immune responses, ensuring restoration of homeostasis, but are already recruited by tissue damage and resulting stress signals early on to regulate the response and contribute to tissue repair (Belkaid et al. 2002; Li et al. 2018). The ratio between effector T cells and Tregs remains stable throughout an immune response despite a rapid influx of effector cells in the early stage of inflammation (Ferreira et al. 2020; Clay et al. 2020), showcasing the role of Tregs in balancing the response to ensure tissue integrity during inflammation (D'Alessio et al. 2009).

Considering the important role of Tregs in immunological homeostasis, it is evident that defects in the mechanisms ensuring their generation and maintenance would materialize in increased incidence of immune-mediated disorders, like autoimmune and allergic diseases. This is exactly what has been experienced for the last half a century in industrialized countries (Pawankar et al. 2011; Lerner et al. 2015; Platts-Mills 2015; Krishna et al. 2020; Miller 2023; Augustine et al. 2023), but without causative mechanisms that could account for the continued increase in disease incidence being pinpointed.

Here we review findings from studies showing associations between environmental factors and increased disease incidence, before presenting recent mouse and human studies that are increasing our knowledge on environmental conditions that may skew early-life immune programming to leave an individual predisposed to noncommunicable diseases. We highlight systems-wide analyses employing multi-omics approaches, which are increasingly being explored to find mechanistic links between microbial metabolites and the generation and maintenance of Tregs. Furthermore, we present multi-omics and epigenomic studies that find potential mechanisms linking autoimmunity to lifestyle-related environmental conditions and discover novel means of promoting Tregs that may be translated into clinical interventions.

2 Regulatory T Cells Protect from Noncommunicable Diseases such as Autoimmune Diseases, Asthma, and Allergies

Regulatory T cells can be broadly divided into two distinct arms. One arm ensures that our immune system does not engage in combatting our own cells as it does for pathogens, while the other arm ensures that we tolerate environmental factors that are not harmful, like innocuous antigens from food and commensal bacteria. These two arms are promoted within the thymus (central tolerance) and in the periphery, respectively.

Tregs undergo vital maturation in the thymus and commence divergence along two separate pathways that ultimately lead to the thymic-derived Tregs (tTregs) and peripheral-derived Tregs (pTregs). As part of their immunosuppressive phenotype, Tregs express forkhead box P3 (Foxp3) as a master transcription factor driving the regulatory phenotype, but several key differences in phenotype and function distinguishes tTregs and pTregs and can be traced back to how they were generated and importantly the epigenetic imprint they obtained during development (Lu et al. 2017). In this context of immune imprinting, epigenetics refers to methylations, acetylations, or phosphorylations of DNA and histones, which control gene expression and define the cellular phenotype.

Within the thymus, tTregs are primed for a Foxp3+ regulatory phenotype upon agonistic recognition of self-antigens and leave as immune-regulating cells to seed secondary lymphoid organs (Rosenblum et al. 2011) and confer tolerance against self-antigens (Thornton et al. 2010; Klein and Jovanovic 2011; Hsieh et al. 2012; Weiss et al. 2012). Foxp3+ pTregs, on the other hand, arise from cells that left the thymus as naïve, self-antigen unresponsive CD4+ T cells and engage in T cell receptor/peptide-major histocompatibility complex interactions with their cognate antigen in secondary lymphoid organs while receiving Treg differentiation signals from a priming (tolerogenic) dendritic cell (Nutsch et al. 2016).

Failure to generate tTregs with specificity against a vital self-antigen results in enhanced risk of development of autoimmune diseases, while failure to generate pTregs leads to impaired peripheral tolerance and increased risk of developing exaggerated reactions against harmless external antigens, such as pollen or food allergens, resulting in allergies, or commensal microorganisms, causing, e.g., inflammatory bowel disease (IBD; Sakaguchi et al. 2020). Thus, Tregs are important cellular regulators of exaggerated immune responses that may otherwise lead to a broad range of noncommunicable diseases of different inflammatory origin (Fig. 1).

2.1 Noncommunicable Diseases Are on the Rise

For half a century, we have now witnessed dramatic increases in incidences of allergic diseases worldwide, especially in children (Pawankar et al. 2011; Platts-Mills 2015; Krishna et al. 2020; Augustine et al. 2023), with strong ties to the concurrent changes in lifestyle and reduced exposure to diverse microbial environments (Pawankar et al. 2011; Platts-Mills 2015; Renz and Skevaki 2021; Augustine et al. 2023).

Increased incidence of autoimmune diseases is being reported as well (Lerner et al. 2015; Miller 2023), but data are scarcer, inconsistent (Gutierrez-Arcelus et al. 2016; Conrad et al. 2023), and on small study groups (Conrad et al. 2023). However, a recent population-based study of 22 million UK citizens by Conrad et al. (2023) supported evidence for an increasing incidence of autoimmune diseases and found

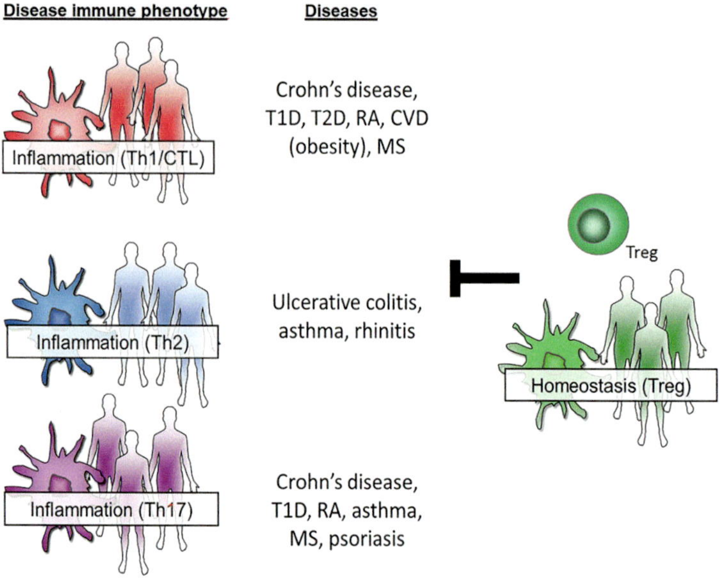

Fig. 1 Tregs mediate their effector function by regulating inflammatory responses of different kinds that otherwise may result in development of various non-communicable diseases. The diseases are here grouped depending on their underlining immunopathology. Some diseases (Crohn's disease, asthma, MS, T1D, and RA) are represented in two groups, as different individuals may present with one or more of the underlying immunopathologies. *T1D* type 1 diabetes; *T2D* type 2 diabetes; *RA* Rheumatoid arthritis; *CVD* cardiovascular disease; *MS* multiple sclerosis

that especially celiac disease, Graves' disease, and rheumatic disorders increased in incidence from 2000 to 2019.

Alongside the rise in allergic and autoimmune diseases, an increasing body of evidence is indicating that the gut microbial composition in early life has a lasting impact on immune development and priming of immunological tolerance to prevent development of both autoimmune and allergic diseases. These findings are especially owing to advances in sequencing technologies within the past decade which have continually improved how well we can interrogate the gut microbiota of both humans and mice.

3 A Network of Environmental Factors Link Perturbed Treg-Based Tolerance in Early Life to Later Disease Incidence

3.1 Associations Between Gut Microbial Dysbiosis in Infancy and Later Disease

Gut microbial dysbiosis in infancy, here used to define a microbial state that diverge from the usual age-dependent composition, has been linked with increased risk of asthma development (Abrahamsson et al. 2014; Arrieta et al. 2015; Stokholm et al. 2020), allergic disease (Arrieta et al. 2015; Fujimura et al. 2016), celiac disease (Olivares et al. 2015), IBD (Hviid et al. 2011; Gevers et al. 2014), and type 1 diabetes (Gevers et al. 2014; Vatanen et al. 2016).

Such dysbiosis may be promoted by early environmental events, like caesarean section (Dominguez-Bello et al. 2010; Stokholm et al. 2016, 2020), which itself is linked to childhood asthma (Sevelsted et al. 2016; Stokholm et al. 2020) and several other chronic immune disorders, like IBD and juvenile arthritis (Sevelsted et al. 2015).

While researchers have found that, e.g., increased proteobacterial load associates with gut microbial dysbiosis and disease (Shin et al. 2015), it has been demonstrated that other taxa, like bifidobacteria and *Bacteroides*, may represent keystone taxa of the infant gut ecosystem promoting healthy immune development (Laforest-Lapointe and Arrieta 2017). Especially the health-promoting role of bifidobacteria in relation to early-life gut microbial influence on tolerance induction has been the focus of several studies.

Vatanen et al. (2016) found an association between lacking bifidobacteria in early life and increased risk of the autoimmune disease type 1 diabetes, while other studies found low bifidobacterial abundance to associate with atopic wheeze (Arrieta et al. 2018), colic, and gut inflammation (Rhoads et al. 2018). Bifidobacteria and *Bacteroides* species usually become the most abundant genera in the gut within the first weeks of life (Underwood et al. 2014) and are promoted by breastfeeding as they can directly consume the human milk oligosaccharides (HMOs) therein (Marcobal and Sonnenburg 2012). The most adept bifidobacterium at HMO utilization, *Bifidobacterium longum* subsp. *infantis* (*B. longum* ssp. *infantis*), is associated with anti-inflammatory responses and decreased intestinal permeability (Underwood et al. 2014). *B. longum* ssp. *infantis* is rare in breastfed infants in Europe (Grzeskowiak et al. 2012; Jost et al. 2012; Roos et al. 2013; Avershina et al. 2014; Abrahamsson et al. 2014) and North America (Casaburi et al. 2021; Azad et al. 2013; Lewis et al. 2015), where prevalence of immune-mediated disorders is especially high, and conversely common in breastfed infants of countries with low incidence of immune-mediated disorders (Grzeskowiak et al. 2012; Huda et al. 2014).

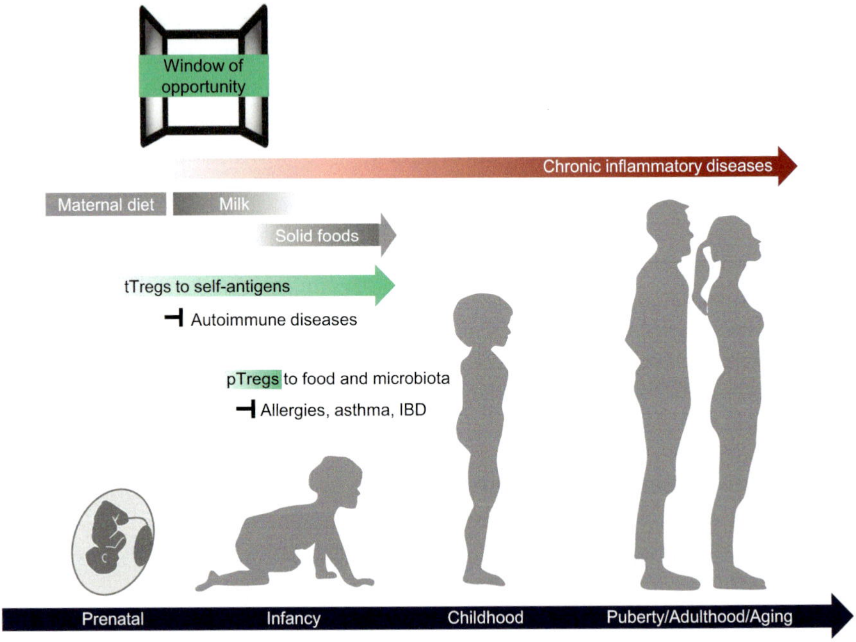

Fig. 2 Early life window for immune programming of central and peripheral tolerance—an event that influences later development of chronic inflammatory diseases. *tTregs* thymic-derived regulatory T cells; *pTreg* peripheral-derived regulatory T cells

3.2 A Window of Opportunity for Inducing Immunological Tolerance

Alongside the discoveries of early-life gut microbial dysbiosis associating with increased risk of noncommunicable diseases in humans, mouse studies supporting the hygiene hypothesis that decreased or lacking early-life exposure to microbes disturbs immune development and increases risk of immune-mediated disorders gained ground. Antibiotic treatment before weaning, but not in adult mice, increased food antigen-specific immunoglobulin E (IgE) response, allergic symptoms (Bashir et al. 2004), and allergic asthma (Russell et al. 2012). Germfree mice and mice with low-diversity microbiota developed increased serum IgE levels and increased risk of exaggerated systemic anaphylaxis (Cahenzli et al. 2013), but germfree mice could be rescued from increased susceptibility to chemically induced colitis and allergic asthma if they were recolonized during the neonatal period (Olszak et al. 2012). These and other mouse and human studies found that bacteria and their metabolic products have a modulating effect on the immune system between birth and weaning (Thorburn et al. 2015; Gensollen et al. 2016), laying the foundation for an arising hypothesis of a temporal neonatal window of opportunity in which microbial exposure has a defining influence on imprinting peripheral immunological tolerance in pTregs (Fig. 2).

Tissue-resident tTreg populations, on the other hand, may primarily originate from perinatal seeding (Yang et al. 2015), and from mouse studies, we know that tTregs seed many tissues, including the skin, lungs, visceral adipose tissue (VAT), liver, and colon during the first 2 weeks of life (Haxhinasto et al. 2008; Kolodin et al. 2015; Sefik et al. 2015; Kalekar et al. 2019). In skin and VAT, these tTregs are proven minimally migratory and inadequately replenished if depleted (Feuerer et al. 2009; Kolodin et al. 2015). They also contribute to tolerance toward the microbiota (Cebula et al. 2013), especially in the early postnatal period where the microbiota establishes (Torow et al. 2015) and pTregs have not been generated (Weiss et al. 2012).

4 Systems Immunology-Based Studies Unravel Mechanisms Involved in Early-Life Priming of Tolerance

Early-life priming of immune tolerance is promoted via the interplay of several complex biological systems that are not easily disentangled, and therefore mechanistic understanding has largely been missing. However, as new experimental omics-based technologies are developed and existing technologies become cheaper and more sensitive, we are improving our understanding of these networks and how we may be able to intervene and prevent disease development. Approaching the system from a systems-wide point of view is essential to move from associations to causation and to build the mechanistic understanding that may lead to translation of findings into clinically relevant interventions or prevention. Such investigations have generally been performed as mouse studies, human cohorts, or intervention studies.

4.1 An Early Window of Opportunity for Priming Peripheral Immunological Tolerance in Mice

Knoop et al. (2017) expanded on the hypothesis of an early-life window of opportunity for priming of peripheral tolerance by demonstrating three distinct phases of antigen encounter and immune imprinting of tolerance in mice: (1) an initial 10-day postnatal phase in which antigens in the intestine are not encountered by the immune system, (2) a following phase lasting until weaning where antigens are primarily encountered by the colonic immune system, and (3) a postweaning phase where antigens are primarily encountered in the small intestine. They found that epidermal growth factor, which is delivered at high levels via breastmilk at birth, suppresses antigen passage across the intestinal epithelium via goblet cell-associated passages and hereby controls the temporal phases via its waning levels from birth until weaning. The temporally controlled antigen delivery allowed for generation of pTregs with affinity for the gut commensals at what appears a crucial time window for tolerance induction, and shutting down antigen passaging or delaying it until after weaning abolished their generation and led to exaggerated colitis in response to epithelial damage and pro-inflammatory responses toward gut commensals in later life.

Al Nabhani et al. (2019b) went on to demonstrate what they referred to as "a weaning reaction" in mice, where the immune system first elicits a strong response to the gut-colonizing microbes, resulting in transient high levels of pro-inflammatory cytokines, particularly tumor necrosis factor-alpha (TNF-α) and interferon-gamma (IFN-γ), which recede again within a week. The authors demonstrate that this phenomenon is a response to the rapid increase in bacterial numbers and species richness accompanying the transition from mother's milk only to a diet containing solid foods and that it associates with opening of goblet cell-associated cell passages and results in generation of pTregs with affinity for the gut commensals. Inhibiting the reaction with antibiotic treatment resulted in increased susceptibility to allergic inflammation and colitis in later life, and the reaction did not occur if adult germfree mice were colonized, supporting the existence of an early-life window of opportunity for priming of peripheral tolerance to gut commensals and external allergens.

Additionally, it was demonstrated that this weaning reaction could be perturbed by neonatal excess calorie intake, as achieved by feeding the mothers a high-fat diet, force-feeding the neonates with lipid-rich food, or removing competition for mother's milk. This resulted in an increase in microbial production of hydrogen sulfide, intestinal permeability, and pro-inflammatory cytokines which led to increased susceptibility to colitis in adulthood. The increased susceptibility was evident even if the high-calorie intake was restricted to the weaning period only and was not induced if the high-fat diet was introduced after weaning (Al Nabhani et al. 2019a), corroborating the importance of peripheral immune imprinting before weaning.

In above studies, the researchers used a range of experimental designs in various mouse models to test their hypotheses and employed various experimental techniques to obtain data. Among these were flow cytometry, fluorescent microscopy, spectrophotometry, time-lapse imaging by two-photon microscopy, ELISA, qPCR, total RNA sequencing, and 16S rRNA sequencing of microbiota.

While studies like these have contributed greatly to the field, the murine immune system and its temporal reactions cannot necessarily be directly translated to humans. Moreover, the murine studies insufficiently account for the genetic heterogeneity, environmental influence, lifespan, natural course of disease, and bacterial colonization setting apart mouse models and man. Determining how the temporal dynamics may work in humans necessitates human studies, but these have been difficult to come by, though, in part because infant samples have been scarce and small sample sizes of especially blood have required processing that only recently became possible (Davis and Brodin 2018; Olin et al. 2018). With new technologies like mass cytometry emerging, existing technologies maturing, and sequencing costs dropping, systems immunology-based studies in humans are becoming attainable and human infant studies more frequent. Typically, these studies are in the form of interventions or longitudinal birth cohorts.

4.2 *Early-Life Administration of* B. longum *ssp.* infantis *and Tolerance-Promoting Mechanisms in Early Life of Humans*

In 2021, a systems-wide study was published, comprising both an intervention in Californian children fed the probiotic *B. infantis* EVC001 strain and a longitudinal cohort analysis of Swedish infants (Henrick et al. 2021). Their cohort analysis comprised analyses of circulating blood by mass cytometry targeting 44 cell markers to characterize 64 immune cell populations, Olink assays quantifying 355 unique circulating proteins, as well as fecal samples analyzed by shotgun sequencing for metagenomic profiling. The intervention study sequenced fecal samples for metagenomic profiling, quantified fecal cytokine levels, performed targeted combined transcriptomics (259 molecules) and protein expression (10 markers) in T cells polarized with fecal water from test subjects, and analyzed for metabolites within feces using metabolomics.

The study showed that a low abundance of bifidobacterial species, and a limited content of genes for HMO utilization within the fecal microbiota, is associated with intestinal inflammation and specifically pinpointed Th2- and Th17-based inflammatory processes. Oppositely, high levels of bifidobacteria capable of HMO utilization associated with elevated levels of the anti-inflammatory cytokine interleukin-10 (IL-10). In their intervention study, the authors were able to specifically demonstrate these findings upon in vitro incubation of naïve T cells with fecal water from children who were not colonized with *B. infantis* EVC001 which led to polarization of the T cells into Th2 or Th17 cells, while fecal water from those fed with *B. infantis* EVC001 generated Th1 cells. Interestingly, activation of the immune system in this cohort was registered around 3 months of age, where most infants had not initiated weaning, leading the authors to the conclusion that the window of opportunity is likely uncoupled from weaning in humans.

They additionally found that *B. infantis* EVC001 is associated with higher fecal IFN-β levels, which may too be of relevance for host health as this cytokine is crucial for antiviral immunity. Henrick et al. (2021) hereby expand on recent studies showing that gut commensals, like *Bacteroides*, induce IFN-β production by colonic dendritic cells, fortifying the gut against viral infections (Schaupp et al. 2020; Stefan et al. 2020). A fortification that does require tolerogenic balancing, though, if T cells are to be prohibited from being primed for aberrant reactions toward innocuous antigens (Schaupp et al. 2020). Such balance might, e.g., be achieved by the ability of *B. longum* ssp. *infantis* to promote upregulation of the regulatory genes *FOXP3*, *GITR*, *IL10*, and *CTLA4* in CD4+ T cells (Vogel et al. 2023).

Henrick et al. (2021) also found that fecal levels of the aromatic lactic acid indole-3-lactic acid (ILA) were tenfold increased in children colonized with *B. infantis* EVC001 and that Th2 and Th17 cells stimulated with ILA strongly upregulated transcription of chemokine receptor 3 (CXCR3), classically associated with Th1 immunity, and galectin-1, a negative regulator of T cell proliferation. Adding to this, Vogel et al. (2023) found that antigens from *B. longum* ssp. *infantis* presented by APCs promoted significant increase in galectin-1 release in both neonatal and adult naïve T cells as well as in memory T cells.

4.3 Aromatic Lactic Acids and Early-Life Tolerance-Inducing Mechanisms in Humans

Other research groups had earlier reported that bifidobacteria-derived ILA possesses anti-inflammatory properties (Meng et al. 2020; Ehrlich et al. 2020; Laursen et al. 2021).

In a subset of a Bangladeshi infant cohort, Ehrlich et al. (2020) studied whether metabolites from *B. longum* ssp. *infantis* grown on HMOs might have a beneficial effect on the host. They collected fecal samples to determine bifidobacterial abundance and quantify metabolite levels and cultured *B. infantis* ATCC 15697 on different substrates to analyze metabolite differences by NMR spectroscopy and their effects on cultivated cell lines using ELISA, Western blot, and qPCR. Production of ILA by *B. longum* ssp. *infantis* was found to be significantly increased in the presence of HMOs as compared to lactate, and the metabolite itself was found to attenuate pro-inflammatory responses in in vitro-cultivated macrophages and intestinal epithelial cells. The authors found ILA to mediate its effect via the transcription factors nuclear factor erythroid 2-related factor 2 (Nrf2) and aryl hydrocarbon receptor (AhR), of which the latter has been associated with homeostasis and intestinal barrier integrity (Metidji et al. 2018; Natividad et al. 2018).

Meng et al. (2020) carried out similar investigations but instead performed the stimulations using human organoids and cell cultures, mouse organ cultures, and bacterial cultures followed by analyses using mass spectrometry, ELISA, and qPCR to answer their questions. They too found ILA to possess anti-inflammatory properties mediated via AhR, specifically inhibition of IL-8 expression, but only in the immature intestine of both mouse and human enterocytes with surface toll-like receptor 4 expression.

Investigating two Danish birth cohorts, Laursen et al. (2021) aimed to improve our understanding of the symbiotic relationship between bifidobacteria promoted by breastmilk in early life. They discovered an unrecognized bifidobacterial aromatic lactate dehydrogenase (*ALDH*) to be responsible for converting aromatic amino acids (which are rich in breastmilk) into aromatic lactic acids, including ILA, and found that the *ALDH* enzyme was only present in infant-type bifidobacteria, such as *B. longum* ssp. *infantis*. Additionally, they demonstrated by using ex vivo cultures of human cells that ILA, dose-dependently via AhR, stimulated secretion of the gut homeostasis-associated cytokine IL-22 in Th17-polarized CD4+ T cells. As part of this study, we performed investigations showing that ILA dose-dependently inhibited monocyte secretion of the pro-inflammatory cytokine IL-12p70 in a pro-inflammatory environment, via activation of both the hydroxycarboxylic acid receptor 3 (HCAR3) and AhR. These data were obtained using 16S rRNA sequencing, bifidobacterial cultivation, cloning and recombinant protein expression, enzyme assays, a mouse monocolonization model, metabolomics, rat and human reporter gene assays, and ex vivo stimulation of isolated human immune cells.

Alongside the above study, we have investigated if the aromatic lactic acids produced by infant bifidobacteria possessing the *ALDH* genes may associate with reduced susceptibility to allergic sensitization in later life (Myers and Dehli et al., unpublished). These studies were performed using a Swedish birth cohort followed longitudinally until 5 years of age, employing metagenomic profiling of fecal samples, targeted metabolomics in feces and blood, Phadiatop measurements of allergen-specific IgE in circulating blood, and primary cell stimulations. We find that relative fecal abundance of *ALDH+* bifidobacteria strongly correlates with fecal levels of aromatic lactic acids and that fecal levels of ILA in turn correlate with serum levels of ILA, indicating that gut bacterially produced ILA may contribute to immune priming beyond the epithelium. We additionally show that high relative abundance of *ALDH+* bifidobacteria in the first 6 months of life significantly associates with reduced risk of developing IgE sensitization to food allergens and that children sensitized to food allergens have significantly lower fecal levels of aromatic lactic acids at 2 months of life. Finally, we demonstrate that monocytes stimulated ex vivo with ILA in a pro-inflammatory environment, in addition to decreased IL-12p70 levels, significantly increase IL-10 levels, potentially contributing to homeostatic or anti-inflammatory properties.

The studies pinpoint important mechanisms and roles of certain bifidobacteria and their metabolic products, but as the gut ecological niche is rich in complex interactions, other metabolites should be investigated too, as Vatanen et al. (2022) allude to in their multi-omics study, warranting further investigation of, e.g., pipecolic acid.

Nevertheless, the above studies, among numerous others, are bringing us closer to understanding the dynamics of neonatal priming of immunological tolerance in humans. By doing so they are beginning to unveil mechanisms that may be responsible for associations between early-life disappearance of certain microbial species, like the aromatic lactic acid-producing bifidobacteria, and increased susceptibility to immunopathologies. However, the gut milieu's influence on regulatory T cell development and stability is multifaceted. Therefore, we need to investigate not only how the abovementioned microbes and metabolic products affect it, but also how other lifestyle-related determinants of mucosal milieus independent of the routes discussed above may influence metabolism and epigenetic imprinting that controls the regulatory phenotype, within and beyond the window of opportunity for early-life priming of immunological tolerance.

5 Metabolic and Epigenetic Pathways Influencing Foxp3 Expression, Regulatory Phenotype, and Disease Outcome

Systems-wide studies linking metabolism and epigenetic modifications with transcription, expression, phenotype, and function are required to fully grasp how Treg development and stability are ensured or perturbed, and a combination of disease models and human studies may help us understand which environments can be causative. Mechanistic studies doing just so are emerging and may pave the way for improved intervention or prevention in the future.

5.1 Dietary Components May Affect Treg Metabolism and Control of Foxp3 Expression

While naïve CD4+ T cells are especially fueled by ATP from oxidative phosphorylation (OXPHOS) driven by fatty-acid β-oxidation (FAO), activation and differentiation into effector CD4+ T cells (like Th1, Th2, and Th17) necessitate a switch to aerobic glycolysis and scavenging of glucose from the environment, as a swift expansion of the effector population comes with a much higher energy expenditure. Interestingly, despite being activated CD4+ T cells, Tregs to a much lesser degree use aerobic glycolysis and predominantly employ FAO-driven OXPHOS instead (Michalek et al. 2011; Procaccini et al. 2016; Gerriets et al. 2016; Geltink et al. 2018). This metabolism is further promoted by Foxp3 inducing OXPHOS and suppressing glycolysis (Gerriets et al. 2016; Angelin et al. 2017).

Sparked by these metabolic differences, Pompura et al. (2021) investigated whether metabolic sensing of the most abundant long-chain fatty acids preserved suppressive Treg function in tissue. They isolated cells from multiple sclerosis (MS) patients and controls and employed suppression assays, metabolic assays, a proliferation assay, phosphorylation staining, lentiviral gene silencing, flow cytometry, mass spectrometry-based quantitation of fatty acids, transcriptomics, and qPCR for gene expression and epigenetic modification and methylation of Foxp3 demethylation. They determined oleic acid to be the most abundant free fatty acid in adipose tissue and found that it increased the suppressive capacity of Tregs isolated from adipose tissue while also being the only studied long-chain fatty acid to promote in vitro Foxp3 expression. They demonstrated that oleic acid exerted its effect via FAO-OXPHOS and that it primed Tregs for a transcriptomic signature similar to that of blood and adipose tissue Tregs from healthy individuals, while Tregs exposed to arachidonic acid resembled those of MS patients. Arachidonic acid is an essential precursor for production of C20 omega-6-derived eicosanoids and is pro-inflammatory of nature, and excessive activation of arachidonic acid pathways is suggested to play a role in the pathogenesis of MS (Palumbo 2017; Broos et al. 2023). Finally, Pompura et al. (2021) found that MS patients had a lower concentration of oleic acid in adipose tissue and that culturing defective Tregs from untreated MS patients with oleic acid partially restored their suppressive function.

A high-salt diet has also been associated with autoimmune disease (Kleinewietfeld et al. 2013; Wu et al. 2013) and is proposed to contribute to increased incidence (Wilck et al. 2017) by affecting Th17 development and Treg balance (Hernandez et al. 2015). Combining mouse models and studies of isolated human Tregs, Sumida et al. (2018) took on the task of finding a direct link and showed that Tregs of MS patients have an IL-10 versus IFN-γ imbalance under high-salt conditions. By performing unbiased transcriptomics on Treg populations, they identified β-catenin to be central in regulation of this cytokine balance, as well as for regulation of the Treg function, and proceeded to show that prostaglandin E receptor 2 (PTGER2) was responsible for the imbalance at high-salt concentrations through a positive feed forward loop with β-catenin. Notably, the authors also showed a positive correlation between IFN-γ, PTGER2, and β-catenin levels in Tregs from MS patients, but not in those of healthy individuals.

Findings like these suggest mechanisms by which the environment can be linked to stability of Tregs and increased risk of autoimmunity. More specifically, they exemplify mechanisms by which a Mediterranean diet rich in oleic acid-containing vegetable oils, low in meat consumption, and with only moderate salt content could favor anti-inflammatory over pro-inflammatory processes as compared to diets in many European and North American countries rich in salt, glucose, animal fats, and meat, of which the latter two are dietary sources of arachidonic acid.

Bridging the early-life microbial composition with these metabolic findings, Vogel et al. (2023) found that *B. longum* ssp. *infantis* stimulated a significant increase in OXPHOS-driven metabolism in neonatal T cells.

5.2 *Genetic Predisposition to Disease and Influence on Epigenetics*

Despite emerging links to environmental influence, there is no doubt that genetics play a crucial role in the development of autoimmune diseases, as evidenced, e.g., by the 200 risk loci for MS development in the human genome (Hafler et al. 2007; Patsopoulos et al. 2019). It is thus important to consider individual genetic predisposition in analyses of how the environment may affect onset and progression of disease, especially in studies including few patients. Axisa et al. (2022) demonstrated how a genetic defect can influence Treg function and might contribute to MS development in a more direct and defining manner than environmental factors were capable of.

While epigenetics by definition is "on top of" genetics, controlling expression through covalent modification of DNA and histones, Axisa et al. (2022) excellently demonstrated that the tables can be turned, in the sense that a genetic defect may alter the epigenetic landscape. In relation to MS development, Axisa et al. (2022)

investigated the role of histone deacetylase 7 (HDAC7), an epigenetic modifier interacting with many central transcription factors (TFs), like Foxo1, Foxp3, STAT3, MEF2D, and Bcl6, making it capable of affecting Treg function in multiple ways. They uncovered that HDAC7 increased IL-10 and Foxp3 expression and that a partial loss of HDAC7 led to defective Treg activation, making mice susceptible to experimental autoimmune encephalitis. By knockin of an HDAC7 variant orthologous to an MS-protective human HDAC7 variant, the mice exhibited decreased disease severity. This study bridges genotype to phenotype variation at a single locus, without interference from environmental cues but warrants further mechanistic studies to decipher how HDAC7 may interact with TFs.

5.3 Potential Therapeutic Avenues Based on Epigenetic Findings

As technologies for studying the epigenome matures, our understanding of how epigenetic imprinting dictates phenotype and function matures with it and allows us to consider clinical interventions based on epigenetic discoveries.

Wagner et al. (2021) developed Compass, an algorithm relying on single cell RNA-seq data and flux balance analysis, and predicted how pathogenicity of Th17 cells was driven by a switch from FAO to aerobic glycolysis and the polyamine pathway. They validated the findings by showing that disruption of the polyamine pathway led to Foxp3 expression and an epigenetic and transcriptomic signature within the Th17 cells similar to that of Tregs, attenuating central nervous system (CNS) autoimmunity. For this purpose, they combined metabolomics, transcriptomics, and epigenomics, the latter by means of assay for transposase-accessible chromatin using sequencing (ATAC-seq), and used both mouse models and isolated human cells.

Regulation of the Foxp3 locus is especially important for Treg development and function, and several TFs are able to regulate *FOXP3* transcription and the regulatory phenotype in Tregs (Fig. 3), as reviewed by Trujillo-Ochoa et al. (2023). Conserved noncoding sequence 2 (CNS2) of the *FOXP3* locus controls expression of the gene (Feng et al. 2014) and needs to be completely demethylated for Foxp3 stability and expression (Floess et al. 2007). CNS2 can be considered a sensor of IL-2, as the downstream signal transducer of this cytokine, STAT5, binds CNS2 to promote *FOXP3* transcription. IL-6-STAT3-mediated activation of the methyltransferase Dnmt3a results in methylation of the CNS2 and likely prevents STAT5 binding and *FOXP3* transcription if IL-2 levels are limited (Feng et al. 2014). Pro-inflammatory IL-6 may thus be able to destabilize Tregs and promote Th17 cells by IL-6-STAT3-mediated activation of Dnmt3a in an IL-2 limited environment. Garg et al. (2019) found that the protein Blimp1 inhibits Dnmt3a upregulation and maintains CNS2 demethylation and stable *FOXP3* expression under inflammatory conditions, demonstrating a Blimp1-dependent antagonism of

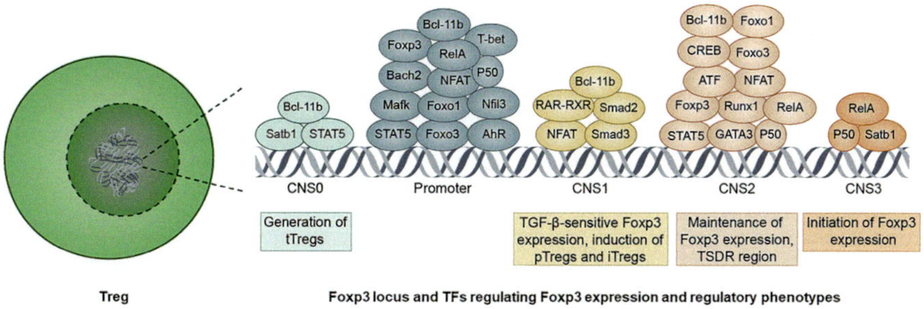

Fig. 3 TFs regulating *FOXP3* expression as reviewed by Trujillo-Ochoa et al. (2023). Environmental cues are relayed via cell-surface receptors and TFs which integrate with other TFs at Conserved Non-coding Sequences (CNS) at the *FOXP3* locus to regulate protein expression and the regulatory phenotype. Complete demethylation of the Treg-specific demethylated region (TSDR) at CNS2 is required for maintaining *FOXP3* expression and a stable regulatory phenotype. Demethylation of CNS0 is required for tTreg generation while demethylation of CNS1 appears crucial for inducing pTregs and iTregs. *Bcl-11b* B-cell leukemia/lymphoma 11b; *Satb1* Special AT-rich sequence binding protein 1; *STAT5* Signal Transducer and Activator of Transcription 5; *Foxp3* Forkhead box p3; *T-bet* T-box transcription factor; *RelA* NF-κB p65; *Bach2* BTB domain and CNC homolog 2; *NFAT* Nuclear Factor of Activated T cells; *Mafk* MAF BZIP Transcription Factor K; *Nfil3* Nuclear factor interleukin-3; *AhR* Aryl hydrocarbon Receptor; *RAR-RXR* Retinoid Acid Receptor (RAR) and Retinoid X Receptor (RXR); *Smad2* Small mothers against decapentaplegic homolog 2; *CREB* cAMP Response Element Binding protein; *ATF* Activating Transcription Factor; *Runx1* Runt-related transcription factor 1; *GATA3* GATA binding protein 3; *TGF-β* Transforming Growth Factor-β; *TF* Transcription Factor; *tTregs* thymic-derived regulatory T cells; *pTregs* peripheral-derived regulatory T cells; *iTregs* induced regulatory T cells

the IL-6-STAT3-mediated abrogation of *FOXP3* expression. Their findings were based on mouse models and employed RNA-seq, CHIP-seq, and ATAC-seq for transcriptomic and epigenetic analyses.

Ramos et al. (2023) discovered that inhibition of the methyltransferase and epigenetic repressor G9a triggers spontaneous induced Treg (iTreg) generation in vitro and in vivo. Combining RNA-seq, ATAC-seq, mass cytometry, and metabolomics in analyses of primary cell cultures and mouse models, the authors found that G9a was able to regulate genes involved in cholesterol biosynthesis and that inhibition of G9a resulted in an increased level of membrane-bound and intracellular cholesterol, supporting cholesterol-based energy derivation and increased cholesterol membrane composition, yielding iTreg development. Following these findings, the authors demonstrated that mice immunized in presence of a G9a inhibitor generated Tregs in vivo capable of protecting from TNBS colitis and that Tregs without G9a expression remain functional and able to suppress chronic intestinal inflammation.

Studies like these are bringing insights into pathways involved in epigenetic regulation of Tregs and lead to formulation of new strategies that may improve the generation of Tregs, maintain established tolerance, or revoke breakdown of established tolerance, thereby preventing trajectories toward development of noncommunicable diseases.

6 Future Studies to Further the Field

ATAC-seq interrogates chromatin accessibility epigenome-wide and has been one of the most prominent advances in the field of epigenetics. Moreover, the opportunity for single cell and multi-omics studies, combining ATAC-seq with, e.g., transcriptomics, provides researchers with unprecedented insights into intracellular regulation of gene expression and intercellular heterogeneity. Satpathy et al. (2018) showcased the technological and biological opportunities, combining ATAC-seq with T cell clonotyping. Such combination enables phenotypic studies of memory T cells (whose phenotypes cannot be determined from their quiescent protein expression but can be interpreted from their epigenome) to be matched with antigen specificity, as determined by their clonotype. Despite the possibilities, such single cell multi-omics technologies have not yet been widely utilized to address early-life programming of immune tolerance; however, we foresee that they will further the field by, e.g., aiding discovery of phenotypically distinct, disease-associated clonotypes, and antigens driving aberrant T cell responses.

In addition to a continued focus on environmental perturbations of modern society, we also advocate for more studies focusing on human biology when addressing questions related to development and plasticity of Tregs, and the use of multi-omics, single cell, and advanced in situ methodologies, such as spatial single-cell technologies, which are advantageous for such studies.

7 Conclusion

We here presented associations between microbial dysbiosis in early life and the increasing incidence in autoimmune and allergic diseases in industrialized countries and highlighted recent studies starting to unveil mechanisms responsible for the associations. Studies in both mice and humans point to an early-life window for priming of immunological tolerance that ensures lifelong protection from disease. Mouse studies have demonstrated specific temporal mechanisms controlling when the window is open, suggesting a crucial timepoint at weaning for imprinting of peripheral tolerance to external antigens, while central tolerance to self-antigens is primed via perinatal infant-maternal interplays. The temporal dynamics in humans is likely different, though, and immune priming of peripheral tolerance is suggested to occur before weaning, while insights into the time window for central tolerance imprinting are limited.

Specific bacteria in the human infant gut appear to promote peripheral tolerogenic priming, and *B. longum* ssp. *infantis* is the most prominent of these. High abundance of the bacterium associates with enhanced circulating levels of anti-inflammatory cytokines, while low abundance associates with pro-inflammatory T cell subsets. It promotes upregulation of regulatory genes in CD4+ T cells and metabolizes aromatic amino acids from mother's milk into aromatic lactic acids, among which ILA appears to be important as several studies have demonstrated its anti-inflammatory effect.

Recent advances in omics technologies, including epigenomics, have also fostered studies discovering mechanisms that may be causative in autoimmune disease development, especially MS. Oleic acid may change CD4+ T cell metabolism to favor Tregs and promote health, while excess arachidonic acid, salt, and glucose oppositely may promote pro-inflammatory processes. Finally, disruption of the polyamine pathway promoting aerobic glycolysis, inhibition of Dnmt3a methylating CNS2, and inhibition of G9a regulating cholesterol biosynthesis all promoted Treg generation or maintenance and prevented autoimmune disease, constituting potential therapeutic avenues to pursue in clinical interventions.

Acknowledgments We acknowledge support from the CAG Greater Copenhagen Research Centre for Systemic Low-Grade Inflammation (LOGINFLAM), Copenhagen, Denmark, Technical University of Denmark (DTU), and FII Institute. S.B. is the incumbent of the FII Institute research chair at DTU in immune-based prediction of disease (2021–2023).

References

Abrahamsson TR, Jakobsson HE, Andersson AF et al (2014) Low gut microbiota diversity in early infancy precedes asthma at school age. Clin Exp Allergy 44:842–850

Al Nabhani Z, Dulauroy S, Lécuyer E et al (2019a) Excess calorie intake early in life increases susceptibility to colitis in adulthood. Nat Metab 1:1101–1109

Al Nabhani Z, Dulauroy S, Marques R et al (2019b) A weaning reaction to microbiota is required for resistance to Immunopathologies in the adult. Immunity 50:1276–1288.e5

Angelin A, Gil-De-Gó Mez L, Wallace DC et al (2017) Foxp3 reprograms T cell metabolism to function in low-glucose, high-lactate environments. Cell Metab 25:1282–1293.e7

Arrieta MC, Arévalo A, Stiemsma L et al (2018) Associations between infant fungal and bacterial dysbiosis and childhood atopic wheeze in a nonindustrialized setting. J Allergy Clin Immunol 142:424–434.e10

Arrieta MC, Stiemsma LT, Dimitriu PA et al (2015) Early infancy microbial and metabolic alterations affect risk of childhood asthma. Sci Transl Med 7:307ra152

Augustine T, Kumar M, Al Khodor S et al (2023) Microbial Dysbiosis tunes the immune response towards allergic disease outcomes. Clin Rev Allergy Immunol 65:43–71

Avershina E, Storrø O, Øien T et al (2014) Major faecal microbiota shifts in composition and diversity with age in a geographically restricted cohort of mothers and their children. FEMS Microbiol Ecol 87:280–290

Axisa PP, Yoshida TM, Lucca LE et al (2022) A multiple sclerosis-protective coding variant reveals an essential role for HDAC7 in regulatory T cells. Sci Transl Med 14:eabl3651

Azad MB, Konya T, Maughan H et al (2013) Gut microbiota of healthy Canadian infants: profiles by mode of delivery and infant diet at 4 months. Can Med Assoc J 185:385–394

Barros L, Ferreira C, Veldhoen M (2021) The fellowship of regulatory and tissue-resident memory cells. Mucosal Immunol 15:64–73

Bashir MEH, Louie S, Shi HN, Nagler-Anderson C (2004) Toll-like receptor 4 signaling by intestinal microbes influences susceptibility to food allergy. J Immunol 172:6978–6987

Belkaid Y, Piccirillo CA, Mendez S et al (2002) CD4+CD25+ regulatory T cells control Leishmania major persistence and immunity. Nature 420:502–507

Broos JY, Loonstra FC, de Ruiter LR et al (2023) Association of Arachidonic Acid-Derived Lipid Mediators with Disease Severity in patients with relapsing and progressive multiple sclerosis. Neurology 101:e533–e545

Cahenzli J, Kö Y, Wyss M et al (2013) Intestinal microbial diversity during early-life colonization shapes long-term IgE levels. Cell Host Microbe 14:559–570

Casaburi G, Duar RM, Brown H et al (2021) Metagenomic insights of the infant microbiome community structure and function across multiple sites in the United States. Sci Rep 11:1472

Cebula A, Seweryn M, Rempala GA et al (2013) Thymus-derived regulatory T cells contribute to tolerance to commensal microbiota. Nature 497:258–262

Clay SL, Bravo-Blas A, Wall DM et al (2020) Regulatory T cells control the dynamic and site-specific polarization of total CD4 T cells following salmonella infection. Mucosal Immunol 13: 946–957

Conrad N, Misra S, Verbakel JY et al (2023) Incidence, prevalence, and co-occurrence of autoimmune disorders over time and by age, sex, and socioeconomic status: a population-based cohort study of 22 million individuals in the UK. Lancet 401:1878–1890

D'Alessio FR, Tsushima K, Aggarwal NR et al (2009) CD4+CD25+Foxp3+ Tregs resolve experimental lung injury in mice and are present in humans with acute lung injury. J Clin Invest 119: 2898–2913

Davis MM, Brodin P (2018) Rebooting Human Immunology. Annu Rev Immunol 36:843–864

Dominguez-Bello MG, Costello EK, Contreras M et al (2010) Delivery mode shapes the acquisition and structure of the initial microbiota across multiple body habitats in newborns. Proc Natl Acad Sci USA 107:11971–11975

Ehrlich AM, Pacheco AR, Henrick BM et al (2020) Indole-3-lactic acid associated with Bifidobacterium-dominated microbiota significantly decreases inflammation in intestinal epithelial cells. BMC Microbiol 20:357

Feng Y, Arvey A, Chinen T et al (2014) Control of the inheritance of regulatory T cell identity by a cis element in the Foxp3 locus. Cell 158:749–763

Ferreira C, Barros L, Baptista M et al (2020) Type 1 Treg cells promote the generation of CD8+ tissue-resident memory T cells. Nat Immunol 21:766–776

Feuerer M, Herrero L, Cipolletta D et al (2009) Lean, but not obese, fat is enriched for a unique population of regulatory T cells that affect metabolic parameters. Nat Med 15:930–939

Floess S, Freyer J, Siewert C et al (2007) Epigenetic control of the foxp3 locus in regulatory T cells. PLoS Biol 5:e38

Fujimura KE, Sitarik AR, Havstad S et al (2016) Neonatal gut microbiota associates with childhood multisensitized atopy and T cell differentiation. Nat Med 22:1187–1191

Garg G, Muschaweckh A, Moreno H et al (2019) Blimp1 prevents methylation of Foxp3 and loss of regulatory T cell identity at sites of inflammation. Cell Rep 26:1854–1868.e5

Geltink RIK, Kyle RL, Pearce EL (2018) Unraveling the complex interplay between T cell metabolism and function. Annu Rev Immunol 36:461–488

Gensollen T, Iyer SS, Kasper DL et al (2016) How colonization by microbiota in early life shapes the immune system. Science 352:539–544

Gerriets VA, Kishton RJ, Johnson MO et al (2016) Foxp3 and toll-like receptor signaling balance Treg cell anabolic metabolism for suppression. Nat Immunol 17:1459–1466

Gevers D, Kugathasan S, Denson LA et al (2014) The treatment-naive microbiome in new-onset Crohn's disease. Cell Host Microbe 15:382–392

Grzeskowiak L, Collado MC, Mangani C et al (2012) Distinct gut microbiota in southeastern African and northern European infants. J Pediatr Gastroenterol Nutr 54:812–816

Gutierrez-Arcelus M, Rich SS, Raychaudhuri S (2016) Autoimmune diseases—connecting risk alleles with molecular traits of the immune system. Nat Rev Genet 17:160–174

Hafler DA, Compston A, Sawcer S et al (2007) Risk alleles for multiple sclerosis identified by a Genomewide study. N Engl J Med 357:851–862

Haxhinasto S, Mathis D, Benoist C (2008) The AKT–mTOR axis regulates de novo differentiation of CD4+Foxp3+ cells. J Exp Med 205:565–574

Henrick BM, Rodriguez L, Lakshmikanth T et al (2021) Bifidobacteria-mediated immune system imprinting early in life. Cell 184:3884–3898.e11

Hernandez AL, Kitz A, Wu C et al (2015) Sodium chloride inhibits the suppressive function of FOXP3+ regulatory T cells. J Clin Invest 125:4212–4222

Hsieh CS, Lee HM, Lio CWJ (2012) Selection of regulatory T cells in the thymus. Nat Rev Immunol 12:157–167

Huda MN, Lewis Z, Kalanetra KM et al (2014) Stool microbiota and vaccine responses of infants. Pediatrics 134:e362–e372

Hviid A, Svanström H, Frisch M (2011) Antibiotic use and inflammatory bowel diseases in childhood. Gut 60:49–54

Jost T, Lacroix C, Braegger CP et al (2012) New insights in gut microbiota establishment in healthy breast fed neonates. PLoS One 7:e44595

Kalekar LA, Cohen JN, Prevel N et al (2019) Regulatory T cells in skin are uniquely poised to suppress Profibrotic immune responses. Sci Immunol 4:eaaw2910

Klein L, Jovanovic K (2011) Regulatory T cell lineage commitment in the thymus. Semin Immunol 23:401–409

Kleinewietfeld M, Manzel A, Titze J et al (2013) Sodium chloride drives autoimmune disease by the induction of pathogenic Th17 cells. Nature 496:518–522

Knoop KA, Gustafsson JK, Mcdonald KG et al (2017) Microbial antigen encounter during a preweaning interval is critical for tolerance to gut bacteria. Sci Immunol 2:eaao1314

Kolodin D, Van Panhuys N, Li C et al (2015) Antigen- and cytokine-driven accumulation of regulatory T cells in visceral adipose tissue of lean mice. Cell Metab 21:543–557

Krishna MT, Mahesh PA, Vedanthan PK et al (2020) The burden of allergic diseases in the Indian subcontinent: barriers and challenges. Lancet Glob Health 8:e478–e479

Laforest-Lapointe I, Arrieta MC (2017) Patterns of early-life gut microbial colonization during human immune development: an ecological perspective. Front Immunol 8:788

Laursen MF, Sakanaka M, von Burg N et al (2021) Bifidobacterium species associated with breastfeeding produce aromatic lactic acids in the infant gut. Nat Microbiol 6:1367–1382

Lerner A, Jeremias P, Matthias T (2015) The world incidence and prevalence of autoimmune diseases is increasing. Int J Celiac Dis 3:151–155

Lewis ZT, Totten SM, Smilowitz JT et al (2015) Maternal fucosyltransferase 2 status affects the gut bifidobacterial communities of breastfed infants. Microbiome 3:13

Li J, Tan J, Martino MM et al (2018) Regulatory T-cells: potential regulator of tissue repair and regeneration. Front Immunol 9:585

Lu L, Barbi J, Pan F (2017) The regulation of immune tolerance by FOXP3. Nat Rev Immunol 17: 703–717

Marcobal A, Sonnenburg JL (2012) Human milk oligosaccharide consumption by intestinal microbiota. Clin Microbiol Infect 18(Suppl 4):12–15

Meng D, Sommella E, Salviati E et al (2020) Indole-3-lactic acid, a metabolite of tryptophan, secreted by Bifidobacterium longum subspecies infantis is anti-inflammatory in the immature intestine. Pediatr Res 88:209–217

Metidji A, Omenetti S, Crotta S et al (2018) The environmental sensor AHR protects from inflammatory damage by maintaining intestinal stem cell homeostasis and barrier integrity. Immunity 50:1542

Michalek RD, Gerriets VA, Jacobs SR et al (2011) Cutting edge: distinct glycolytic and lipid oxidative metabolic programs are essential for effector and regulatory CD4+ T cell subsets. J Immunol 186:3299–3303

Miller FW (2023) The increasing prevalence of autoimmunity and autoimmune diseases: an urgent call to action for improved understanding, diagnosis, treatment, and prevention. Curr Opin Immunol 80:102266

Natividad JM, Agus A, Planchais J et al (2018) Impaired aryl hydrocarbon receptor ligand production by the gut microbiota is a key factor in metabolic syndrome. Cell Metab 28:737–749.e4

Nikolova M, Lelievre JD, Carriere M et al (2009) Regulatory T cells differentially modulate the maturation and apoptosis of human CD8+ T-cell subsets. Blood 113:4556–4565
Nutsch K, Chai JN, Ai TL et al (2016) Rapid and efficient generation of regulatory T cells to commensal antigens in the periphery. Cell Rep 17:206–220
Olin A, Henckel E, Chen Y et al (2018) Stereotypic immune system development in newborn children. Cell 174:1277–1292.e14
Olivares M, Neef A, Castillejo G et al (2015) The HLA-DQ2 genotype selects for early intestinal microbiota composition in infants at high risk of developing coeliac disease. Gut 64:406–417
Olszak T, An D, Zeissig S et al (2012) Microbial exposure during early life has persistent effects on natural killer T cell function. Science 336:489–493
Palumbo S (2017) Pathogenesis and progression of multiple sclerosis: the role of arachidonic acid–mediated Neuroinflammation. Perspectives in Treatment and Pathogenesis, Multiple Sclerosis, pp 111–124
Patsopoulos NA, Baranzini SE, Santaniello A et al (2019) Multiple sclerosis genomic map implicates peripheral immune cells and microglia in susceptibility. Science 365:eaav7188
Pawankar R, Canonica G, Holgate S et al (2011) The WAO white book on. Allergy:2011–2012
Platts-Mills TAE (2015) The allergy epidemics: 1870–2010. J Allergy Clin Immunol 136:3–13
Pompura SL, Wagner A, Kitz A et al (2021) Oleic acid restores suppressive defects in tissue-resident FOXP3 Tregs from patients with multiple sclerosis. J Clin Invest 131:e138519
Procaccini C, Carbone F, Di SD et al (2016) The proteomic landscape of human ex vivo regulatory and conventional T cells reveals specific metabolic requirements. Immunity 44(2):406–421
Ramos GP, Bamidele AO, Klatt EE et al (2023) G9a modulates lipid metabolism in CD4 T cells to regulate intestinal inflammation. Gastroenterology 164:256–271.e10
Renz H, Skevaki C (2021) Early life microbial exposures and allergy risks: opportunities for prevention. Nat Rev Immunol 21:177–191
Rhoads JM, Collins J, Fatheree NY et al (2018) Infant colic represents gut inflammation and Dysbiosis. J Pediatr 203:55–61.e3
Roos S, Dicksved J, Tarasco V et al (2013) 454 pyrosequencing analysis on Faecal samples from a randomized DBPC trial of colicky infants treated with lactobacillus reuteri DSM 17938. PLoS One 8:e56710
Rosenblum MD, Gratz IK, Paw JS et al (2011) Response to self antigen imprints regulatory memory in tissues. Nature 480:538–542
Russell SL, Gold MJ, Hartmann M et al (2012) Early life antibiotic-driven changes in microbiota enhance susceptibility to allergic asthma. EMBO Rep 13:440–447
Sakaguchi S, Mikami N, Wing JB et al (2020) Regulatory T cells and human disease. Annu Rev Immunol 38:541–566
Sakaguchi S, Miyara M, Costantino CM et al (2010) FOXP3+ regulatory T cells in the human immune system. Nat Rev Immunol 10:490–500
Satpathy AT, Saligrama N, Buenrostro JD et al (2018) Transcript-indexed ATAC-seq for precision immune profiling. Nat Med 24:580–590
Schaupp L, Muth S, Rogell L et al (2020) Microbiota-induced type I interferons instruct a poised basal state of dendritic cells. Cell 181:1080–1096.e19
Sefik E, Geva-Zatorsky N, Oh S et al (2015) Individual intestinal symbionts induce a distinct population of RORγ+ regulatory T cells. Science 349:993–997
Sevelsted A, Stokholm J, Bisgaard H (2016) Risk of asthma from cesarean delivery depends on membrane rupture. J Pediatr 171:38–42.e1-4
Sevelsted A, Stokholm J, Bønnelykke K et al (2015) Cesarean section and chronic immune disorders. Pediatrics 135:e92–e98
Shin NR, Whon TW, Bae JW (2015) Proteobacteria: microbial signature of dysbiosis in gut microbiota. Trends Biotechnol 33:496–503
Stefan KL, Kim MV, Iwasaki A et al (2020) Commensal microbiota modulation of natural resistance to virus infection. Cell 183:1312–1324.e10

Stokholm J, Thorsen J, Blaser MJ et al (2020) Delivery mode and gut microbial changes correlate with an increased risk of childhood asthma. Sci Transl Med 12:eaax9929

Stokholm J, Thorsen J, Chawes BL et al (2016) Cesarean section changes neonatal gut colonization. J Allergy Clin Immunol 138:881–889.e2

Sumida T, Lincoln MR, Ukeje CM et al (2018) Activated β-catenin in Foxp3+ regulatory T cells links inflammatory environments to autoimmunity. Nat Immunol 19:1391–1402

Suvas S, Kumaraguru U, Pack CD et al (2003) CD4+CD25+ T cells regulate virus-specific primary and memory CD8+ T cell responses. J Exp Med 198:889–901

Thorburn AN, McKenzie CI, Shen S et al (2015) Evidence that asthma is a developmental origin disease influenced by maternal diet and bacterial metabolites. Nat Commun 6:7320

Thornton AM, Korty PE, Tran DQ et al (2010) Expression of Helios, an Ikaros transcription factor family member, differentiates Thymic-derived from peripherally induced Foxp3+ T regulatory cells. J Immunol 184:3433–3441

Torow N, Yu K, Hassani K et al (2015) Active suppression of intestinal CD4+TCRαβ+ T-lymphocyte maturation during the postnatal period. Nat Commun 6:7725

Trujillo-Ochoa JL, Kazemian M, Afzali B (2023) The role of transcription factors in shaping regulatory T cell identity. Nat Rev Immunol 23:842–856

Underwood MA, German JB, Lebrilla CB et al (2014) Bifidobacterium longum subspecies infantis: champion colonizer of the infant gut. Pediatr Res 77:229–235

Vatanen T, Ang QY, Siegwald L et al (2022) A distinct clade of Bifidobacterium longum in the gut of Bangladeshi children thrives during weaning. Cell 185:4280–4297.e12

Vatanen T, Kostic AD, D'Hennezel E et al (2016) Variation in microbiome LPS immunogenicity contributes to autoimmunity in humans. Cell 165:842–853

Vogel K, Arra A, Lingel H et al (2023) Bifidobacteria shape antimicrobial T-helper cell responses during infancy and adulthood. Nat Commun 14:5943

Wagner A, Wang C, Fessler J et al (2021) Metabolic modeling of single Th17 cells reveals regulators of autoimmunity. Cell 184:4168–4185.e21

Weiss JM, Bilate AM, Gobert M et al (2012) Neuropilin 1 is expressed on thymus-derived natural regulatory T cells, but not mucosa-generated induced Foxp3+ T reg cells. J Exp Med 209:1723–1742

Wilck N, Matus MG, Kearney SM et al (2017) Salt-responsive gut commensal modulates TH17 axis and disease. Nature 551:585–589

Wu C, Yosef N, Thalhamer T et al (2013) Induction of pathogenic Th17 cells by inducible salt sensing kinase SGK1. Nature 496:513–517

Yang S, Fujikado N, Kolodin D et al (2015) Regulatory T cells generated early in life play a distinct role in maintaining self-tolerance. Science 348:589–594